LONDON MATHEMATICAL SOCIETY LECTURE NOTE SERIES

T0296682

Managing Editor: Professor I.M.James,
Mathematical Institute, 24-29 St Giles, Oxford

1. General cohomology theory and K-theory, P.HILTON
4. Algebraic topology: a student's guide, J.F.ADAMS
5. Commutative algebra, J.T.KNIGHT
8. Integration and harmonic analysis on compact groups, R.E.EDWARDS
9. Elliptic functions and elliptic curves, P.DU VAL
10. Numerical ranges II, F.F.BONSALL & J.DUNCAN
11. New developments in topology, G.SEGAL (ed.)
12. Symposium on complex analysis, Canterbury, 1973, J.CLUNIE & W.K.HAYMAN (eds.)
13. Combinatorics: Proceedings of the British combinatorial conference 1973, T.P.McDONOUGH & V.C.MAVRON (eds.)
14. Analytic theory of abelian varieties, H.P.F.SWINNERTON-DYER
15. An introduction to topological groups, P.J.HIGGINS
16. Topics in finite groups, T.M.GAGEN
17. Differentiable germs and catastrophes, Th.BROCKER & L.LANDER
18. A geometric approach to homology theory, S.BUONCRISTIANO, C.P.ROURKE & B.J.SANDERSON
20. Sheaf theory, B.R.TENNISON
21. Automatic continuity of linear operators, A.M.SINCLAIR
23. Parallelisms of complete designs, P.J.CAMERON
24. The topology of Stiefel manifolds, I.M.JAMES
25. Lie groups and compact groups, J.F.PRICE
26. Transformation groups: Proceedings of the conference in the University of Newcastle upon Tyne, August 1976, C.KOSNIOWSKI
27. Skew field constructions, P.M.COHN
28. Brownian motion, Hardy spaces and bounded mean oscillation, K.E.PETERSEN
29. Pontryagin duality and the structure of locally compact abelian groups, S.A.MORRIS
30. Interaction models, N.L.BIGGS
31. Continuous crossed products and type III von Neumann algebras, A.VAN DAELE
32. Uniform algebras and Jensen measures, T.W.GAMELIN
33. Permutation groups and combinatorial structures, N.L.BIGGS & A.T.WHITE
34. Representation theory of Lie groups, M.F.ATIYAH et al.
35. Trace ideals and their applications, B.SIMON
36. Homological group theory, C.T.C.WALL (ed.)
37. Partially ordered rings and semi-algebraic geometry, G.W.BRUMFIEL
38. Surveys in combinatorics, B.BOLLOBAS (ed.)
39. Affine sets and affine groups, D.G.NORTHCOTT
40. Introduction to H_p spaces, P.J.KOOSIS
41. Theory and applications of Hopf bifurcation, B.D.HASSARD, N.D.KAZARINOFF & Y-H.WAN
42. Topics in the theory of group presentations, D.L.JOHNSON
43. Graphs, codes and designs, P.J.CAMERON & J.H.VAN LINT
44. Z/2-homotopy theory, M.C.CRABB
45. Recursion theory: its generalisations and applications, F.R.DRAKE & S.S.WAINER (eds.)
46. p-adic analysis: a short course on recent work, N.KOBLITZ
47. Coding the Universe, A. BELLER, R. JENSEN & P. WELCH
48. Low-dimensional topology, R. BROWN & T.L. THICKSTUN (eds.)
49. Finite geometries and designs, P. CAMERON, J.W.P. HIRSCHFELD & D.R. HUGHES (eds.)
50. Commutator Calculus and groups of homotopy classes, H.J. BAUES
51. Synthetic differential geometry, A. KOCK
52. Combinatorics, H.N.V. TEMPERLEY (ed.)
53. Singularity theory, V.I. ARNOLD
54. Markov processes and related problems of analysis, E.B. DYNKIN
55. Ordered permutation groups, A.M.W. GLASS
56. Journees arithmetiques 1980, J.V. ARMITAGE (ed.)
57. Techniques of geometric topology, R.A. FENN
58. Singularities of differentiable functions, J. MARTINET
59. Applicable differential geometry, F.A.E. PIRANI and M. CRAMPIN
60. Integrable systems, S.P. NOVIKOV et al.

London Mathematical Society Lecture Note Series. 47

Coding the Universe

A. BELLER
Department of Computer Science, Temple University and
Department of Decision Science, The Wharton School,
University of Pennsylvania

R. JENSEN
Mathematisches Institut, University of Freiburg

P. WELCH
Mathematical Institute, University of Oxford

CAMBRIDGE UNIVERSITY PRESS

CAMBRIDGE

LONDON NEW YORK NEW ROCHELLE

MELBOURNE SYDNEY

CAMBRIDGE UNIVERSITY PRESS
Cambridge, New York, Melbourne, Madrid, Cape Town, Singapore, São Paulo

Cambridge University Press
The Edinburgh Building, Cambridge CB2 8RU, UK

Published in the United States of America by Cambridge University Press, New York

www.cambridge.org
Information on this title: www.cambridge.org/9780521280402

First published 1982
Re-issued in this digitally printed version 2008

A catalogue record for this publication is available from the British Library

Library of Congress Catalogue Card Number: 81–2663

ISBN 978-0-521-28040-2 paperback

CONTENTS

0 . An introduction

In this book we prove some results about Zermelo-Fraenkel (ZF) set theory. Before discussing the main result (Theorem 0.1) below we introduce some concepts, and notation, together with just a few remarks to try and put the results in an historical setting as well as illuminating, we hope, their contemporary context for someone who is not a set-theorist.

The Constructible Universe of sets, L , was invented by Kurt Gödel in 1938 in order to prove the consistency of the Axiom of Choice (AC) and the Generalised Continuum Hypothesis (that $2^\kappa = \kappa^+$ for all infinite cardinals, the "GCH") with the other axioms of set theory, i.e. ZF . L , the smallest model of ZF containing all the ordinals, is constructed by starting with the empty set, $\emptyset$, and putting in L only what must be there by virtue of the axioms of ZF . At each stage of the construction, any newly constructed element can be defined by a first-order formula using previously defined elements, thus giving L its well-defined structure. Indeed this method implicitly contains the well-ordering of the universe that provides for the truth of AC (and ultimately the GCH in L .

It is possible to inspect the internal structure of this universe more closely, as has recently been done, and the fine structure that arises enables one to have a better "grip" on L than on a general model of ZF or ZF + GCH ; this has led to several results of interest about L .

If we turn our attention away from L to V , by which we mean the universe of all sets of mathematical and set-theoretical discourse, we may ask: "to what extent does L capture all of V". Indeed, the statement that $V = L$ may be posited, and used as an extra "axiom" with ZF, but it is generally assumed that $L \subset V$. Then of course one has the question concerning the different degrees of V not equalling L.

For example, let us suppose that there were an embedding of L into itself, via some map j. Since L is the smallest model of ZF containing the ordinals (the smallest 'inner' model) j takes L into L, written $j: L \longrightarrow L$. But it is a theorem of ZF + AC that there is no such embedding of V to V, the only possibility being that of some $i: V \longrightarrow M$ where $M \subset V$. Thus the existence of j above implies $V \neq L$, and in fact from such a j we can infer the existence of a set of integers in V called $0^{\#}$, one interpretation of which is as a sort of truth definition for L, and which cannot be a member of L. Thus to say "$0^{\#}$ does not exist" (or equivalently that there is no such j as above) implies a restriction on the size of V, and one may obtain as a theorem of ZF : if $0^{\#}$ does not exist ($\neg 0^{\#}$) then every uncountable set of ordinals in V is a subset of one in L of the same cardinal power. This "Covering Theorem" in a very broad sense causes the cardinality structure of L to be not radically different from that of V. One may weaken the restriction $\neg 0^{\#}$ to that of the nonexistence of other specific sets and obtain analogous results for other kinds of "minimal" inner models than L, constructible in a broader sense.

In a different direction one may "force" a model M of $ZF + V = L$ to become one of $ZF + V \neq L$. This method of forcing was invented in 1963 by Paul Cohen. Originally used by him to demonstrate that the AC and GCH were independent of the other axioms of ZF, it has subsequently become much developed. It enables one to start with a model of ZF and construct, in the case that M is countable, an extension N of M with certain desired properties (in case M is not countable, the *consistency* of the existence of N is assured). This generic extension N of M has had the desired properties "forced" by a set of conditions in M or some suitable class of M . The generic set or class so produced, G say, is such that N is the result of adjoining G to M and we may write $N = M[G]$. In the case M also modelled $V = L$, and G is a set, we may write that N models "$V = L[G] \neq L$" . L[G] is the constructible closure of L with G and may be regarded as the universe of sets obtained by starting with the set G instead of the empty set in the definition of L above. How much L[G] will differ from L will, of course, depend on G , but in any case G can only alter the properties of L in a local way: only the bottom part of the universe will be affected whilst the top part will exhibit the same kind of structure and behaviour that L has. For example the GCH will continue to hold 'above G ', as will the fine structural properties of L .

Theorem 0.1 below is in fact an essay in forcing which utilises some of the aforementioned ideas. It says that in any model M of ZF + GCH one can define a class of conditions, $\mathbb{P}$, such that if N is an extension of M as determined by $\mathbb{P}$, then N models "$ZF + GCH + V = L[a]$" where 'a' is a set of integers or, equivalently, a real. This subset of ω predetermines the cardinality structure of N all the way through the ordinals of N in such a way that the cardinals of N are precisely those of M as are cofinalities. This even if M contains $0^{\#}$ or is even a very large model compared to L . N behaves as if it were the constructible closure from this real a . Indeed other properties of the cardinals of M are preserved into N , namely all those that are consistent with N being of the form L[a] . Thus the least possible "distortion" of the cardinality structure is introduced in the transition from M to N, and the conditions ensure that this information can be retrieved from the set a in which it has been "encoded".

Originally the theorem had been proven by Jensen under the strong assumption that the set $0^{\#}$ mentioned earlier did not exist in M. As already stated this implied that the cardinality structure of M already bore some resemblance to that of L. Later this assumption was seen by Jensen to be unnecessary.

Theorem 0.1 *Let* $\langle M,A\rangle$ *be a transitive model of* ZF + GCH, *with* $A \subseteq On \cap M$. *Then there is an* $\langle M,A\rangle$*-definable class of conditions* $\mathbb{P}$ *such that if* G *is* $\mathbb{P}$*-generic over* $\langle M,A\rangle$, *and* $N = M[G]$ *then we have:*

(I) $N \models$ "ZF + GCH + V = L[a]" *for a set* $a \subseteq \omega$, $a \in N$, *where* A *and* G *are* N*-definable from the parameter* a.

(II) The cardinals of N *are precisely those of* M *as are cofinalities and the following large cardinal properties are also preserved: Mahlo, weakly compact,* Π^1_m*-indescribable, subtle, ineffable and* α*-Erdös* $(\alpha < \omega_1)$.

The proof of this theorem is the greater part of the book. It sheds light on a well known conjecture of Solovay. Call a set $X \subseteq On$ *set generic* iff $L[X] = L[G]$ for some G which is $\mathbb{P}$-generic over L for some $\mathbb{P}$ a set in L. Solovay's conjecture then reads:

SC If $a \subseteq On$ is a set such that $L[a] \models \neg 0^{\#}$ then a is set generic.

Using the theorem we construct a model in which SC fails as follows. Let M' be a model of ZF + V = L. For M' cardinals γ, let $\mathbb{Q}_\gamma$ be the set of Cohen conditions of size less than γ^+ for adding a generic subset of $[\gamma,\gamma^+)$. By

Easton forcing [EA] , add a $\mathbf{Q}_\gamma$-generic $A_\gamma \subseteq [\gamma,\gamma^+)$ for all such γ . Set $A = \bigcup_\gamma A_\gamma$ and $M = M'[A]$. Using $\langle M,A\rangle$ as the ground model apply the theorem to get N and a such that $V = L[a]$ in N . Suppose that a were set generic by a set $\mathbf{R}$ of conditions in L . Let $\kappa = \mathrm{card}(\mathbf{R})$. Let B and C be the complete Boolean algebras (in L) generated by $\mathbf{R}$ and $\mathbf{Q}_\kappa$ respectively. Since $A_\kappa \in L[a]$ it follows by standard methods that C is embeddable in B . But $\mathrm{card}(B) < \mathrm{card}(C)$. A contradiction.

There is a weaker form of SC which reads:

<u>WSC</u> If $a \in L[0^\#]$ and $0^\# \notin L[a]$, then a is set generic.

We derive a counterexample to WSC by taking L as the ground model $(A = \emptyset)$ and using $0^\#$ to construct a $\mathbf{P}$-generic class G ($\mathbf{P}$ as stated in Theorem 0.1). We then collapse G to an a by the theorem. It is then easy to show that a is a counterexample to WSC . The above is embodied in:

Theorem 0.2 *Assume* $0^\#$ *exists. Then there is a set* a *such that*

(I) a *is not set generic.*

(II) $L[a]$ *has the preservation properties of Theorem 0.1 with respect to* L *(hence* $0^\# \notin L[a]$*)* .

(III) $0^\# = a^\#$ *as Turing degrees.*

We have thus shown that SC is not provable in ZFC even from $\neg 0^\#$, and that WSC + "$0^\#$ exists" is provably false.

Chapters 1-4 prove Theorem 0.1 under the assumption that $\langle M,A\rangle \models \neg 0^\#$. The first chapter contains the fundamental ideas about coding, and begins with a plan for the complete coding method. In Chapter 2 the conditions for coding are defined and their extendability properties are proven. The third chapter deals with the distributivity of the conditions and in Chapter 4 the theorem is finally proven. The real a of Theorem 0.2 is also constructed at the end of this chapter. Chapter 5 includes some applications of the coding method. Some properties of the sets of conditions which go to make up the class $\mathbb{P}$, and the models in which they exist were assumed in Chapter 2; these properties require fine-structural arguments for their justification and these appear in Chapters 6 and 7. Chapter 8 shows how to dispense with the condition of $\neg 0^{\#}$ and reworks the necessary parts of the earlier proofs. Chapter 9 has some more applications in this context and the last chapter is an appendix of facts not directly related to the coding but which are utilised there. Excepting for Chapter 5 and the preservation of α-Erdös cardinals for $\alpha < \omega_1$, which are due to Beller, the results are all those of Jensen.

The set theoretical notation and conventions are standard. There is a small glossary of notation and abbreviations together with an index of notational definitions. The forcing is modelled on Shoenfield forcing [SH] and in general the notation of [BU] is used. The set of Cohen conditions for adding a subset to a cardinal κ is the set of functions f such that $f: \kappa \longrightarrow 2$ and $\text{card}(\text{dom}(f)) < \kappa$.

Except for Chapters 6 and 7 fine structure is not used although many of its methods implicitly are. We shall try to state explicitly when such methods are employed. If the reader wishes to read Chapters 6 and 7 a knowledge of the fine structure of L (as in [J1]) is really required. At certain points a knowledge of the implications of the Covering Lemma for L, [DJ], would clarify matters slightly, although the implications and their source are usually stated, albeit without going into any details.

1 . The building blocks

1.1 THE PLAN

The basic step of the coding process will be coding a $B \subseteq \kappa^+$ by a $D \subseteq \kappa$ for a cardinal κ. (We understand D codes B to mean than $B \in L[D]$.) The conditions, $\mathbb{P}$, we use are of the 'Easton' type which are 'essentially forward' i.e. the coding step for larger cardinals must be performed before the smaller ones. For a cardinal τ, we can define an upper (or class) part of the forcing conditions, $\mathbb{P}_\tau$ (we have assigned the lower index for the class case; thinking of it as going from τ to ∞). $\mathbb{P}_\tau$ codes from ∞ to τ^+ in the sense that if G is $\mathbb{P}_\tau$-generic, we have a $B \subseteq \tau^+$ which codes G (G is $L[B]$ definable) and $V = L[B]$. (We also call such a B $\mathbb{P}_\tau$-generic.) Using a $\mathbb{P}_\tau$-generic B we can define a set of 'lower' conditions, $\mathbb{P}^B$, which can code from τ^+ to ω. Forcing first with $\mathbb{P}_\tau$ and then with $\mathbb{P}^B$ is the same as forcing with all of $\mathbb{P}$. This factoring gives $\mathbb{P}$ its 'Easton'-like properties, but unlike regular Easton forcing the conditions are 'connected' (i.e. the lower conditions can only be defined in terms of the upper conditions). For $p \in \mathbb{P}_\tau$, we can define a 'lower' set of conditions $\mathbb{P}^p$, so if $p \in \mathbb{P}$, $p/\!\!\upharpoonright[\tau,\infty) \in \mathbb{P}_\tau$ and $p\,[\omega,\tau) \in \mathbb{P}^{p/\!\upharpoonright[\tau,\infty)}$.

It is instructive to see how V is reconstructed by a generic subset $a_0 \subseteq \omega$. a codes a subset $a_1 \subseteq \aleph_1$ s.t. $H_{\aleph_1} = L_{\aleph_1}[a_1]$ and in turn $a_i \subseteq \aleph_i$ codes a subset $a_{i+1} \subseteq \aleph_{i+1}$ s.t. $H_{\aleph_{i+1}} = L_{\aleph_{i+1}}[a_{i+1}]$ for $i < \omega$. At $\aleph_\omega$ (a limit stage) the forcing is arranged so that $\bigcup_{i<\omega} a_i$ codes $a_{\omega+1} \subseteq \aleph_{\omega+1}$ s.t. $H_{\aleph_{\omega+1}} = L_{\aleph_{\omega+1}}[a_{\omega+1}]$, etc. A condition $p \in \mathbb{P}$ must reflect the above process. Since a condition must be a set it can only code up to some cardinal, but then room must be left to code the rest of the way up to ∞. Therefore $p/\aleph_0$ must be bounded in ω and code $p/[\aleph_0,\aleph_1)$ which is also bounded in $\aleph_1$ etc. As before, each $p/[\aleph_i,\aleph_{i+1})$ must contain some information for coding $p/[\aleph_\omega,\aleph_{\omega+1})$.

For each cardinal, τ, we define a set of conditions $\mathbb{S}_\tau$ which are bounded subsets of $[\tau,\tau^+)$ that are potential initial segments of conditions in $\mathbb{P}_\tau$. For $s \in \mathbb{S}_\tau$ we define our basic set of lower conditions $\mathbb{P}^s$, from which we can obtain a subset of ω that codes $s \subseteq [\tau,\tau^+)$. (Note that s potentially codes the rest of V.)

There are three types of forcing used to build up $\mathbb{P}^s$. At successor cardinals the coding will be done by Solovay's almost disjoint method (the method is actually good at any regular cardinal), at limit cardinals a new coding method is introduced and in addition at all cardinals τ we have to 'reshape' the interval $[\tau,\tau^+)$ by conditions of type $\mathbb{S}_\tau$ mentioned above. These three types of forcing are the building blocks of $\mathbb{P}$ and putting them together will be intricate.

The above discussion suggests that the following problem be solved first.

Problem Assume G.C.H., let γ be a cardinal and $B \subseteq \gamma^+$ s.t. $V = L[B]$. Can we generically add a $D \subseteq \gamma$ s.t.
(i) $B \in L[D]$
(ii) Cardinalities and cofinalities are preserved in $L[D]$?
For short we will write 'D codes B' for (i) and (ii).

1.2 ALMOST DISJOINT FORCING

Solovay's almost disjoint forcing appears in its purest form in the following theorem:

Theorem 1.1 (Solovay) *Let* $\kappa \geq \omega$ *be regular s.t.* $2^{\underset{\smile}{\kappa}} = \kappa$ *Let* $B \subseteq \kappa^+$ *s.t. for all* $\delta < \kappa^+$ $L[B \cap \delta] \models \text{'card}(\delta) \leq \kappa\text{'}$ *; then we can generically add a* $D \subseteq \kappa$ *that codes* B .

Proof We will not give the simplest proof, but rather a proof which corresponds to later use.

We can assume w.l.o.g. that $B \subseteq [\kappa, \kappa^+)$. Since $2^{\underset{\smile}{\kappa}} = \kappa$ we can find an $A \subseteq \kappa$ s.t. $H_\kappa = L_\kappa[A]$. For $b \subseteq \kappa$, let $S(b)$ be the set of sequence numbers of b (i.e. the set of codes (in L[A]) of $\chi_b \restriction \delta$ where χ_b is the characteristic function of b and $\delta < \kappa$. We require that the code of $\chi_b \restriction \delta$ be less than the code of $\chi_b \restriction \eta$ for $\eta > \delta$.) Then if $b_1 \neq b_2$, $b_1, b_2 \subseteq \kappa$ there is a $\tau < \kappa$ s.t. $S(b_1) \cap S(b_2) \subseteq \tau$ (i.e. $S(b_1)$ and $S(b_2)$ are almost disjoint). Therefore in order to generate a family of κ^+ almost disjoint sets in κ , it is sufficient to define $b_\xi \subseteq \kappa$ for $\xi \in [\kappa, \kappa^+)$ and take $\langle S(b_\xi) : \xi \in [\kappa, \kappa^+) \rangle$ as the family. We select our b_ξ so that each b_ξ is definable from $B \cap \xi$. Define a sequence μ_ξ for $\xi \in [\kappa, \kappa^+)$ by:

$$\mu_\xi = \text{the least } \mu > \kappa \text{ s.t. } \mu > \sup_{\nu \in [\kappa, \kappa^+)} \mu_\nu \text{ and}$$

$$L_\mu[A, B \cap \xi] \models ZF^- + \text{card}(\xi) = \kappa .$$

Since $L[A, B \cap \xi] \models \text{card}(\xi) = \kappa$, it is easy to see that $\mu_\xi < \kappa^+$ exists for all $\xi \in [\kappa, \kappa^+)$.

Lemma 1.2 *Every* $x \in L_{\mu_\xi}[A, B \cap \xi]$ *is definable in* $L_{\mu_\xi}[A, B \cap \xi]$ *from ordinals less than* $\kappa, A, B \cap \xi$.

Proof

<u>Claim 1</u> The sequence $\langle \mu_\xi : \eta < \xi \rangle$ is definable in $L_{\mu_\xi}[A, B \cap \xi]$ from $\kappa, A, B \cap \xi$. We can define $\langle \mu_\xi : \eta < \xi \rangle$ in

$L_{\mu_\xi}[A,B\cap\xi]$ in exactly the same way we did in $L[A,B\cap\xi]$, using $\kappa,A,B\cap\xi$.

<u>Claim 2</u> $L_{\mu_\xi}[A,B\cap\xi] \models$ 'κ is the last cardinal'.
If not, let $\alpha < \mu_\xi$ be such that

$$L_{\mu_\xi}[A,B\cap\xi] \models \text{'}\alpha = \kappa^{+}\text{'} .$$

Then $L_\alpha[A,B\cap\xi] \models ZF^- + \text{card}(\xi) = \kappa$ and $\alpha > \sup_{\delta<\xi} \mu_\delta$, since $L_{\mu_\xi}[A,B\cap\xi] \models$ '$\exists f \in L_{\kappa^+}[A,B\cap\xi](f\colon \kappa \xrightarrow{\text{On}} \xi)$ and the sequence $\langle \mu_\xi\colon \eta < \xi\rangle$ is bounded under κ^+ and $\alpha = \kappa^+$'. But this contradicts the minimality of μ_ξ.

Q.E.D. (Claim 2)

Now κ is definable in $\langle L_{\mu_\xi}[A,B\cap\xi],A,B\cap\xi\rangle$; hence also $\langle \mu_\delta\colon \delta < \xi\rangle$. Then ξ is also definable as the length of the sequence $\langle \mu_\delta\colon \delta < \xi\rangle$ (i.e., repeating the definition of μ_δ, ξ is the first s.t. μ_ξ is undefined).
Set: $\mathfrak{A}_\xi = \langle L_{\mu_\xi}[A,B\cap\xi],A,B\cap\xi\rangle$ and
X_0 = the smallest $X \prec \mathfrak{A}_\xi$ s.t. $X \supseteq \kappa$.

<u>Claim 3</u> $X_0 = \mathfrak{A}_\xi$.
Let $\mu = \sup_{\delta<\xi} \mu_\delta$ then $X_0 \supseteq \mu$ since $X \models \forall\delta < \xi \exists f\, (f\colon \kappa \xrightarrow{\text{Onto}} \mu_\delta)$.
Let $\pi\colon \langle L_\delta[A,B\cap\xi],A,B\cap\xi\rangle \cong X_0$. $\pi{\restriction}\mu = \text{id}{\restriction}\mu$ and $\pi(\xi) = \xi$ so $\langle L_\delta[A,B\cap\xi],A,B\cap\xi\rangle \models ZF^- + \text{card}(\xi) = \kappa$. Therefore $\mu_\xi = \delta$ and $X_0 = \mathfrak{A}_\xi$.

Q.E.D. (Lemma 1.2)

<u>Conclusion (from Claim 2)</u> $L_{\mu_{\xi+1}}[A,B\cap\xi] \models$ '$\text{card}(\mu_\xi) = \kappa$'.

Now we define

$$b_\xi = \text{the } L[A,B\cap\xi]\text{-least } b \subseteq \kappa \text{ s.t. } b \notin L_{\mu_\xi}[A,B\cap\xi] .$$

By the above conclusion $b_\xi \in L_{\mu_{\xi+1}}[A,B\cap\xi]$ since in $L_{\mu_{\xi+1}}[A,B\cap\xi] \models \text{card}(\xi) = \kappa$ and we have not yet exhausted all the subsets of κ .

Set $\tilde{B} = \{b_\xi : \xi \in B\}$.

We define our conditions $\mathbb{P}$ as follows:

$p \in \mathbb{P}$ iff $p = \langle \dot{p}, |p| \rangle$ and

(i) $\mathrm{card}(\dot{p}), \mathrm{card}(|p|) < \kappa$

(ii) $|p| \subseteq \kappa$, $\dot{p} \subseteq \tilde{B} \times \kappa$

(iii) $\langle b,\eta \rangle \in \dot{p} \to |p| \cap (S(b) \setminus \eta) = \emptyset$

We set $q \le p$ iff $\dot{q} \supseteq \dot{p}$ and $|q| \supseteq |p|$.

$\mathbb{P}$ satisfies the κ^+-A.C., since if $|q| = |p|$ then $r \le p,q$ where $r = \langle \dot{p} \cup \dot{q}, |p| \rangle$. Hence $p/q \to |p| \ne |q|$, but there are only κ many $|p|$ $(2^{\kappa} = \kappa)$. $\mathbb{P}$ is also τ-closed for $\tau < \kappa$, since if $p_0 \ge p_1 \ge \ldots \ge p \ge \ldots$ $(r < \tau)$ then $p \le p_r$ $(r < \tau)$, where $p = \langle \bigcup_i \dot{p}_i, \bigcup_i |p_i| \rangle$. So forcing with $\mathbb{P}$ preserves cardinalities and cofinalities. Now let G by $\mathbb{P}$-generic and set: $D = \bigcup_{P \in G} |p|$.

<u>Claim</u> $B \in L[A,D]$.

By genericity $\xi \in B$ iff $S(b_\xi) \cap D$ is bounded in κ . We define $B \cap \delta$ by induction on $\delta \in [\kappa,\kappa^+)$ in $L[A,D]$ as follows

$B \cap \kappa = 0$

Let $\delta = \xi + 1$; $B \cap \xi$ is already defined. Define μ_ξ , $\mathfrak{N}_\xi$ and b_ξ as before from $B \cap \xi$ $\langle \mu_\eta : \eta < \xi \rangle$ and A

$$B \cap \xi + 1 = B \cap \delta = \begin{cases} (B \cap \xi) \cup \{\xi\} & \text{if } D \cap S(b_\xi) \text{ is bounded in } \kappa \\ B \cap \xi & \text{otherwise} \end{cases}$$

For $\lim(\lambda)$ set: $B \cap \lambda = \bigcup_{\xi < \lambda} B_\xi$.

Clearly this definition can be carried out in $L[A,D]$; hence $B = \bigcup_{\xi \in [\kappa,\kappa^+)} B \cap \xi \in L[A,D]$. So the final D' that codes B will be satisfied by a D' that recursively codes A and D .

Q.E.D. (Theorem 1.1).

As we mentioned in 1.1, we will have recourse to code $s \in \mathbb{S}_\kappa$ (where s is a bounded subset in $[\kappa,\kappa^+)$) and in the case where κ is a successor cardinal we will use almost disjoint forcing. We need a variation of almost disjoint forcing for bounded subsets of $[\kappa,\kappa^+)$ that has the following continuity property: Let $s,r \in \mathbb{S}_\kappa$ and $s = r \cap \xi$. If $D \subseteq \kappa$ is a generic that codes r , then D is also a generic that codes s . For the case $\kappa = \aleph_0$, Solovay devised such a variation by redefining the b_ξ ($\xi \in [\aleph_0,\aleph_1)$) as follows:

b_ξ = the $L[A,B\cap\xi]$-least $b \subseteq \omega$ s.t. b is Cohen generic over $\mathfrak{A}_\xi$.

In order to generalize the above for $\kappa > \omega$ we need:

<u>Assumption</u> There is a sequence $\langle b_\xi : \kappa \leq \xi < \kappa^+ \rangle$ s.t.

(i) $b_\xi \subseteq \kappa$.

(ii) Let $\xi \leq \xi_0 < \ldots < \xi_\nu < \ldots$ $(\nu < \delta < \kappa)$. Then $\langle b_{\xi_\nu} : \nu < \delta \rangle$ is Cohen generic over $\mathfrak{A}_\xi$.

(iii) $b \in L_{\mu_{\xi+1}}[A,B\cap\xi]$ and is uniformly $\langle L_{\mu_{\xi+1}}[A,B\cap\xi]\ A,B\cap\xi \rangle$ - definable.

We shall be able to construct such a sequence and for the moment we assume that such a sequence exists. Given a $B \subseteq [\kappa,\kappa^+)$, we wish to code $B \cap \xi$ (for $\xi \leq \kappa^+$) with the above continuity property. Set $\tilde{B}_\xi = \{b_\nu : \nu \in B\cap\xi\}$, where the b_ν are those mentioned in the above assumption.

Define $\mathbb{P}_\xi$ as follows:

$p \in \mathbb{P}_\xi$ iff $p = \langle \dot{p}, |p| \rangle$ and

(i) $\mathrm{card}(\dot{p})$, $\mathrm{card}(|p|) < \kappa$.

(ii) $|p| \leq \kappa$, $\dot{p} \subseteq (P_{<\kappa}(\kappa) \cup \tilde{B}_\xi) \times \kappa$

(iii) $\langle b,\eta \rangle \in \dot{p} \rightarrow |p| \cap (S(b) \setminus \eta) = \emptyset$.

The addition of $P_{<\kappa}(\kappa)$ is to facilitate obtaining the continuity property. Note that $\mathbb{P}_\xi \in \mathfrak{A}_\xi$ and $\mathbb{P}_\xi = \mathbb{P}_{\kappa^+} \cap \mathfrak{A}_\xi$. We can then carry out the proof of the theorem, noting $b_\delta \in L_{\mu_{\delta+1}}[A, B\cap\xi]$ (*not* $B \cap (\delta+1)$) so the inductive definition of $B \cap \xi$ can be carried out. Also note that to define $\langle b_\nu : \nu \le \xi\rangle$ it will be sufficient to have $\forall\delta \le \xi \;\; L[B\cap\delta] \models \text{card}(\delta) \le \kappa$.

The following lemma will show that if $D \subseteq \kappa$ is $\mathbb{P}_{\kappa^+}$-generic then D is $\mathbb{P}_\xi$-generic over $\mathfrak{A}_\xi$.

Lemma 1.3 *Let* $\mathbb{P}_{\kappa^+}$ *and* $\mathbb{P}_\xi$ *be as above. If* $\Delta \subseteq \mathbb{P}_\xi$ *, and* Δ *is predense in* $\mathbb{P}_\xi$ *then* Δ *is predense in* $\mathbb{P}_{\kappa^+}$.

Proof (Notation as in Theorem 1.2 and set $\mathbb{P}_{\kappa^+} = \mathbb{P}$.) Suppose not. Then there is a $p \in \mathbb{P}$ incompatible with every member of Δ (i.e. no $q \le p$ meets Δ). Let $\dot{p} = v \cup u$ where $v \subseteq (P_{<\kappa}(\kappa) \cup \tilde{B}_\xi) \times \kappa$ and $u \subseteq \{b_{\xi_1}, \ldots, b_{\xi_\nu}, \ldots\}$ for $\nu < \delta < \kappa$ and $\xi \le \xi_1 < \ldots < \xi_\nu < \ldots$. Note that $\langle v, |p|\rangle \in \mathbb{P}_\xi$. By the above assumption it follows that $\langle b_{\xi_\nu} : \nu < \delta\rangle$ is Cohen generic over $\mathfrak{A}_\xi$. Let $\mathbb{Q}$ be a set of Cohen conditions for adding such a sequence. Let $\mathring{b}_\nu$ $(\nu < \delta)$ denote the ν-th subset out of the δ subsets of κ added in the forcing language for $\mathbb{Q}$.

Since no $q \le p$ meets Δ , there must be an $r \in \mathbf{Q}$ s.t.

$$r \Vdash_{\mathbb{Q}} \forall q \in \check{P}(q \le \langle \check{v} \cup \mathring{u}, |\check{p}|\rangle \rightarrow q \text{ does not meet } \check{\Delta}) ,$$

where $\mathring{u}$ is given by

$$\mathring{u} = \bigcup_{\nu<\delta} \{\langle \mathring{b}_\nu, \check{\eta}\rangle : \langle \check{\nu}, \check{\eta}\rangle \in \check{\bar{u}}\}, \quad \bar{u} = \{\langle \nu, \eta\rangle : \langle b_{\xi_\nu}, \eta\rangle \in u\} .$$

Let $\gamma = \max(\sup\{\eta : \exists\, \nu<\delta(\langle\nu,\eta\rangle \in \bar{u})\}, \sup\{\eta : \exists\, \nu<\delta(r \Vdash \check{\eta} \in \mathring{b}_\nu)\})$.

Then we can assume w.l.o.g. that $r \Vdash \mathring{b}_\nu \cap \check{\gamma} = (b_{\xi_\nu} \cap \gamma)\check{}$ and that r does not decide $\check{\beta} \in \mathring{b}_\nu$ for $\beta \ge \gamma$ and $\nu < \delta$.

Set: $u^* = \{\langle b_{\xi_\nu} \cap \gamma, \eta\rangle : \langle \nu, \eta\rangle \in \bar{u}\}$.
$b_{\xi_\nu} \cap \gamma \in P_{<\kappa}(\kappa)$ so $p^* = \langle v \cup u^*, |p|\rangle \in \mathbb{P}_\xi$. Δ is predense in $\mathbb{P}_\xi$ so there is a $q \in \mathbb{P}_\xi$ s.t. $q \leq p^*$ and q meets Δ . Let $\beta > \gamma$ be s.t. $\beta > \sup(|q|)$. Clearly there is an $s \in \mathbb{Q}$, $s \leq r$ s.t. $s \Vdash \check{j} \notin \mathring{b}_r$ for $\gamma \leq j < \beta$. But then $s \Vdash \mathring{b}_\nu \cap \check{\beta} = \widecheck{b_{\xi_\nu} \cap \gamma}$ and in particular $s \Vdash |\check{q}| \cap S(\mathring{b}_\nu) = |\check{q}| \cap S(\widecheck{b_{\xi_\nu} \cap \gamma})$. We wish to show that $s \Vdash \langle \mathring{u}, |\check{q}|\rangle \in \check{P}$. Let $\langle \mathring{b}_\nu, \check{\eta}\rangle$ be a constraint in $\mathring{u}$. Since $q \leq p^*$ and $q \in \mathbb{P}_\xi$ we have that $q \cap (S(b_{\xi_\nu} \cap \gamma)\setminus\eta) = \emptyset$ where we interpret $S(b_{\xi_\nu} \cap \gamma)$ to be a sequence number of $b_{\xi_\nu} \cap \gamma$ as a subset of κ . But $s \Vdash \mathring{b}_\nu \cap [\gamma, \beta) = \emptyset$, so $\mathring{b}_\nu$ is forced to act like $b_{\xi_\nu} \cap \gamma$ up to β . Then $s \Vdash \langle \check{q} \cup \mathring{u}, |\check{q}|\rangle \leq q$, hence $s \Vdash \langle q \cup \mathring{u}, |q|\rangle \leq \langle v \cup \mathring{u}, |p|\rangle$ and $s \Vdash \langle \check{q} \cup \mathring{u}, |\check{q}|\rangle$ meets Δ . Thus we have reached a contradiction.

Q.E.D.

1.3 RESHAPING

Theorem 1.1 almost solves the regular case of the problem in §1.1; the only thing missing is that Theorem 1.1 has the assumption

$$\forall \delta < \kappa^+ \; L[B \cap \delta] \models \mathrm{card}(\delta) \leq \kappa .$$

Taking care of this does not involve coding, but only 'reshaping' of B .

Theorem 1.4 *Let* γ *be an infinite cardinal and let* $B \subseteq \gamma^+$ s.t. $V = L[B]$. *Then we can generically add a* $D \subseteq \gamma$ s.t.

(i) $B \in L[D]$.

(ii) $\forall \delta < \gamma^+ \; L[D \cap \delta] \models \mathrm{card}(\delta) \leq \gamma$.

(iii) $L[D]$ *preserves cardinals and cofinalities.*

Proof We define our conditions, $\mathbb{P}$, as the set of $p: |p| \to 2$ s.t. $|p| \in On$, $|p| < \gamma^+$ and $\forall \delta \leq |p| \ (L[B \cap \delta, p \upharpoonright \delta] \models card(\delta) \leq \gamma)$. $p \leq q$ iff $p \supseteq q$. We call $|p|$ the height of p. We have to show that any $p \in \mathbb{P}$ can be extended to any arbitrary height, i.e. if $|p| = \eta$ and $\delta \in (\eta, \gamma^+)$ then there is a $q \leq p$ s.t. $|q| \geq \delta$. If $\delta \leq \eta + \gamma$ then any extension of p to a function q where $dom(q) = \delta$ will do since $L[B \cap \eta, p] \models card(\delta) \leq \gamma$. If $\delta > \eta + \gamma$ we can code (in $L[B]$) a well-ordering of δ of order type γ by a $d \subseteq [\eta, \eta+\gamma)]$. Any extension, q, of $p \cup \chi_d$ s.t. $dom(q) = \delta$ will do since

$$\forall \xi (\eta+\gamma < \xi \leq \delta \longrightarrow L[B \cap \eta, p \cup \chi_d] \models card(\xi) \leq \gamma)$$

(note that $card(\xi + \omega)^L \leq \xi$ for discussion after the theorem).

We now prove γ-distributivity. Let Δ_i be open dense in $\mathbb{P}$ for $i < \gamma$. We must show that $\bigcap_i \Delta_i$ is dense in $\mathbb{P}$. Let $p \in \mathbb{P}$, we must find a $q \leq p$ s.t. $q \in \bigcap_i \Delta_i$. Set $b = \langle L_{\gamma^{++}}[B], B \rangle$ and construct a sequence $X_i \prec b$ $(i \leq \gamma)$ as follows:

X_0 = the smallest $X \prec b$ s.t. $\gamma \cup \{p, \langle \Delta_i : i < \gamma \rangle\} \subseteq X$,

X_{i+1} = the smallest $X \prec b$ s.t. $X_i \cup \{X_i\} \subseteq X$,

$X_\lambda = \bigcup_{i<\lambda} X_i$ for $\lim(\lambda)$.

Here we are using $V = L[B]$, because $card(P) = \gamma^+$ and any $\Delta \subseteq \mathbb{P}$ appears before $L_{\gamma^{++}}[B]$. Set $\sigma_i : b_i \cong X_i$ the unique transitive collapse. Then $b_i = \langle L_{\delta_i}[B \cap \alpha_i], B \cap \alpha_i \rangle$, where $\alpha_i = X_i \cap \gamma^+ = \sigma_i^{-1}(\gamma^+)$. Note that $X_i \cap \gamma^+$ is transitive, because $\gamma \in X_i$ and $X_i \models \forall \delta < \gamma^+ (\exists f : \gamma \xrightarrow{\text{Onto}} \delta)$. It is readily seen that:

(1) $\sigma_i^{-1}(\mathbb{P}) = \mathbb{P} \cap L_{\alpha_i}[B]$,

(2) $\sigma_i^{-1}(\langle \Delta_j : j < \gamma \rangle) = \langle \Delta_j \cap L_{\alpha_i}[B] : j < \gamma \rangle$,

(3) $\langle \alpha_j : j < \gamma \rangle$ is definable from $b, p \ \langle \Delta_j : j < \gamma \rangle$.

Applying σ_i^{-1} we get that $\langle\alpha_j: j < i\rangle$ is definable from $b_i, p, \sigma_i^{-1}(\langle\Delta_j: j < \gamma\rangle)$ by the same definition since $\sigma_i \restriction \alpha_i = \mathrm{id} \restriction \alpha_i$ and up to α_i everything is the same.

Now define a sequence $p_i \in \mathbb{P}$ by:

$$p_0 = p$$

$$p_{i+1} = \text{the } L_{\gamma^+}[B]\text{-least } p' \leq p_i \text{ s.t. } |p'| \geq \alpha_i \text{ and } p' \in \Delta_i$$

$$p_\lambda = \begin{cases} \bigcup_{i<\lambda} p_i & \text{if } \bigcup_{i<\lambda} p_i \in \mathbb{P} \\ \text{undefined otherwise} \end{cases}$$

<u>Claim</u> p_i is defined for $i \leq \gamma$.

Proof (of claim) By induction on i simultaneously showing: $p_i \in b_{i+1}$ for $i < \gamma$ (hence $|p_i| < \alpha_{i+1}$). The successor case is easy because we have already shown that a condition can be extended to an arbitrary height. Now let $i = \lambda$ be a limit ordinal. Set $q = \bigcup_{i<\lambda} p_i$. We must show $q \in \mathbb{P}$ and $q \in b_{\lambda+1}$. Since $\alpha_i \leq |p_i| < \alpha_{i+1}$ for $i < \lambda$ we have $\mathrm{dom}(q) = [\gamma, \alpha_\lambda)$. For $\gamma < \alpha_\lambda$ we have by the induction hypothesis

$$\delta < \alpha_i < \alpha_\lambda, \quad q \restriction \alpha_i = p \restriction \alpha_i \text{ and } L[B \cap \delta, q \restriction \delta] \models \mathrm{card}(\delta) \leq \gamma .$$

So the only problem is $L[B \cap \alpha_\lambda, q] \models \mathrm{card}(\alpha_\lambda) \leq \gamma$ and $q \in b_{\lambda+1}$. Set $\sigma = \sigma_\lambda$; by the method of (3) above $\langle p_i: i < \lambda\rangle$ is definable from b_λ, p and $\sigma^{-1}(\langle\Delta_j: j < \gamma\rangle)$ as it was defined from $b, p\ \langle\Delta_j: j < \gamma\rangle$. Then $\langle p_i: i < \lambda\rangle \in b_{\lambda+1}$, since $b_\lambda \in b_{\lambda+1}$; hence $q \in b_{\lambda+1}$. We have $L_{\delta_\lambda}[B \cap \alpha_\lambda] \models \forall \delta < \alpha_\lambda\ (\mathrm{card}(\delta) \leq \gamma)$ since $\sigma(\alpha_\lambda) = \gamma^+$. By (3) above we have that $\langle\alpha_i: i < \lambda\rangle$ is definable in $L_{\delta_\lambda}[B \cap \alpha_\lambda]$ from parameters, so $\langle\alpha_i: i < \lambda\rangle \in L_{\delta_\lambda + \delta_\lambda}[B \cap \alpha_\lambda]$ and $L_{\delta_\lambda+\delta_\lambda}[B \cap \alpha_\lambda] \models \mathrm{card}(\alpha_\lambda) \leq \gamma$ since $\alpha_\lambda = \sup_{i<\lambda} \alpha_i \leq \lambda.\gamma = \gamma (\lambda < \gamma)$.

We have in fact shown that $L[B \cap \alpha_\lambda] \models \mathrm{card}(\alpha_\lambda) \leq \gamma$

(without the help of q) and it is important to note that this is true by $L_{\delta_\lambda+\delta_\lambda}[B\cap\alpha_\lambda]$ for the discussion after the theorem.

Q.E.D. (claim)

Then $p_\gamma \le p$ and $p_\gamma \in \bigcap_{i\in\gamma} \Delta_i$ so $\mathbb{P}$ is γ-distributive. Together with γ^{++}-A.C. this shows that $\mathbb{P}$ preserves cardinalities and cofinalities. Let G be $\mathbb{P}$-generic and set:

$$D' = \{\xi: (\exists p \in G)\ p(\xi) = 1\} .$$

Then $(\forall\delta < \gamma^+)(L[B\cap\delta_\lambda D'\cap\delta] \models \text{card}(\delta) \le \gamma)$ and the D that satisfies the theorem is one which recursively codes B and D' .

Q.E.D.

When we actually use the reshaping conditions we will demand more from them. If $\neg 0^\#$ and $\gamma \ge \omega_2$ we may impose on the conditions of the above theorem the additional requirement that, if $\delta \le |p|$ then $L_\mu[B\cap\delta, p\restriction\delta] \models \text{card}(\delta) \le \gamma$ for a μ s.t. $\text{card}(\mu)^L \le \delta$. The proof still goes through noting the following:

(1) When we showed that $p \in \mathbb{P}$ can be arbitrarily extended to any height, we already showed that the above constraint causes no problem.

(2) To show that the γ-distributivity goes through the only problem is showing that $p_\lambda \in \mathbb{P}$ for $\lim(\lambda)$ and the only problem there is showing $L_\mu[B\cap\alpha_\lambda, p_\lambda] \models \text{card}(\alpha_\lambda) \le \gamma$ for $\text{card}(\mu)^L \le \alpha_\lambda$. We have already noted above that $L_{\delta_\lambda+\delta_\lambda}[B\cap\alpha_\lambda] \models \text{card}(\alpha_\lambda) \le \gamma$. Hence it is sufficient to show that $\text{card}(\delta_\lambda+\delta_\lambda)^L = \text{card}(\delta_\lambda)^L \le \alpha_\lambda$. Suppose not. Then some $\eta \in [\alpha_\lambda, \delta_\lambda]$ is regular in L since $\alpha_\lambda < \text{card}(\delta_\lambda)^L$. Then by Marganalia [DJ] $\text{cf}(\delta) = \text{card}(\delta) = \gamma > \omega$. (Here we use $\neg 0^\#$ and $\gamma \ge \omega_2$.) Let $\delta = \sup(\sigma_\lambda{}''\delta)$ and $\sigma' = \sigma_\lambda \restriction L_\delta$. Then it is easy to see that $\sigma': L_\delta \xrightarrow[\Sigma_1]{} L_{\delta'}$ cofinally. δ is regular in L , $\text{cf}(\delta) > \omega$

and $\sigma' \neq id$, hence by Marginalia $0^{\#}$ exists. Thus we have reached a contradiction.

We will have to use the reshaping conditions, $\mathbb{P}$, when there is a $X \subseteq [\gamma,\gamma^{+})$ s.t. $p \in \mathbb{P}$ is restricted to $dom(p) \subseteq [\alpha,|p|) \cap X$. An inspection of the proof yields that if $\forall \xi \in [\gamma,\gamma^{+}) \exists \eta \in [\gamma,\gamma^{+})$ ($\eta > \xi$ and $L \models card(\eta) \leq \xi$ and $o.t.(X \cap [\xi,\eta)) \leq \gamma$), the theorem goes through.

Definition: The Gödel Pairing Function. A well-ordering on $On \times On$ is defined as follows:

$$(\alpha_1,\alpha_2) \prec (\beta_1,\beta_2) \text{ iff } [\max(\alpha_1,\alpha_2) < \max(\beta_1,\beta_2) \text{ or}$$
$$[\max(\alpha_1,\alpha_2) = \max(\beta_1,\beta_2) \text{ and } \alpha_1 < \beta_1] \text{ or}$$
$$[\max(\alpha_1,\alpha_2) = \max(\beta_1,\beta_2) \text{ and } \alpha_1 = \beta_1 \text{ and } \alpha_2 < \beta_2]] .$$

Set $ot\langle(\alpha,\beta),\prec\rangle = \{\alpha,\beta\} \in On$ and $Z_\alpha = \{\{\xi,\alpha\} : \xi \in On\}$. It is easy to see that Z_α satisfies the above condition.

1.4 LIMIT CARDINALS

The problem in §1.1 is now solved for regular cardinals, but we will use almost disjoint forcing only for successor cardinals, for all limit cardinals we will use the following method.

Theorem 1.5 *Assume G.C.H. Let* β *be a limit cardinal. Let* $B \subseteq \beta^{+}$ s.t. $V = L[B]$ *and* $\forall \delta < \beta^{+}\ L[B \cap \delta] \models card(\delta) \leq \beta$. *Then we can generically add a* $D \subseteq \beta$ *that codes* B.

Proof We may assume w.l.o.g. that $B \subseteq [\beta,\beta^{+})$. Let $A \subseteq \beta$ s.t. $H_\tau = L_\tau[A]$ for all infinite cardinals τ. (We can do this because of the G.C.H.) Set

$$I = \{\tau < \beta : \tau \text{ is an infinite cardinal}\}$$

The coding will use a variation of the almost disjoint method. Our conditions will be functions $f: \beta \longrightarrow 2$ s.t. for $\tau \in I$,

$$\mathrm{card}(\mathrm{dom}(p) \cap [\tau,\tau^+)) \le \tau .$$

For each $\xi \in [\beta,\beta^+)$ we will define functions $\rho_\xi: I \longrightarrow \beta$ s.t. $\forall \tau \in I(\rho_\xi(\tau) \in (\tau,\tau^+))$ (they are analogous to the b_ξ of Theorem 1.1). A condition p will decide $\xi \in B$ or $\xi \notin B$ if an end segment of $\langle \rho_\xi(\tau): \tau \in I\rangle$ is in $\mathrm{dom}(p)$ by:

$$p \text{ decides } \xi \in B \text{ iff } \exists \tau \in I\, \forall \xi \in I(\nu \ge \tau \longrightarrow p(\rho_\xi(\nu)) = 1) .$$

While defining the conditions and ρ_ξ we have to take care that

1) ρ_ξ and ρ_δ (for $\xi \ne \delta$) must be almost disjoint or differ from some point on.

2) A condition p does not decide too much.

With this in mind we define μ_ξ and $\mathfrak{N}_\xi$ exactly as in Theorem 1.1. Set $X_{\xi,\tau}$ = the smallest $X \prec \mathfrak{N}_\xi$ s.t. $\tau + 1 \subseteq X$ and $\Pi_{\xi\tau}: \mathfrak{N}_{\xi\tau} \cong X_{\xi\tau}$. Let $\rho_\xi(\tau)$ be the L[A]-code of $\mathfrak{N}_{\xi\tau}$. Note that $\mathrm{code}(x) \ge \mathrm{rank}(x)$ and $\forall \tau \in I(\rho_\xi(\tau) \in (\tau,\tau^+))$ since $H_{\tau^+} = L_{\tau^+}[A]$.

We define our conditions $\mathbb{P}$ to be the set of $p: \mathrm{dom}(p) \to 2$ s.t.

a) $\mathrm{dom}(p) \subseteq [\omega,\beta)$

b) $\tau \in I \longrightarrow \mathrm{card}(\mathrm{dom}(p) \cap [\tau,\tau^+)) \le \tau$

Set $|p|$ = the least $\xi \in [\beta,\beta^+)$ s.t. $p \in \mathfrak{N}_\xi$

c) If $\xi < |p|$, then there is a $\tau \in I$ s.t. for all $\nu \in I$,

$$p(\rho_\xi(\nu)) = \begin{cases} 1 & \text{if } \xi \in B \\ 0 & \text{if not.} \end{cases}$$

Finally set $p \leq q$ iff $p \supseteq q$.

Remarks

1) (c) says that for $\xi < |p|$, p decides if $\xi \in B$ or not.

2) Let $p \in \mathbb{P}$. We shall show that if $\xi \geq |p|$ then p does not decide $\xi \in B$, i.e. there is a $\tau \in I$ s.t. $\forall \nu \in I$ $(\nu \geq \tau \longrightarrow \rho_\xi(\nu) \not\in \mathrm{dom}(p))$.

Proof By Lemma 1.3 $\bigcup_{\nu \in I} X_{\xi\nu} = \mathfrak{A}_\xi$. Choose a τ s.t. $p \in X_{\xi\tau}$ (note that $p \in \mathfrak{A}_\xi$). For $r \in I$ we have $\nu \in X_{\xi\nu}$ and hence $\nu^+ \in X_{\xi\nu}$. In addition $X_{\xi\nu} \cap \nu^+$ is transitive since $\mathfrak{A}_\xi \models \forall \eta < \gamma^+ (\exists f: \nu \xrightarrow{\text{Onto}} \eta)$. So for $\nu \geq \tau$, $\Pi_{\xi\nu}^{-1} \upharpoonright \nu^+ = \mathrm{id} \upharpoonright \nu^+$, $p \upharpoonright \nu^+ \in X_{\xi\nu}$, $\Pi_{\xi\nu}^{-1}(p \upharpoonright \nu^+) = p \upharpoonright \nu^+ \subseteq \mathfrak{A}_{\xi\nu}$ and $\Pi_{\xi\nu}^{-1}(p \upharpoonright \nu^+) = p \upharpoonright \nu^+ \in \mathfrak{A}_{\xi\nu}$. Hence $\mathrm{dom}(p \upharpoonright \nu^+) \subseteq \mathfrak{A}_{\xi\nu}$. But $\rho_\xi(\nu) \not\in \mathfrak{A}_{\xi\nu}$ because as mentioned above $\rho_\xi(\nu) \geq \mathrm{rank}(\mathfrak{A}_{\xi\nu})$ (or we cannot define the truth of $\mathfrak{A}_{\xi\nu}$ inside $\mathfrak{A}_{\xi\nu}$).

We define $\mathbb{P}_\tau$ $(\tau \in I)$ exactly as $\mathbb{P}$ using τ in the place of ω (then $\mathbb{P} = \mathbb{P}_\omega$). We let $\mathbb{P}^\tau$ $(\tau \in I)$ be the set $p: \mathrm{dom}(p) \longrightarrow 2$ s.t.

a) $\mathrm{dom}(p) \subseteq [\omega, \tau)$

b) $\nu \in I \longrightarrow \mathrm{card}(\mathrm{dom}(p) \cap [\nu, \nu^+)) \leq \nu$

For $p \in \mathbb{P}$, $\tau \in I$ set

$$(p)_\tau = p \upharpoonright [\tau, \beta) \quad \text{and} \quad (p)^\tau = p \upharpoonright [\omega, \tau)$$

Then $\mathbb{P}_\tau = \{(p)_\tau : p \in \mathbb{P}\}$, $\mathbb{P}^\tau = \{(p)^\tau : p \in \mathbb{P}\}$ and $\mathbb{P} = \mathbb{P}_\tau \times \mathbb{P}^\tau$.

Lemma 1.6 *Let* $p \in \mathbb{P}$, $\xi \geq |p|$. *Then* $\exists q \leq p(|q| = \xi)$.

Proof By induction on ξ . For $\xi = |p|$ there is nothing to prove. Now let $\xi = \gamma + 1$. We may assume w.l.o.g. that $|p| = \gamma$. Pick a $\eta \in I \setminus \tau$ s.t. $\nu \geq \eta \longrightarrow \rho_\gamma(\nu) \notin \mathrm{dom}(p)$. Set

$$q = p \cup \{\langle \rho_\gamma(\nu), \chi_B(\gamma) \rangle : \nu \in I \setminus \eta\} .$$

We carry out this definition in $\mathfrak{A}_\xi$; so $q \in \mathfrak{A}_\xi$, $q \subseteq p$, and $|q| = \xi$.

Now let $\lim(\xi)$, $\xi > |p|$. We know that $\Box_\beta$ holds since

$$\forall \delta < \beta^+ (L[B \cap \delta] \models \mathrm{card}(\delta) \leq \beta) \quad \text{(see Chapter 6)}$$

The proof of $\Box_\beta$ shows that there is a cub $C \subseteq [\beta, \xi)$ s.t.

1) $C \in \mathfrak{A}_\xi$.

2) $\mathrm{o\,t}(C) \leq \beta$.

3) If r_γ is a limit point of C then $C \cap \gamma \in \mathfrak{A}_\gamma$.

Let $\mathrm{o\,t}(C) = \epsilon$. Let $\langle \gamma_i : i < \epsilon \rangle$ be a monotone enumeration of C . We may assume w.l.o.g. that $\gamma_{i_0} = |p|$. Let $\langle \eta_i : i \leq \epsilon \rangle$ be a canonical normal function s.t. $\eta_i \in I \setminus \tau$. Define p_i by induction on $i < \epsilon$ as follows:

$p_0 = p$

$p_{i+1} = p \cup q$ where $q =$ the $\mathfrak{A}_{\gamma_{i+1}}$-least q s.t. $q \leq (p_i)_{\eta_{i+1}}$ in $\mathbb{P}_{\eta_{i+1}}$ and $|q| = \gamma_{i+1}$.

$p_\lambda = \bigcup_{i<\lambda} p_i$ for $\lim(\lambda)$.

We have to show that $p_\lambda \in \mathbb{P}_\tau$ for $\lim(\lambda)$. Let $\gamma \in I$, we must show that $\mathrm{card}(\mathrm{dom}(p_\lambda) \cap [\gamma, \gamma^+)) \leq \gamma$. p_i contributes to $\mathrm{dom}(p_\lambda) \cap [\gamma, \gamma^+)$ only if $\eta_i \leq \gamma$ but $\mathrm{card}(i) \leq \eta_i \leq \gamma$ so only $\leq \gamma$ of the p_i contribute.

$$\mathrm{card}(\mathrm{dom}(p) \cap [\gamma, \gamma^+)) = \mathrm{card}(\mathrm{dom}(\bigcup_i p_i) \cap [\gamma, \gamma^+)) \leq \gamma . \gamma = \gamma .$$

We also have to show that $p_\lambda \in \mathfrak{A}_{\gamma_\lambda}$. $\sup_{i<\lambda} \gamma_i = \gamma_\lambda$ is a limit point of C, so $C \cap \gamma_\lambda \in \mathfrak{A}_{\gamma_\lambda}$ and $p_\lambda \in \mathfrak{A}_{\gamma_\lambda}$ and $|p_\lambda| = \gamma_\lambda$, since $\langle p_i : i < \lambda\rangle$ is definable from $C \cap \gamma_\lambda$. The same is true for p_ϵ, so $p_\epsilon \le p$ and $|p_\epsilon| = \xi$.

Q.E.D.

Lemma 1.7 $\mathbb{P}_\tau$ *is* τ*-distributive for* $\tau \in I$.

Proof We rather closely imitate the proof of the corresponding assertion of Theorem 1.4.

Let Δ_i be open dense in $\mathbb{P}_\tau$ for $i < \tau$. We have to show that $\bigcap_i \Delta_i$ is dense in $\mathbb{P}_\tau$. Let $p \in \mathbb{P}_\tau$ and set:

$$b = \langle L_{\beta^{++}}[A,B],A,B\rangle .$$

Construct $X_i \prec b \ (i \le \tau)$ as follows:

X_0 = the smallest $X \prec b$ s.t. $B \cup \{p, \langle \Delta_j : j < i\rangle\} \subseteq X$.

X_{i+1} = the smallest $X \prec b$ s.t. $X_i \cup \{X_i\} \subseteq X$.

$X_\lambda = \bigcup_{i<\lambda} X_i$ for $\lim(\lambda)$.

Set: $\sigma_i : b \cong X_i$; then $b_i = \langle L_{\delta_i}[A, B \cap \alpha_i], A, B \cap \alpha_i\rangle$, where $\alpha_i = \beta^+ \cap X_i = \sigma_i^{-1}(\beta^+)$. The sequence $\langle \alpha_i \rangle$ is normal.

Remember $\mathfrak{A}_{\alpha_i} = \langle L\mu_{\alpha_i}[A, B \cap \alpha_i], A, B \cap \alpha_i\rangle$ and note $\delta_i < \mu_{\alpha_i}$ because α_i is regular in b_i but not in $\mathfrak{A}_{\alpha_i}$. So $b_i \in \mathfrak{A}_{\alpha_i}$. Now we define $p_i \in \mathbb{P}_\tau$ by

$p_0 = 0$

p_{i+1} = the $L_{\beta^+}[A,B]$-least $q \le p_i$ s.t. $|q| \ge \alpha_i$ and $q \in \Delta_i$.

$$p_\lambda = \begin{cases} p & \text{if } p = \bigcup_{i<\lambda} p_i \in \mathbb{P} \\ \text{undefined otherwise.} \end{cases}$$

We must show that p_i is defined for $i \le \tau$. The only problematic case is p_λ for $\lim(\lambda)$. Exactly as in the case of Theorem 1.4 we have: $\langle p_i : i < \lambda \rangle$ is definable from b_λ, p, $\langle \Delta_j \cap b_\lambda : j < \tau \rangle$ as it was definable from b, p, $\langle \Delta_j : j < \tau \rangle$. Hence $\langle p_i : i < \lambda \rangle \in \mathfrak{A}_{\alpha_\lambda}$, since $b_\lambda \in \mathfrak{A}_{\alpha_\lambda}$. Hence $q \in \mathfrak{A}_{\alpha_\lambda}$ where $q = \bigcup_{i<\lambda} p_i$.

We must show that $q \in \mathbb{P}_\tau$, q obviously satisfies a) and b) in the definition of $\mathbb{P}_\tau$, because $\lambda \le \tau$. If $\xi \in [\beta, \alpha_\lambda)$, then $\xi < |p_i|$ for some $i < \lambda$ and $\exists \eta \in I \, \forall \nu \in I \, (\nu \ge \eta \to q(\rho_\xi(\nu)) = \chi_B(\xi))$. Hence p_λ is defined, $p_\tau \le p$ and $p_\tau \in \bigcap_i \Delta_i$.

Q.E.D.

Lemma 1.8 *Let* Δ_ν *be open dense in* $\mathbb{P}_\nu$ *for* $\nu \in I \backslash \tau$. *Then* Δ *is dense in* $\mathbb{P}_\tau$, *where*

$$\Delta = \{p \in \mathbb{P}_\tau : \forall \nu \ge \tau \, ((p)_\nu \in \Delta_\nu)\} .$$

Proof The proof is similar to Lemma 1.7 and we only point out the difference.

Let $I \backslash \nu = \langle \eta_i : i < \delta \le \beta \rangle$. Construct $X_i \prec b$ $(i \le \delta)$ as before starting with X_0 = the smallest $X \prec b$ s.t. $B \cup \{p, \langle \Delta_{\eta_i} : i < \delta \rangle\} \subseteq X$. $p_{i+1} = p_i \cup q$, where q is the $L_{\beta^+}[A,B]$-least $q \le (p_i)_{\eta_i}$ s.t. $|q| \ge \alpha_i$ and $q \in \Delta_{\eta_i}$. Then one must show that p_i is defined for $i \le \delta$; clause b) going through as in Lemma 1.6.

Q.E.D.

Lemma 1.9 $\mathbb{P}$ *preserves cardinals and cofinalities.*

Proof It suffices to prove that $\mathbb{P}$ preserves cofinalities. We shall show that for every regular γ and cardinal α s.t. $\mathrm{cf}(\alpha) > \gamma$, forcing with $\mathbb{P}$ will not cause $\mathrm{cf}(\alpha)$ to become $\le \gamma$. Since $\mathrm{card}(\mathbb{P}) = \beta^+$ this holds for every $\gamma \ge \beta^+$. For $\gamma \le \beta$ we shall prove the preservation by induction.

Case 1 $\gamma < \beta$ and γ is a successor, i.e. $\gamma = \eta^+$. Then $card(\mathbb{P}^\gamma) = \gamma$ and hence satisfies the γ-A.C. We also have that $\mathbb{P}_\gamma$ satisfies γ-distributivity so we can complete this case by standard methods.

Case 2 $\gamma < \beta$ and γ is inaccessible. The proof of Case 2 is similar and simpler than Case 3, so we will return to Case 2 after Case 3.

Case 3 $\gamma = \beta$. Since we are assuming that γ is regular, we get $\gamma = \beta$ is inaccessible. Let $f: \beta \longrightarrow On$ be in a generic extension by $\mathbb{P}$, i.e. for some q in the generic filter and the term $\mathring{f}$ we have: $q \Vdash \mathring{f}: \check{\beta} \longrightarrow \check{On}$. We claim that there is an $X \in L[B]$ (the ground model) $card(X) = \beta$ and a $p \leq q$, p in the generic filter s.t. $p \Vdash rng(\mathring{f}) \subseteq \check{X}$. This clearly will show that if $cf(\alpha) > \beta$ then $cf(\alpha)$ will remain so in a generic extension. For $\nu < \beta$, $p \in \mathbb{P}_{\omega_{\nu+1}}$ set

$$\Delta_\nu^p = \{r \in \mathbb{P}^{\omega_{\nu+1}} \; r \cup p/q \; \vee \exists \delta (r \cup p \Vdash \mathring{f}(\check{\nu}) = \check{\delta})\} \text{ and}$$

$$\Delta_\nu = \{p \in \mathbb{P}_{\omega_{\nu+1}} : \Delta_\nu^p \text{ is dense in } \mathbb{P}^{\omega_{\nu+1}}\}$$

Claim Δ_ν is dense in $\mathbb{P}_{\omega_{\nu+1}}$.

Proof (of Claim) The facts that $card(\mathbb{P}^{\omega_{\nu+1}}) = \omega_{\nu+1}$, $\{p \in \mathbb{P}: p/q \text{ or } p \| \mathring{f}(\check{\nu})\}$ is open dense in $\mathbb{P}$ and $\mathbb{P}_{\omega_{\nu+1}}$ is $\omega_{\nu+1}$-distributive allows us to carry out the following.

Let $\mathbb{P}^{\omega_{\nu+1}} = \langle r_i : i < \omega_{\nu+1} \rangle$ and set

$$\Delta_i = \{p \in \mathbb{P}_{\omega_{\nu+1}} : r_i \cup p/q \; \vee \exists r \leq r_i \; (r \cup p \Vdash \mathring{f}(\check{\nu}))\} \text{ for } i < \omega_{\nu+1}$$

Δ_i is clearly dense in $\mathbb{P}_{\omega_{\nu+1}}$ and $\bigcap_{i<\omega_{\nu+1}} \Delta_i \subseteq \Delta_\nu$ so Δ_ν is dense in $\mathbb{P}_{\omega_{\nu+1}}$.

Q.E.D. (claim)

Now by Lemma 1.8 $\Delta = \{p \in \mathbb{P}: \forall \nu < \beta((p)_{\omega_{\nu+1}} \in \Delta_\nu)\}$ is dense in $\mathbb{P}$. Let p be in the generic filter and $p \in \Delta$. Then $p \leq q$. For $\gamma < \beta$, $r \in \mathbb{P}^{\omega_{\nu+1}}$ set

$$\xi_{\pi_\nu} = \text{that } \xi \text{ s.t. } r \cup (p)_{\omega_{\nu+1}} \Vdash \mathring{f}(\check{\nu}) = \check{\xi} \text{ if such a } \xi \text{ exists.}$$

Let $X = \{\xi_{\pi_\nu} : \gamma < \beta \text{ and } r \in \mathbb{P}^{\omega_{\nu+1}}$. Clearly $p \Vdash \text{rng}(\mathring{f}) \subseteq \check{X}$ and $\text{card}(X) = \beta$.

Q.E.D. (Case 3)

Proof (of Case 2) $\mathbb{P} \cong \mathbb{P}_\gamma \times \mathbb{P}^\gamma$ and $\mathbb{P}_\gamma$ is γ-distributive so it is sufficient to show that $\mathbb{P}^\gamma$ has the desired properties. Define $\mathbb{P}^\gamma_\tau$ as $\mathbb{P}^\gamma$ except that we replace ω by τ. Then $\mathbb{P}^\gamma = \mathbb{P}^\tau \times \mathbb{P}^\gamma_\tau$ and we can prove lemmas like 1.7 and 1.8 with respect to $\mathbb{P}^\gamma$ in place of $\mathbb{P}$ (here the proofs are trivial). Now proceed as in Case 3.

Q.E.D. (Lemma 1.9)

Now let G be a $\mathbb{P}$-generic and set

$$D' = \{r < \beta: \exists p \in G(p(r) = 1)\}.$$

Lemma 1.10 $B \in L[A,D']$.

Proof Exactly as in Theorem 1.1 using

$$\xi \in B \text{ iff } \exists \tau \in I\, \forall \eta \in I(\eta \geq \tau \longrightarrow \rho_\xi(\eta) \in D')$$

But then Theorem 1.5 holds with a D that recursively codes A and D'.

Q.E.D.

2 . The conditions

2.1 INTRODUCTION

We are now ready to begin the proof of the main theorem (Theorem 0.1). From now on we will work in a transitive model

$$\langle M,A\rangle \models ZF + GCH + \neg 0^{\#} , \quad \text{where } A \subseteq On .$$

We perform a preliminary forcing with the class of Easton conditions, $\mathbb{Q}$ for adding a Cohen generic subset $A'_\alpha \subseteq [\alpha+\alpha , \alpha^+)$ for every cardinal α . Set: $A^o_\alpha = \{\alpha+\nu: \nu \in A\cap\alpha\}$, $A^*_\alpha = A^o_\alpha \cup A'_\alpha$, $A' = \bigcup_{\alpha\in On} A^*_\alpha$ and $M' = M[A']$

$$\langle M',A'\rangle \models ZF + GCH + \neg 0^{\#} \text{ and}$$

1. A is definable in $\langle M',A'\rangle$.

2. $\langle M',A'\rangle$ = 'global axiom of choice' (this was shown by Easton [EA]).

3. $\langle M',A'\rangle \models$ 'there is a $\diamond$-sequence $\langle S_\alpha: \alpha < \beta^+\rangle$ for β^+ which is uniformly $\langle H_{\beta^+},A\cap\beta^+\rangle$-definable for β a cardinal ($\langle S_\alpha: \alpha < \beta^+\rangle$ is a $\diamond$-sequence for β^+ iff whenever $X \subseteq \beta^+$, the set $\{\alpha: X\cap\alpha = S_\alpha\}$ is stationary). (3. is needed for the proof of Lemma 2.5.)

4. $\langle M',A'\rangle$ also satisfies properties which we will discuss in Chapter 4 and which we need for some cardinal preservation theorems.

Levy [LE] has shown that if $\langle M',A'\rangle \models$ 'global choice' then there is a $B \subseteq On$ s.t. $\langle M',A'\rangle \models "V = L[B]"$. In addition, since we have GCH we may assume w.l.o.g. that $H_\alpha = L_\alpha[B]$ for every infinite cardinal α .

Let $\mathbb{P}$ be the conditions which we are defining in this chapter and do the coding. Since Theorem 0.1 is stated in terms of a class generic extension, we may take the conditions of Theorem 0.1 to be the pairs $\langle q,p\rangle$ s.t. $q \in \mathbb{Q}$ and $q \Vdash_{\mathbb{Q}} p \in \mathring{\mathbb{P}}$. Noting that A is definable from A' in $\langle M',A'\rangle$ we can assume w.l.o.g. that our original A has the following properties:

$\langle M,A\rangle \models ZF+GCH+\neg 0^{\#}+"V=L[A]"+"H_\alpha=L_\alpha[A]$ for every infinite cardinal α^{+}" + the properties 3. and 4. above.

When the ground model is L (i.e. $M = L$, $A = \emptyset$) we do not need the preliminary forcing since L already satisfies all of the above properties.

We define the class card by:

$\alpha \in$ card iff $\alpha = \emptyset$ or α is an infinite cardinal

and adopt the convention that $\aleph_0 = \emptyset^{+}$; thus in our terms $\aleph_0$ becomes a successor cardinal.

2.2 THE RESHAPING CONDITIONS $\mathbb{S}_\alpha$

We define conditions similar to those of Theorem 1.4.

Definition 1

For $\alpha \in$ card , we define a set of conditions $\mathbb{S}_\alpha$ as follows:

$p \in \mathbb{S}_\alpha$ iff $p: [\alpha,|p|) \longrightarrow 2$ where $|p| \in [\alpha,\alpha^{+})$ and

1) $\xi \in [\alpha,|p|] \longrightarrow L[A\cap\alpha, p\restriction\xi] \models "card(\xi) = \alpha"$.

2) If $\alpha \geq \omega_2$ then

$(\forall\xi \in [\alpha, |p|])(\exists\mu_\xi < \alpha^+)(L_{\mu_\xi}[A\cap\alpha, p/\xi] \models$ "$\mathrm{card}(\xi) = \alpha$" and $\mathrm{card}(\mu_\xi)^L \leq \xi)$.

Notation We denote by $\emptyset_\alpha$ the null function of $\mathbf{S}_\alpha$.

We are going to define μ_p^i similar to the μ_ξ^i of Chapter 1, except that μ_p^i will depend on $p \in \mathbf{S}_\alpha$; not just $|p|$.

Definition 2

Let $\alpha \geq \omega$, $\alpha \in$ card . By induction on $|p|$ $(p \in \mathbf{S}_\alpha)$ define μ_p^i $(i \leq \omega)$ as follows:

$\mu_p^0 = \sup(\alpha \cup \{\mu_{p/\xi}^\omega : \xi \in [\alpha, |p|)\})$

μ_p^{i+1} = the least $\mu > \mu_p^i$ s.t. setting $b = L_\mu[A\cap\alpha, p]$ we have

a) $b \models$ "$ZF^- + \mathrm{card}(|p|) = \alpha$"

b) $\exists\mu' < \mu$ s.t. $L_{\mu'}[A\cap\alpha, p] \models$ "$\mathrm{card}(p) = \alpha$" and
$L_\mu \models$ "$\mathrm{card}(\mu') = \mathrm{card}(|p|)$"

c) $\alpha \geq \omega_2 \rightarrow b \models$ "$\forall x \subseteq \alpha(\mathrm{card}(x) \geq \omega_2 \rightarrow \exists y \in L(x \subseteq y \wedge \mathrm{card}(x) = \mathrm{card}(y)))$"

d) $\mathrm{cf}(\mu) = \alpha$ if α is a successor cardinal

Remarks It is easily shown that μ_p^{i+1} exists and $\mu_p^{i+1} < \alpha^+$; just note that a) and b) can be satisfied since $p \in \mathbf{S}_\alpha$. c) can be satisfied since we assume $\neg 0^\#$ and hence, by the Marginalia, we have the covering lemma and d) can be trivially satisfied. b) will play an important role in guaranteeing that μ_p^1 is "big enough".
Finally define $\mu_p = \mu_p^\omega = \sup_{i<\omega} \mu_p^i$.

Notation For a fixed p , we shall often write μ_ξ^i in place of $\mu_{p/\xi}^i$ $(\xi \in [\alpha, |p|])$. In particular, we may write μ_α^i for $\mu_{\emptyset_\alpha}^i$.

Definition 3

$\mathfrak{A}^i_p = \langle L_{\mu^i_p}[A\cap\alpha,p^*],A\cap\alpha,p^*\rangle$ for $i \le \omega$ where $p^* = \{\mu^0_\xi : p(\xi)=1\}$.

Claim 2.1

For $i > 0$, $\mathfrak{A}^i_p$ is essentially the same as $\langle L_{\mu^i_p}[A\cap\alpha,p],A\cap\alpha,p\rangle$; i.e. $L_{\mu^i_p}[A\cap\alpha,p^*] = L_{\mu^i_p}[A\cap\alpha,p]$ and $p \in \mathfrak{A}^1_p$.

Proof We define $p(\xi)$ by induction for $\xi \in \mathrm{dom}(p)$

$$p(\alpha) = 1 \text{ iff } \alpha = \mu^0_\alpha \in p^*$$

μ^i_α $(i \le \omega)$ can be defined in $\mathfrak{A}^1_p$ and $\mu^0_{p\restriction\alpha+1} = \mu^\omega_\alpha$ (remember that $L_{\mu^i_p}[A\cap\alpha,p\restriction\alpha+1] \models ZF^-$ so such a construction can be carried out in $\mathfrak{A}^1_p$). Then

$$p(\alpha+1) = 1 \text{ iff } \mu^0_{p\restriction\alpha+1} \in p^*$$

Limit points are just as easy and assuming the case that $\lim(|p|)$ we can define $p = \bigcup_{\xi<|p|} p\restriction\xi$ in $\mathfrak{A}^1_p$ since $\mu^1_p > \mu^0_p \ge |p|$. Recovering p^* from p is similar.

Q.E.D. (Claim)

For $i = 0$ the above claim is not true and we used p^* in the definition of $\mathfrak{A}^1_p$ so that

$$\mathfrak{A}^0_p = \bigcup_{\xi<|p|} \mathfrak{A}^0_{p\restriction\xi} \quad \text{for } \lim(|p|)$$

Note that if $\mu^0_p > |p|$, then in general $L_{\mu^0_p}[A\cap\alpha,p] \ne \bigcup_{\xi<|p|} L_{\mu^0_p}[A\cap\alpha,p\restriction\xi]$ since we cannot use p as a parameter on the right hand side. This problem does not occur with p^* since $\sup(p^*) = \mu^0_p$. Summing up: $\mathfrak{A}^0_p = \bigcup_{\xi<|p|} \mathfrak{A}^0_\xi$ for $\lim(|p|)$ and

$$\mathfrak{A}^i_p = \langle L_{\mu^i_p}[A\cap\alpha, p], A\cap\alpha, p\rangle \quad \text{for} \quad i > 0$$

Notation $\mathfrak{A}_p = \mathfrak{A}^\omega_p$ and we often write $\mathfrak{A}^i_\xi$ for $\mathfrak{A}^i_{p/\xi}$ $(\xi \in [\alpha, |p|])$.

Claim 2.2

$\langle \mu^i_{p/\xi} \mid \xi < |s|$ or $(\xi = |s|$ and $i < n)\rangle$ is uniformly $\Sigma_1(\mathfrak{A}^n_s)$ in α for $s \in \mathbf{S}_\alpha$ and hence:

$\langle \mathfrak{A}^i_{s/\xi} \mid \xi < |s|$ or $(\xi = |s|$ and $i < n)\rangle$ is uniformly $\Sigma_1(\mathfrak{A}^n_s)$ in α for $s \in \mathbf{S}_\alpha$.

Proof By standard methods (cf. Jensen [J1]).

2.3 THE AUXILIARY LEMMAS

We now state three lemmas whose proofs require fine structure. The proofs appear in Chapters 6 and 7. The first lemma is a generalization of $\square$. Instead of indexing the Cs on $\beta \in [\alpha, \alpha^+)$ we index them on $s \in \mathbf{S}_\alpha$ and $C_s \subseteq [\alpha, \mu^0_s)$.

Lemma 2.3 There exists a sequence $\langle C_p : p \in \mathbf{S}_\alpha, |p| > \alpha\rangle$ s.t.

(i) $C_p \subseteq [\alpha, \mu^0_p)$ and C_p is cub in μ^0_p.

(ii) $\text{ot}(C_p) \leq \alpha$

(iii) If $\lambda \in (C_p)'$ then $\lambda = \mu^0_\xi$, where $C_{p/\xi} = \lambda \cap C_p$.

(iv) $C_p \in \mathfrak{A}^1_p$ is uniformly $\Sigma_1(\mathfrak{A}^1_p)$.

(v) Let α be a limit cardinal. Let $\bar{\alpha} \in \text{card} \cap \alpha$ and let

$$\pi : \langle \bar{\mathfrak{A}}, \bar{C}\rangle \xrightarrow[\Sigma_1]{} \langle \mathfrak{A}^0_p, C_p\rangle \quad \text{s.t.}$$

$\bar{\mathfrak{A}}$ is transitive, $\pi/\bar{\alpha} = \text{id}/\bar{\alpha}$, $\pi(\bar{\alpha}) = \alpha$.
Let $\bar{p} = (\pi^{-1})''(p)$, then $\bar{p} \in \mathbf{S}_{\bar{\alpha}}$, $\bar{\mathfrak{A}} = \mathfrak{A}^0_p$ and $\bar{C} = C_{\bar{p}}$.

The $\mathfrak{A}_s$, $s \in \mathbb{S}_\alpha$ will play a role similar to the L_{μ_ξ}, $\xi \in [\alpha,\alpha^+)$ of the building blocks. In particular it was not trivial to show in Theorem 1.5 that a condition can be extended. In this chapter it will be much more difficult to prove this. When we wish to extend a condition from height $|s/\xi|$ to $|s|$ where

$$\mathfrak{A}_{s/\xi} \models \text{"}\alpha \text{ is regular" and } \mathfrak{A}_s \models \alpha \text{ is singular},$$

"continuity" difficulties arise and the next lemma helps solve them.

Definition 1 Let α be a singular cardinal. $\mathbb{S}^*_\alpha = \{p \in \mathbb{S}_\alpha : (\mathfrak{A}_p \models (\alpha \text{ is singular})) \text{ and } (|p| = \alpha \text{ or } \mathfrak{A}^o_p \models \alpha$ is regular).

Lemma 2.4 There exist functions $p \longrightarrow D_p$, $p \longrightarrow \langle p_\beta : \beta \in D_p \rangle$ and $p \longrightarrow \langle \pi^p_\beta : \beta \in D_p \rangle$ defined on $\mathbb{S}^*_\alpha$ s.t.

(i) $D_p \subseteq \alpha$ is a closed set of singular cardinals.

(ii) $\text{ot}(D_p) < \alpha$; either $\sup(D_p) = \alpha$ or $\text{cf}(\alpha) = \omega$.

(iii) If $\beta \in D_p$ then:

- (a) $p_\beta \in \mathbb{S}^*_\beta$
- (b) $D_{p_\beta} = \beta \cap D_p$
- (c) $(p_\beta)_\gamma = p_\gamma$ for $\gamma \in D_{p_\beta}$
- (d) $\pi^p_\beta : \langle \mathfrak{A}^o_{p_\beta}, C_{p_\beta} \rangle \xrightarrow{\Sigma_o} \langle \mathfrak{A}^o_p, C_p \rangle$, where the Cs are those provided by Lemma 2.3, $\pi^p_\beta / \beta = \text{id}/\beta$ and $\pi^p_\beta(\beta) = \alpha$ if $\alpha < |p|$.
- (e) $\pi^p_\beta \pi^{p_\beta}_\gamma = \pi^p_\gamma$ for $\gamma \in D_{p_\beta}$.

(iv) $\sup D_p = \alpha \longrightarrow \mathfrak{A}^o_p = \bigcup_{\beta \in D_p} \pi^{p\,\prime\prime}_\beta \mathfrak{A}^o_{p_\beta}$.

(v) D_p, $\langle p_\beta : \beta \in D_p \rangle$, $\langle \pi^p_\beta : \beta \in D_p \rangle$ and are uniformly $\Sigma_1(\mathfrak{A}_p)$.

Remarks

1. Set $\lambda = \sup(\mathrm{rng}(\pi^p_\beta))$; by (iii)(d) we have that $\lambda \in (C_p)'$ (C_{p_β} is cub in $\mu^0_{p_\beta}$ by Lemma 2.3(i)). Then $\lambda = \mu^0_\xi$ for $\xi \le |p|$ by Lemma 2.3(iii) and

$$\pi^p_\beta : \langle \mathfrak{A}^0_{p_\beta}, C_{p_\beta} \rangle \xrightarrow[\Sigma_1]{} \langle \mathfrak{A}^0_{p \restriction \xi}, C_{p \restriction \xi} \rangle \quad \text{cofinally.}$$

Hence in particular $(\pi^p_\beta)''C_{p_\beta} = C_{p\restriction\xi}$.

2. By Lemma 2.4(iv) if $\sup D_p = \alpha$ then $\mathfrak{A}^0_p = \bigcup_{\beta \in D_p} \pi^{p}_\beta{}''\mathfrak{A}^0_{p_\beta}$. If in addition $\mathrm{cf}(\mu^0_p) < \mathrm{cf}(\alpha)$ then there must be a $\beta \in D_p$ s.t. π^p_β is cofinal in $\mathfrak{A}^0_p$; otherwise $\mathrm{cf}(\mu^0_p) = \mathrm{cf}(\mathrm{ot}(D_p))$ $= \mathrm{cf}(\alpha) > \omega$ contradicting $\mathrm{cf}(\mu^0_p) < \mathrm{cf}(\alpha)$. In particular the above holds if $\mathrm{cf}(\mu^0_p) > \omega$ and $|p|$ is a successor ordinal (since then $\mathrm{cf}(\mu^0_p) = \omega$).

3. $\pi^p_\beta \restriction \beta = \mathrm{id} \restriction \beta$, $\pi^p_\beta(\beta) = \alpha$ and $\mathrm{ot}(C_{p_\beta}) \le \beta$ so $[\alpha, \alpha+\beta) \subseteq \mathrm{rng}(\pi^p_\beta)$ and $(\pi^p_\beta)''C_{p_\beta}$ is an initial segment of C_p .

The next lemma is the promised assumption from lemma 1.3.

Lemma 2.5 Let α be a successor cardinal. There is a sequence $\langle b_p : p \in \mathbf{S}_\alpha \rangle$ s.t.

(i) $b_p \subseteq \alpha$.

(ii) $b_p \in \mathfrak{A}^2_p$ is uniformly $\Sigma_1(\mathfrak{A}^2_p)$ in α .

(iii) Let $\delta < \alpha$ and let f inject δ into $[\xi, |p|]$. Then $\langle b_{p \restriction f(i)} : i < \delta \rangle$ is Cohen generic over $\mathfrak{A}^1_{p\restriction\xi}$.

2.4 $\mathbb{R}^s$ - THE SUCCESSOR STAGE

In this section we will define the conditions used at successor cardinals. We use the Gödel pairing function to factor out ordinals for different types of coding.

2.4.1 Definition 1

(a) $Z_\delta = \{\{\xi,\delta\} : \xi \in On\}$, where $\{,\}$ is the Gödel pairing function defined in Chapter 1.

(b) $V_\delta = \{\{\{\xi,\delta\},0\} : \xi \in On\}$; so $V_\delta \subseteq Z_0$.

(c) For $b \subseteq \alpha$, let $\mathbf{S}(b) = \mathbf{S}_\alpha(b)$ = the set of $\{\eta,1\}$ s.t. η is a sequence number of an initial segment of b (i.e. η = the L[A]-code of $\chi_{b/\delta}$ for some $\delta < \alpha$) . Then $\mathbf{S}(b) \subseteq \alpha^+ \cap Z_1$ for $\alpha \in$ card .

2.4.2

For $s \in \mathbf{S}_{\alpha^+}$ we construct a set of conditions, $\mathbf{R}^s \in \mathfrak{N}^1_s$ s.t. :

Assertion Forcing with $\mathbf{R}^s$ over $\mathfrak{N}^1_s$ satisfies:

(a) $\mathbf{R}^s$ is α^{++}-A.C. and α-distributive; in particular cardinals and cofinalities are preserved by $\mathbf{R}^s$.

(b) $\mathbf{R}^s$ adds a $D \subseteq [\alpha,\alpha^+)$ s.t.

(i) $A \cap \alpha^+, s \in L_{\mu^1_s}[D]$.

(ii) If $\delta \in [\alpha,\alpha^+)$ then $L_{\mu^1_s}[A\cap\alpha, D\cap\delta] \models$ "card(δ) $\leq \alpha$" and $L_\mu[A\cap\alpha, D\cap\delta] \models$ "card(δ) $\leq \alpha$" for a μ s.t. $L_{\mu^1_s} \models$ "card(μ) $\leq \delta$" if $\alpha \geq \omega_2$.

Z_0, Z_1 are reserved for specific codings leaving $\bigcup_{m>1} Z_n$ free for use in reshaping. This does not weaken the reshaping powers of $\mathbf{S}_\alpha$ as was noted in the remarks after Theorem 1.4.

2.4.3 Definition of $\mathbf{R}^s$:

Let $s \in \mathbf{S}_{\alpha^+}$.
$\mathbf{R}^s$ = the set of pairs $\langle \dot{r}, r\rangle$ s.t.

(i) $r \in \mathbf{S}_\alpha$.

(ii) $\dot{r} = \dot{r}^0 \times \{0\} \cup \dot{r}^1 \times \{1\}$ where

$\dot{r}^0 \subseteq (A \cap \alpha^+) \times [\alpha,\alpha^+)$ and

$\dot{r}^1 \subseteq (P_{<\alpha^+}(\alpha^+) \cup \{b_{s\restriction\xi}: \xi \in \tilde{s}\}) \times [\alpha,\alpha^+)$

(where $\tilde{s} = \{\xi: s(\xi) = 1\}$ for $s \in \mathbf{S}_\beta$, $\beta \in$ card) .

(iii) $\mathrm{card}(\dot{r}) \leq \alpha$.

(iv) $\langle \delta,\eta \rangle \in \dot{r}^0 \longrightarrow (V_\delta \setminus \eta) \cap \tilde{r} = 0$.

(v) $\langle b,\eta \rangle \in \dot{r}^1 \longrightarrow (\mathbf{S}(b) \setminus \eta) \cap \tilde{r} = 0$.

$\langle \dot{p},p \rangle \leq \langle \dot{r},r \rangle$ iff $\dot{p} \supseteq \dot{r}$ and $p \supseteq r$.

Remarks

1. $\langle b_r: r \in \mathbf{S}_{\alpha^+} \rangle$ are those that exist by virtue of Lemma 2.5.

2. $\mathbb{R}^s$ trivially codes $A \cap \alpha^+$ on Z_0 .

3. $\mathbb{R}^s$ codes s by almost disjoint forcing using Z_1 .

4. $\mathbb{R}^s$ reshapes $[\alpha,\alpha^+)$.

2.4.4 Fact

Let $s \in \mathbf{S}_{\alpha+}$ and let $f: \alpha \longrightarrow \mathfrak{A}^1_s$ then $f \in \mathfrak{A}^1_s$.

Proof From §2.2, by definition 2(d) $\mathrm{cf}(\mu^1) = \alpha^+$ hence: $f \subseteq L_\delta[A\cap\alpha^+,s]$ for $\alpha^+ \leq \delta < \mu^1_s$. $\mathfrak{A}^1_s \models \mathrm{card}(L_\delta[A\cap\alpha^+,s]) = \alpha^+$ so there is a $g \in \mathfrak{A}^1_s$ s.t. $g: L_\delta[A\cap\alpha,s] \longrightarrow \alpha^+$, 1 to 1 . ${}^\alpha(\alpha^+) \subseteq L_{\alpha^+}[A] \subseteq \mathfrak{A}^1_s$ hence: $g \circ f \in \mathfrak{A}^1_s$ and $g^{-1} \circ g \circ f = f \in \mathfrak{A}^1_s$

Q.E.D. (Fact)

We use the above fact when $\mathbb{R} \in \mathfrak{A}^1_s$ is a set of conditions and we wish to show

$\mathfrak{A}^1_s \models$ "$\mathbb{R}$ is α-distributive" is absolute,

i.e., for any $\langle \Delta_i: i < \alpha \rangle$ of open dense sets in $\mathbb{R}$, $\bigcap_{i<\alpha} \Delta_i$ is dense in $\mathbb{R}$ since $\langle \Delta_i: i < \alpha \rangle \in \mathfrak{A}^1_s$ by the above fact.

Lemma 2.6 *Let* $s \in \mathbf{S}_{\alpha^+}$ *then Assertion 2.4.2 holds.*

Proof We can show that $\mathbb{R}^s$ is α^{++}-A.C. as we did in Theorem 1.1

Forcing with $\mathbb{R}^s$ is equivalent to the three following forcing operations:

(1) Forcing over $\mathfrak{A}^1_s$ with the set $\mathbb{R}_0$ of pairs $\langle \dot{r}, r\rangle$ s.t. r is a Cohen condition for adding a subset of $Z_0 \cap [\alpha,\alpha^+)$, $\mathrm{card}(\dot{r}) \leq \alpha$, $\dot{r} \subseteq (A \times \alpha^+) \times [\alpha,\alpha^+)$ and $\langle \delta,\eta\rangle \in \dot{r} \longrightarrow \tilde{r} \cap (V_\delta \setminus \eta) = 0$. $\mathbb{R}_0$ is trivially α-closed (hence α-distributive). Let $D_0 \subseteq Z_0 \cap [\alpha,\alpha^+)$ be $\mathbb{R}_0$-generic over $\mathfrak{A}^1_s$, then $A \cap \alpha^+ \in L_{\mu^1_s}[D_0]$ since:

$$\delta \in A \cap \alpha^+ \text{ iff } \exists \eta < \alpha^+ (D_0 \cap (V_\delta \setminus \eta) = 0) .$$

(2) Forcing over $\mathfrak{A}^1_s[D_0] = \langle L_{\mu^1}[D_0, s], D_0, s\rangle$ with the set $\mathbb{R}_1$ of pairs $\langle \dot{r}, r\rangle$ s.t.

(a) r is a Cohen condition for adding a subset of $Z_1 \cap [\alpha,\alpha^+)$.

(b) $\dot{r} \subseteq (P_{<\alpha^+}(\alpha^+) \cup \{b_\xi : \xi \in \tilde{s}\}) \times [\alpha,\alpha^+)$.

(c) $\mathrm{card}(\dot{r}) \leq \alpha$.

(d) $\langle b,\eta\rangle \in \dot{r} \longrightarrow \tilde{r} \cap (S(b) \setminus \eta) = 0$.

$\mathbb{R}_1$ is almost disjoint forcing on Z_1 and is trivially α-closed. To carry out the proof of Theorem 1.1 we need,

$$\forall \delta < |s| \; (L_{\mu^1_s}[D_0 \restriction \delta, s \cap \delta] \models \mathrm{card}(\delta) \leq \alpha^+)$$

and we have this because $s \in \mathbf{S}_\alpha$. If D_1 is $\mathbb{R}_1$-generic over $\mathfrak{A}^1_s[D_0]$, we can carry out the inductive definition of $\tilde{s} \cap \xi$, $b_{s\restriction\xi}$, $\mathfrak{A}^n_{s\restriction\xi}$ $(\xi < |s|, n < \omega)$ in $L_{\mu^1_s}[D_0 \cup D_1]$, using the fact that $b_{s\restriction\xi} \in \mathfrak{A}^2_{s\restriction\xi} \in \mathfrak{A}^1_s$ is uniformly $\Sigma_1(\mathfrak{A}^2_{s\restriction\xi})$ by Lemma 2.5.

Let $D_1 \subseteq Z_1 \cap [\alpha,\alpha^+)$ by $\mathbb{R}_1$-generic over $\mathfrak{A}^1_s[D_0]$. Then $s \in L_{\mu^1_s}[A \cap \alpha^+, D_1]$ by the argument of Theorem 1.1. Set

$D = D_0 \cup D_1$, then $A \cap \alpha^+ , s \in L_{\mu^1_s}[D]$; hence

$\mathfrak{A}^1_s[D] \models$ "$V = L[D]$" .

(3) Forcing over $\mathfrak{A}^1_s[D]$ with the set $\mathbf{R}_2$ of pairs $\langle r, |r| \rangle$ s.t.

(a) $r: (\bigcup_{\delta>1} Z_\delta \cap [\alpha, |r|]) \longrightarrow 2$.

(b) $|r| \in [\alpha, \alpha^+)$.

(c) $\delta \le |r| \Rightarrow L[A \cap \alpha, D \cap \delta, r \restriction \delta] \models \mathrm{card}(\delta) \le \alpha$ and $L_\mu[A \cap \alpha, D \cap \delta, r \restriction \delta] \models \mathrm{card}(\delta) \le \alpha$ for a μ s.t. $\mathrm{card}(\mu)^L \le \delta$ if $\alpha \ge \omega_2$.

We can carry out the argument of Theorem 1.4 since $\mathfrak{A}^1_s[D] \models$ "$V = L[D]$" and hence $\mathbf{R}_2$ is α-distributive. Let D_2 be $\mathbf{R}_2$-generic over $\mathfrak{A}^1_s[D]$ then $D_2 \cup D$ satisfies our lemma.

Q.E.D. (lemma)

Remark

It must of course be shown that forcing with $\mathbf{R}^s$ is actually equivalent to successive forcing with $\mathbf{R}_0, \mathbf{R}_1, \mathbf{R}_2$. Since $\mathbf{R}_0, \mathbf{R}_1 \in \mathfrak{A}_s$ and $\mathbf{R}_2 \subseteq \mathfrak{A}_s$, it follows by standard arguments that the successive forcing is equivalent to the set $\mathbf{P}_0$ of triples $\langle p_0, p_1, p_2 \rangle$ s.t. $p_i \in \mathbf{R}_i$ $(i = 0,1)$ and $\langle p_0, p_1 \rangle \Vdash p_2 \in \mathbf{R}_3$ (forcing with $\mathbf{R}_0 \times \mathbf{R}_1$). Let $\mathbf{P}_1$ be the set of $p = \langle \langle \dot{p}_0, p_0 \rangle, \langle \dot{p}_1, p_1 \rangle, \langle \dot{p}_2, p_2 \rangle \rangle \in \mathbf{P}_0$ s.t. $|p_2| \subseteq \mathrm{dom}(p_0) \cup \mathrm{dom}(p_1) \cup \mathrm{dom}(p_2)$. It is clear that $\mathbf{P}_1$ is dense in $\mathbf{P}_0$. But p is element of $\mathbf{P}_1$ iff:

(a) $\langle \dot{p}_i, p_i \rangle \in \mathbf{R}_i$ $(i = 0,1)$.

(b) $p_2: \bigcup_{\eta>1} Z_\eta \cap [\alpha, |p_2|) \longrightarrow 2$.

(c) $|p_2| \subseteq \mathrm{dom}(p_0) \cup \mathrm{dom}(p_1) \cup \mathrm{dom}(p_2)$.

(d) $\delta \le |p_2| \rightarrow L_\mu[A \cap \alpha, p' \restriction \delta] \models \mathrm{card}(\delta) \le \alpha$ for a $\mu > \delta$ s.t. $\mathrm{card}(\mu)^L \le \delta$ if $\alpha \ge \omega_2$, where $p' = p_0 \cup p_1 \cup p_2$.

Now let $\mathbb{P}_2$ be the set of $p \in \mathbb{P}_1$ s.t. $|p_2| = \text{dom}(p_0) \cup \text{dom}(p_1) \cup \text{dom}(p_2)$. $\mathbb{P}_2$ is easily seen to be dense in $\mathbb{P}_1$, by the characterization (a)-(d). Hence the successive forcing is equivalent to forcing with $\mathbb{P}_2$. For $p \in \mathbb{P}_2$ set:

$$\sigma(p) = \langle \dot{p}_0 \times \{0\} \cup \dot{p}_1 \times \{1\} , p_0 \cup p_1 \cup p_2 \rangle .$$

By the characterization (a)-(d), it follows that σ is an isomorphism of $\mathbb{P}_2$ and $\mathbb{R}^s$.

Lemma 2.7 *Let* $s \in \mathbb{S}_{\alpha^+}$ *and* $\xi \in [\alpha, |s|)$. *If* $\Delta \in \mathfrak{A}^1_\xi$ *is predense in* $\mathbb{R}^{s \restriction \xi}$ *then* Δ *is predense in* $\mathbb{R}^s$.

Proof Same as Lemma 1.3.

Definition 2 Let $\alpha \in$ Card . Then,

(a) $\mathbb{S}^+_\alpha$ = the set of $D \subseteq [\alpha, \alpha^+)$ s.t.
(i) $\delta \in [\alpha, \alpha^+) \longrightarrow \chi_D \restriction \delta \in \mathbb{S}_\alpha$, where D is not necessarily in $\langle M, A \rangle$.
(ii) $\mathfrak{A}_D \models ZF^-$, where $\mathfrak{A}_D = \langle L_{\alpha^+}[A \cap \alpha, D], A \cap \alpha, D \rangle$.

(b) Let $D \in \mathbb{S}^+_{\alpha^+}$. $\mathbb{R}^D = \bigcup_{\delta \in [\alpha^+, \alpha^{++})} \mathbb{R}^{\chi_D \restriction \delta}$.

For $D \in S^+_{\alpha^+}$, $\mathbb{R}^D$ is a proper class in $\mathfrak{A}_D$

$p \in \mathbb{R}^D$ iff $\exists \delta \, (p \in \mathbb{R}^{\chi_D \restriction \delta})$

and each $\mathbb{R}^s$ is uniformly definable in s ; so $\mathbb{R}^D$ is definable in $\mathfrak{A}_D$.

Lemma 2.8 *Let* $D \in S^+_{\alpha^+}$. *Then* $\mathbb{R}^D$ *satisfies* α^{++}*-A.C. and is* α*-distributive in* $\mathfrak{A}_D$.

Proof $\mathbb{R}^D$ satisfies α^{++}-A.C. means that if Δ is an $\mathfrak{A}_D$-definable class of mutually incompatible conditions, then

$\operatorname{card}(\Delta) \le \alpha^+$ in $\mathfrak{A}_D$ (hence $\Delta \in \mathfrak{A}_D$) .

If $r \in \mathbf{R}^{s\frown\xi}$ then $r \in \mathbf{R}^s$; so α^{++}-A.C. follows by the argument of Theorem 1.1. For α-distributivity we must show that if $\langle \Delta_i : i < \alpha \rangle$ is a definable sequence of definably dense classes in $\mathfrak{A}_D$ (i.e., $\{\langle p,i \rangle : p \in \Delta_i\}$ is $\mathfrak{A}_D$-definable), then $\bigcap_i \Delta_i$ is dense in $\mathfrak{A}_D$.

Let $r \in \mathbf{R}^D$, we must find a $q \le r$, $q \in \bigcap_i \Delta_i$. Let $\Delta_i^* \subseteq \Delta_i$ be a maximal set of mutual incompatible conditions. By α^{++}-A.C., $\operatorname{card}(\Delta_i^*) \le \alpha^+$ and hence $\langle \Delta_i^* : i < \alpha \rangle \in \mathfrak{A}_D$. Clearly Δ_i^* is predense in $\mathbf{R}^D$. Pick $s = \chi_D\frown\xi$, $\xi < \alpha^{++}$ s.t. p , $\langle \Delta_i^* : i < \alpha \rangle \in \mathfrak{A}_s^1$. Then Δ_i^* is predense in $\mathbf{R}^s$. Apply the α-distributivity of $\mathbf{R}^s$ to get a $q \in \mathbf{R}^s$ s.t. $q \le r$ and

$$q \in \{p \in \mathbf{R}^s : \forall i < \alpha \ (p \text{ meets } \Delta_i^*)\} .$$

Q.E.D. (Lemma)

2.5 THE LIMIT CASE AND $\mathbb{P}_\tau$

The $\mathfrak{A}_s^n$ are analogous to the L_{μ_ξ} of Chapter 1 and just as in Lemma 1.2 we can show that every $x \in \mathfrak{A}_s^n$ is definable in $\mathfrak{A}_s^n$ from ordinals less than α (for $s \in \mathbb{S}_\alpha$) . For limit cardinals, α , we define the analogues of the $X_{\xi\tau}$ of Theorem 1.5.

Definition 1 Let α be a limit cardinal, $s \in \mathbb{S}_\alpha$, $1 \le n \le \omega$ and $\tau \in \operatorname{card} \cap \alpha$

$$X_{s\tau}^n = \text{the least } X \prec \mathfrak{A}_s^n \text{ s.t. } \tau \subseteq X \text{ and } \pi_{s\tau}^n : \mathfrak{A}_{s\tau}^n \cong X_{s\tau}^n .$$

Since each $x \in \mathfrak{A}_s^n$ is definable from α , we have

$$\bigcup_{\tau\in\alpha} X_{s\tau}^n = \mathfrak{A}_s^n .$$

We now prove a few facts about the 'fine structure' and collapsing properties of the $\mathfrak{A}^n_s$ that will be used frequently later.

Fact 1 Let $s \in \mathbb{S}_\alpha$, $\bar{s} \in \mathbb{S}_{\bar{\alpha}}$, $\bar{\alpha} < \alpha$, α limit cardinal and

$$\pi : \mathfrak{A}^i_{\bar{s}} \xrightarrow[\Sigma_1]{} \mathfrak{A}^1_s$$

where $\pi(\bar{s}) = s$, $\pi \restriction \bar{\alpha} = id \restriction \bar{\alpha}$ and $\pi(\bar{\alpha}) = \alpha$ (if $\bar{\alpha} \in \mathfrak{A}^i_{\bar{s}}$). Then,

(a) $\pi(\mathfrak{A}^j_{\bar{s}\restriction\xi}) = \mathfrak{A}^j_{s\restriction\xi}$.

(b) $\pi''\mathfrak{A}^j_{\bar{s}\restriction\xi} = X^j_{s\restriction\xi,\bar{\alpha}}$.

(c) $\mathfrak{A}^j_{\bar{s}\restriction\bar{\xi}} = \mathfrak{A}^j_{s\restriction\xi,\alpha}$.

(d) $\pi/\mathfrak{A}^j_{\bar{s}\restriction\bar{\xi}} = \pi^j_{s\bar{\alpha}}$ for $\pi(\bar{\xi}) = \xi$ and $\xi < |s|$ or $\xi = |s|$ and $j < i$.

Proof $\bar{\alpha}$ is a limit cardinal since $\pi : \mathfrak{A}^i_{\bar{s}} \xrightarrow[\Sigma_1]{} \mathfrak{A}^1_s$ and $\pi(\bar{\alpha}) = \alpha$. By Claim 2.2 $\mathfrak{A}^j_{\bar{s}\restriction\bar{\xi}} \in \mathfrak{A}^i_{\bar{s}}$ is uniformly Σ_1 definable in $\mathfrak{A}^i_{\bar{s}}$ from $\bar{s}$, $\bar{\xi}$, j and $\bar{\alpha}$. So we have, $\pi(\mathfrak{A}^j_{\bar{s}\restriction\bar{\xi}}) = \mathfrak{A}^j_{\pi(\bar{s})\restriction\pi(\bar{\xi})} = \mathfrak{A}^j_{s\restriction\xi}$ and (a) holds. To prove (b) we first show $\pi''\mathfrak{A}^j_{\bar{s}\restriction\bar{\xi}} \prec \mathfrak{A}^j_{s\restriction\xi}$. This is true because a Σ_n statement about $\mathfrak{A}^j_{s\restriction\xi}$ is $\Sigma_0(\mathfrak{A}^i_s)$ when the quantifiers are relativized to $\mathfrak{A}^j_{s\restriction\xi}$. Next $\pi''(\mathfrak{A}^j_{\bar{s}\restriction\bar{\xi}}) \supseteq X^j_{s\restriction\xi,\bar{\alpha}}$ since $\pi \restriction \bar{\alpha} = id \restriction \bar{\alpha}$ and $X^j_{s\restriction\xi,\bar{\alpha}}$ is the least such. Finally $\pi''\mathfrak{A}^j_{\bar{s}\restriction\bar{\xi}} = X^j_{s\restriction\xi,\bar{\alpha}}$ since each element of $\mathfrak{A}^i_{\bar{s}\restriction\bar{\xi}}$ is definable from ordinals less than $\bar{\alpha}$. $y \in \mathfrak{A}^j_{\bar{s}\restriction\bar{\xi}}$ is definable in $\mathfrak{A}^j_{\bar{s}\restriction\bar{\xi}}$ from ordinals less than $\bar{\alpha}$, hence $\pi(y) \in \mathfrak{A}^j_{s\restriction\xi}$ is definable in $\mathfrak{A}^j_{s\restriction\xi}$ from ordinals less than $\bar{\alpha}$ ($\pi\restriction\bar{\alpha} = id\restriction\bar{\alpha}$). Hence $\pi(y) \in X^j_{s\restriction\xi,\bar{\alpha}}$. This completes (b). (c) and (d) follow from the uniqueness of the Mostowski collapse.

Q.E.D. (Fact)

For the next facts we assume α is a limit cardinal, $s \in \mathbf{S}_\alpha$, $1 \le n \le \omega$ and $\bar{\alpha} < \alpha$ is a cardinal s.t. $X^n_{s,\bar{\alpha}} \cap \alpha = \bar{\alpha}$.

Fact 2 $(\pi^n_{s\bar{\alpha}})^{-1}(s) = \bar{s} \in \mathbf{S}_{\bar{\alpha}}$.

Proof $n \ge 1$ implies $\pi^n_{s\bar{\alpha}}(\bar{\alpha}) = \alpha$ and $s \in \mathfrak{A}^n_s$. From §2.2, by definition $s \in \mathbf{S}_{\bar{\alpha}}$ is a definable property of $\mathfrak{A}^n_s$ (i.e. $\mathfrak{A}^n_s \models$ "$s \in \mathbf{S}_\alpha$"), hence by the isomorphism $\bar{s} \in \mathbf{S}_{\bar{\alpha}}$.

Q.E.D. (Fact)

For the following facts we assume $\pi^n_{s\bar{\alpha}}(\bar{s}) = s$.

Fact 3 $\pi^n_{s\bar{\alpha}}(\mathfrak{A}^i_{\bar{s}\wedge\bar{\xi}}) = \mathfrak{A}^i_{s\wedge\xi}$ and $\mathfrak{A}^i_{\bar{s}\wedge\bar{\xi}} = \mathfrak{A}^i_{s\wedge\xi,\alpha}$ for $\pi^n_{s\bar{\alpha}}(\bar{\xi}) = \xi$ and $\xi < |s|$ or $\xi = |s|$ and $i < n$.

Proof Direct application of the facts above.

Fact 4 $\mathfrak{A}^n_{s\bar{\alpha}} = \mathfrak{A}^n_{\bar{s}}$.

Proof $\mathfrak{A}^n_{\bar{s}} = \langle L_{\mu^n_{\bar{s}}}[A\cap\bar{\alpha},\bar{s}], A\cap\bar{\alpha},\bar{s}\rangle$ and $\mathfrak{A}^n_{s\bar{\alpha}} = \langle L_{\bar{\mu}}[A\cap\bar{\alpha},\bar{s}], A\cap\bar{\alpha},\bar{s}\rangle$ so it is sufficient to show $\bar{\mu} = \mu^n_{\bar{s}}$.

If $n = \omega$ then

$$\bar{\mu} = \sup_{i<\omega}(\pi^n_{s\bar{\alpha}})^{-1}(\mu^i_s) = \sup_{i<\omega} \mu^i_{\bar{s}} = \mu^\omega_{\bar{s}} .$$

If $n = m+1$ then $\pi^n_{s\bar{\alpha}}(\mu^m_{\bar{s}}) = \mu^m_s$. Since $\mu^{m+1}_s = \mu^n_s =$ the least $\mu > \mu^m_s$ s.t. clauses (a), (b), (c), (d) of the second definition of §2.2 hold; $\bar{\mu} =$ the least $\mu > \mu^m_{\bar{s}}$ satisfying the clauses. Hence $\bar{\mu} = \mu^n_{\bar{s}}$.

Q.E.D. (Fact)

Fact 5 $\pi^n_{s\bar{\alpha}} \supseteq \pi^m_{s\wedge\xi,\bar{\alpha}}$, where $\xi < |s|$ or $\xi = |s|$ and $n < m$.

Proof Use the previous facts noting

$$\pi^n_{s\bar{\alpha}}: \mathfrak{A}^n_{\bar{s}} \longrightarrow \mathfrak{A}^n_s$$

because by Fact 2 $\bar{s} = (\pi^n_{s\bar{\alpha}})^{-1}(s)$ and by Fact 4 $\mathfrak{A}^n_{s\bar{\alpha}} = \mathfrak{A}^n_{\bar{s}}$.

Definition 2 Let α be a limit cardinal, $\beta \in \mathrm{card} \cap \alpha$, and $s \in \mathbb{S}_\alpha$. $\rho^n_{s\beta} = \{\rho', 2\}$, where ρ' = the L[A]-code of $\mathfrak{A}^n_{s\beta^+}$. Hence $\rho^n_{s\beta} \in Z_2 \cap (\beta^+, \beta^{++})$.

Notation We write $X_{s\tau}$, $\pi_{s\tau}$, $\mathfrak{A}_{s\tau}$, $\rho_{s\tau}$ for $X^\omega_{s\tau}$, $\pi^\omega_{s\tau}$, $\mathfrak{A}^\omega_{s\tau}$, $\rho^\omega_{s\tau}$ and $X^n_{\xi\tau}$, $\pi^n_{\xi\tau}$, $\mathfrak{A}^n_{\xi\tau}$, $\rho^n_{\xi\tau}$ for $X^n_{s\wedge\xi,\tau}$, $\pi^n_{s\wedge\xi,\tau}$, $\mathfrak{A}^n_{s\wedge\xi,\tau}$, $\rho^n_{s\wedge\xi,\tau}$.

By [J1] Theorem 5.1 there is a sequence

$$\langle C'_\beta : \beta \text{ is a singular cardinal in } L \rangle \text{ s.t.}$$

(a) $C'_\beta \subseteq \beta$ is closed in β.

(b) $\gamma \in C'_\beta \longrightarrow \gamma$ is singular in L and $C'_\gamma = \gamma \cap C'_\beta$.

(c) $\mathrm{ot}(C'_\beta) < \beta$.

(d) $\sup(C'_\beta) < \beta \longrightarrow \mathrm{cf}(\beta) = \omega$.

(e) If $\mu > \beta$ is p.r. closed and $L_\mu \models$ "β is singular", then $C'_\beta \in L_\mu$ and C'_β is uniformly L_μ definable.

By [DJ] and the fact that we have $\neg 0^\#$, if β is a singular cardinal then β is singular in L. Hence we can define:

Definition 3 $C_\beta = \mathrm{card} \cap C'_\beta$ for singular cardinals β. The sequence, $\langle C_\beta : \beta \text{ is a singular cardinal}\rangle$ satisfies,

(a) $C_\beta \subseteq \beta$ is closed in β .

(b) $\gamma \in C_\beta \longrightarrow \gamma$ is singular and $C_\gamma = \gamma \cap C_\beta$.

(c) $ot(C_\beta) < \beta$.

(d) $\sup(C_\beta) < \beta \longrightarrow cf(\beta) = \omega$.

(e) If $\mu > \beta$ is p.r. closed and $L_\mu[A \cap \beta] \models$ "β is singular" then $C_\beta \in L_\mu[A \cap \beta]$ and C_β is uniformly $L_\mu[A \cap \beta]$ definable. ($H_\beta = L_\beta[A \cap \beta]$, hence $\text{card} \cap \beta \in L[A \cap \beta]$.)

Definition 4 Let β be a singular cardinal

$$\lambda_\beta = \begin{cases} |C_\beta \setminus |C_\beta|| & \text{if } \sup(C_\beta) = \beta \\ \omega & \text{otherwise.} \end{cases}$$

If $\sup(C_\beta) = \beta$, set $\langle \gamma_i^\beta : i < \lambda_\beta \rangle$ = the monotone enumeration of $C_\beta \setminus |C_\beta|$. If not set $\langle \gamma_i^\beta : i < \omega \rangle$ = the $L[A \cap \beta]$-least ω-sequence of successor cardinals converging to β s.t. $\gamma_0^\beta > \sup(C_\beta)$.

Note that by definition 3(b),

$$C_{\gamma_{i+1}^\beta} = \gamma_{i+1}^\beta \cap C_\beta = C_\beta \cap (\gamma_i^\beta) + 1$$

so $C_{\gamma_{i+1}^\beta}$ is bounded in γ_{i+1}^β ; hence by (d) $cf(\gamma_{i+1}^\beta) = \omega$.

Notation When the context is clear we write γ_i for γ_i^β and set $\gamma_{\lambda_\beta} = \beta$.

Definition 5 Let β be a singular cardinal, $s \in \mathbf{S}_\beta$. For $i < \lambda_\beta$ set:

$$\tilde{\rho}_{si} = \left\{\left\{\rho_{s\gamma_i} , |C_\beta|\right\} , 3\right\} \text{ . Hence } \tilde{\rho}_{si} \in Z_3 \cap (\gamma_i^+, \gamma_i^{++}) \text{ .}$$

Notation We write $\tilde{\rho}_{\xi i}$ for $\tilde{\rho}_{s/\xi, i}$.

When defining our conditions we use the codes $\rho^i_{s\gamma}$ when $\mathfrak{A}_s \models \beta$ is regular ($s \in \mathbf{S}_\beta$) and $\tilde{\rho}_{si}$ when $\mathfrak{A}_s \models$ " β is singular".

Definition 6 Let β be a limit cardinal, $s \in \mathbf{S}_\beta$. If $\mathfrak{A}_s \models \beta$ is singular set: $\Omega_\beta = \Omega^s_\beta$ = the least $\xi \le |s|$ s.t. $\mathfrak{A}_\xi \models \beta$ is singular. Otherwise Ω_β is undefined.

Remarks

(1) By definition 3(e), $C_\beta \in \mathfrak{A}_{\Omega_\beta}$ and this implies

$$\langle \gamma_i : i < \lambda_\beta \rangle \in \mathfrak{A}_{\Omega_\beta} .$$

If $cf(\beta) = \omega$ and $\mathfrak{A}_{\Omega_\beta} \models$ " β is singular" then $\mathfrak{A}_{\Omega_\beta} \models$ "$cf(\beta) = \omega$" because $\mathfrak{A}_{\Omega_\beta} \models$ "$cf(\beta)$ is regular" and the regular cardinals of $\mathfrak{A}_{\Omega_\beta}$ are absolute.

(2) If $\xi \in [\Omega_\beta, |s|]$, then $\langle \tilde{\rho}_{\xi i} : i < \lambda_\beta \rangle \in \mathfrak{A}_\xi$ since $\langle \mathfrak{A}^\omega_{s/\xi, \gamma_i^+} : i < \lambda_\beta \rangle \in \mathfrak{A}_\xi$.

2.6 DEFINITION OF $\mathbf{P}^s_\tau$ and $\mathbf{P}_\tau$

We now have all the tools to define $\mathbf{P}^s_\tau$ whose purpose it is to code $s \in S_\alpha$ by a generic subset of $[\tau, \tau^+)$.

Definition of $\mathbf{P}^s_\tau$ Let $s \in \mathbf{S}_\alpha$, $\alpha \in$ card , $\tau \in$ card and $\tau \le \alpha$. We define $\mathbf{P}^s_\tau$ by induction on α as follows: $\mathbf{P}^s_\tau$ = the set of maps $p : \text{card} \cap [\tau, \alpha) \longrightarrow V^2$ s.t. Setting $p(\gamma) = \langle \dot{p}_\gamma, p_\gamma \rangle$ for $\gamma \in dom(p)$ we have:

(A) If $\alpha = \gamma^+$, $\tau \le \alpha$, then $p(\gamma) \in \mathbf{R}^s$.

(B) $\forall \gamma \in \text{card} \cap [\tau, \alpha) \ (p/\gamma \in \mathbf{P}^{p_\gamma}_\tau)$.

Comment (A) and (B) imply that $p_\gamma \in \mathbf{S}_\gamma$ and $p(\gamma) \in \mathbf{R}^{p_{\gamma+}}$

(or if $\alpha = \gamma^+$ $p(\gamma) \in \mathbf{R}^s$) for $\gamma \in \text{card} \cap [\tau,\alpha)$.

(C) If α is a limit cardinal, $\tau < \alpha$, then the following hold:

(i) $p \in \mathfrak{A}_s$; set $|p|$ = the least $\alpha \le \xi \le |s|$ s.t. $p \in \mathfrak{A}_\xi$.

Comment Note that $|\ |$ has two meanings. $|s|$ is $\sup(\text{dom}(s))$ for $s \in \mathbf{S}_\alpha$ and $|p|$ we defined above for $p \in \mathbb{P}_\tau^s$. The dual use is very convenient.

(ii) If $\mathfrak{A}_{|p|} \models$ "α is regular", there is a cub $C \subseteq [\tau,\alpha)$ s.t. $C \in \mathfrak{A}_{|p|}$ and $\gamma \in C \longrightarrow \gamma$ is a limit cardinal, $|p{\restriction}\alpha| = |p_\gamma|$ and $\dot{p}_\gamma = \emptyset$. (Note the dual use of $|\ |$ above.)

(iii) If $\xi < |p|$, $\mathfrak{A}_\xi \models$ (α is regular), there is an $n \ge 1$ s.t. for all $i \ge n$ a $C \in \mathfrak{A}_\xi$ exists s.t. $C \subseteq [\tau,\alpha)$ is cub in α and $\gamma \in C \longrightarrow \widehat{p}(\rho^i_{\xi\gamma}) = s(\xi)$ (where $\widehat{p} = \bigcup_{\gamma \in [\tau,\alpha)} p_\gamma$).

(iv) If $|p| \ge \Omega_\alpha^s$, $\text{cf}(\alpha) > \omega$, there is a cub $D \subseteq [\tau,\alpha)$ s.t. $D \in \mathfrak{A}_{\Omega_\alpha}$ and $\delta \in D \longrightarrow \Omega_\delta^{p\delta}$ exists, $|p{\restriction}\delta| \ge \Omega_\delta$ and $D \cap \delta \in \mathfrak{A}_{\Omega_\delta}$.

(v) If $\xi \in [\Omega_\alpha, |p|)$ then,

$$\exists \eta < \lambda_\alpha \quad \forall i < \lambda_\alpha \quad (i \ge \eta \longrightarrow \widehat{p}(\widetilde{\rho}_{\xi i}) = s(\xi)) .$$

We set $q \le p$ in $\mathbb{P}_\tau^s$ iff $p,q \in \mathbf{P}_\tau^s$ and

$$\forall \gamma \in \text{card} \cap [\tau,\alpha) \ (\dot{q}_\gamma \supseteq \dot{p}_\gamma \text{ and } q_\gamma \supseteq p_\gamma) .$$

Remarks If $\alpha = \gamma^+$ then s is coded by $\mathbf{R}^s$ using almost disjoint forcing. If α is a limit cardinal s is coded by other methods. Set:

$s_{reg} = \{\xi < |p| : \mathfrak{A}_\xi \models \alpha \text{ is regular}\}$ and

$$s_{sing} = \{\xi < |p| : \mathfrak{A}_\xi \models \alpha \text{ is singular}\} = \{\xi < |p| : \xi \geq \Omega_\alpha\} .$$

For $\xi \in s_{reg}$ we decide $\xi \in \tilde{s}$ if there is an n s.t. $\forall i \geq n$ there is a cub $C \in \mathfrak{A}_\xi$, $C \subseteq [\tau,\alpha)$ s.t. $\forall\gamma \in C(\hat{p}(\rho^i_{\xi\gamma}) = 1)$. We require this for each $i \geq n$ to facilitate construction of conditions in $\mathbb{P}^s_\tau$. For $\xi \in [\Omega_\alpha, |p|)$ the coding is similar to Theorem 1.5 and on an end segment of λ_α,

$$\xi \in s \text{ iff } \hat{p}(\tilde{\rho}_{\xi i}) = 1 .$$

(1) By (iii) and (v) if $\xi < |p|$, then $s(\xi)$ is determined by, p, $s \restriction \xi$.

(2) By the methods of Theorem 1.5, Remark (2) we can show that for $\xi \geq |p|$ $(p \in \mathfrak{A}_\xi)$,

(a) If $p \in \mathfrak{A}^n_\xi$ then there is a $\delta \in \text{card} \cap [\tau,\alpha)$ s.t.

$$\gamma \geq \delta \longrightarrow \rho^n_{\xi\gamma} \notin \text{dom}(\hat{p}) .$$

(b) $\xi \geq \Omega_\alpha \longrightarrow \exists n < \lambda_\alpha$ s.t. $i \geq n \longrightarrow \tilde{\rho}_{\xi i} \notin \text{dom}(\hat{p})$. To imitate the proof in Theorem 1.5 one only has to note, $\rho^n_{\xi\gamma} \in (\gamma^+,\gamma^{++})$ $(\tilde{\rho}_{\xi i} \in (\gamma^+_i,\gamma^{++}_i))$ is a code of $\mathfrak{A}^n_{\xi\gamma^+}$ $(\mathfrak{A}_{\xi\gamma^+_i})$ and for large enough γ, $p \in X^n_{\xi\gamma^+}$ $(p \in X_{\xi\gamma^+_i})$ and $\gamma^{++} \in X^n_{\xi\gamma^+}$ $(\gamma^{++}_i \in X_{\xi\gamma^+_i})$ giving

$$\Pi^n_{\xi\gamma^+}(\hat{p} \restriction \gamma^{++}) = \hat{p} \restriction \gamma^{++} \quad (\Pi_{\xi\gamma^+_i}(\hat{p} \restriction \gamma^{++}_i) = \hat{p} \restriction \gamma^{++}_i) ;$$

hence $\rho^n_{\xi\gamma}$ $(\tilde{\rho}_{\xi i}) \notin \text{dom}(\hat{p})$. Hence if $\xi \geq |p|$, $s(\xi)$ is not determined by p, $s \restriction \xi$.

(3) $q \leq p \longrightarrow |q| \geq |p|$ since otherwise q would determine $s(\xi)$ for $\xi \geq |q|$ contradicting (2).

(4) $p \in \mathbb{P}^{s \restriction \xi}_\tau \Rightarrow p \in \mathbb{P}^s_\tau$ (follows from inspection of the definition).

Definition of $\mathbb{P}_\tau$ We define the class $\mathbb{P}_\tau$ as follows:

$p \in \mathbb{P}_\tau$ iff $p: \mathrm{dom}(p) \longrightarrow V^2$ s.t.

(1) $\mathrm{dom}(p) \subseteq \mathrm{card} \cap [\tau,\infty)$.

(2) Setting $\hat{p}(\gamma) = \begin{cases} p(\gamma) & \text{if } \gamma \in \mathrm{dom}(p) \\ \langle \emptyset,\emptyset \rangle & \text{otherwise} \end{cases}$

and writing $\hat{p}(\gamma) = \langle \dot{\hat{p}}_\gamma, \hat{p}_\gamma \rangle$, we have

$$\forall \gamma \in \mathrm{card} \cap [\tau,\infty) \ (\hat{p}\restriction\gamma \in \mathbb{P}^{\hat{p}_\gamma}) .$$

$p \leq q$ in $\mathbb{P}_\tau$ iff $p,q \in \mathbb{P}_\tau$ and

$$\forall \gamma \in \mathrm{card} \cap [\tau,\infty) \ (\hat{p}\restriction\gamma \leq \hat{q}\restriction\gamma \text{ in } \mathbb{P}^{\hat{p}_\gamma}) .$$

Set $\mathbb{P}^s = \mathbb{P}^s_0$, $\mathbb{P} = \mathbb{P}_0$.

$\mathbb{P}$ is our ultimate class of conditions. However before showing that $\mathbb{P}$ does the job, we must prove a great deal of lemmas, most of which deal with the sets $\mathbb{P}^s_\tau$.

Lemma 2.9 *Let* $s \in \mathbb{S}_\alpha$, $\tau \leq \alpha \in \mathrm{card}$. *Then* $\emptyset_{\mathbb{P}^s_\tau} \in \mathbb{P}^s_\tau$, *where* $\emptyset_{\mathbb{P}^s_\tau} = \emptyset\restriction[\tau,\alpha)$ = the constant function $\langle\emptyset,\emptyset\rangle$ on $\mathrm{card} \cap [\tau,\alpha)$.

Proof By induction on α . The only problem is to satisfy (iv) of the definition of $\mathbb{P}^s_\tau$ and we only have to do this when $\mathrm{cf}(\alpha) > \omega$ and $\alpha = \Omega_\alpha$. In such a case we must find a $D \in \mathfrak{A}_\alpha = \mathfrak{A}_{\Omega_\alpha}$ s.t. D is cub in α and $\delta \in D \longrightarrow \Omega_\delta^{p_\delta}$ exists, $|p\restriction\delta| \geq \Omega_\delta$ and $D \cap \delta \in \mathfrak{A}_{\Omega_\delta}$. Since $\mathfrak{A}_{\Omega_\delta} \models$ "α is singular", $\emptyset_\alpha \in \mathbb{S}^*_\alpha$ and we may use Lemma 2.4. Inspection shows that $D_{\emptyset_\alpha}$ is just what is needed.

Notation We write $\mathbb{P}^{\xi}_{\tau}$ in place of $\mathbb{P}^{s/\xi}_{\tau}$ so $\mathbb{P}^{\gamma}_{\tau} = \mathbb{P}^{\emptyset_\gamma}_{\tau}$ for $\gamma \in$ card .

Corollary 2.9.1 If $p \in \mathbb{P}^{s}_{\tau}$, $s \in \mathbf{S}_{\alpha}$, $\gamma \leq \tau$, then

$$p \cup \emptyset_{\mathbb{P}^{\tau}_{\gamma}} \in \mathbb{P}^{s}_{\gamma} .$$

Corollary 2.9.2 $\emptyset \in \mathbf{P}_{\tau}$.

Corollary 2.9.3 If $p \in \mathbf{P}_{\tau}$, $\gamma \leq \tau$, then $p \in \mathbf{P}_{\gamma}$.

Lemma 2.10 *If* $p \in \mathbf{P}^{s}_{\tau}$, $s \in \mathbf{S}_{\alpha}$ *then* $p \cup \{\langle \alpha, \langle \emptyset, s\rangle\rangle\} \in \mathbf{P}_{\tau}$.

Proof Trivial.

Lemma 2.11 *Let* $s \in \mathbf{S}_{\alpha^+}$*,* $\tau \leq \alpha$ *,* $\xi \in [\alpha, |s|]$. *If* $\Delta \in \mathfrak{A}^{1}_{\xi}$ *is predense in* $\mathbb{P}^{\xi}_{\tau}$ *, then* Δ *is predense in* $\mathbb{P}^{s}_{\tau}$.

Proof Let $p \in \mathbf{P}^{s}_{\tau}$. We must find $q \leq p$ s.t. q meets Δ . Set:

$$\Delta' = \{q(\alpha) \in \mathbf{R}^{s/\xi} \mid q \in \mathbf{P}^{\xi}_{\tau} \cap \Delta \text{ and } (q/\alpha \leq p/\alpha \text{ or } p/\alpha \not\Vdash \mathbf{P}^{q_\alpha}_{\tau}) .$$

Δ' is predense in $\mathbf{R}^{s/\xi}$, hence by Lemma 2.7 Δ' is predense in $\mathbf{R}^{s}$. Let $q(\alpha) \in \mathbf{R}^{s}$, $q(\alpha) \leq p(\alpha)$ meet Δ' , then $q \leq p$ and q meets Δ .

Q.E.D.

Definition 1 Let $D \in \mathbf{S}^{+}_{\alpha}$, $\alpha \in$ card , $\tau \leq \alpha$

$$\mathbf{P}^{D}_{\tau} = \bigcup_{\delta \in [\alpha, \alpha^+)} \mathbf{P}^{D/\delta}_{\tau}$$

Corollary 2.11.1 Let $D \in \mathbf{S}^{+}_{\alpha^+}$ and $\tilde{s} = D \cap \delta$ for $\delta < \alpha^{++}$.

If $\Delta \in \mathfrak{A}^1_s$ is predense in $\mathbf{P}^s_\tau$ then Δ is predense in $\mathbf{P}^D_\tau$.

The following lemma is important and will be used frequently in all that follows. It shows that the $s \in \mathbf{S}_\alpha$, $p \in \mathbf{P}^s_\tau$ and $X^n_{s\bar{\alpha}}$ collapse coherently under certain conditions.

Lemma 2.12 The Collapsing Lemma *Let* α *be a limit cardinal,* $s \in \mathbf{S}_\alpha$ *and* $\mathfrak{A}_s \models$ (α is regular). *Let* $p \in \mathbf{P}^s_\tau$ $(\tau < \alpha)$ s.t. $|p| = |s|$. *Let* $1 \le n \le \omega$ s.t. $p \in \mathfrak{A}^n_s$ *and there is a* cub *set* $C \subseteq [\tau,\alpha)$, $C \in \mathfrak{A}^n_s$ s.t.

$$\gamma \in C \longrightarrow \gamma \text{ is a limit cardinal, } \dot{p}_\gamma = 0 \text{ and } |p{\restriction}\gamma| = |p_\gamma| .$$

(This always holds for some $n \le \omega$ by definition.) Let $\bar{\alpha} < \alpha$ be s.t. $\bar{\alpha} = \alpha \cap X^n_{s\bar{\alpha}}$ and $p \in X^n_{s\bar{\alpha}}$, then $\dot{p}_{\bar{\alpha}} = 0$, $p_{\bar{\alpha}} = (\Pi^n_{s\bar{\alpha}})^{-1}(s)$, $|p{\restriction}\bar{\alpha}| = |p_{\bar{\alpha}}|$ and $\mathfrak{A}^n_{p_{\bar{\alpha}}} = \mathfrak{A}^n_{s\bar{\alpha}}$.

Proof Let C = the $\mathfrak{A}^n_s$-least C satisfying the above conditions. $C \in X^n_{s\bar{\alpha}}$ since it is definable from p in $\mathfrak{A}^n_s$. Set $\Pi = \Pi^n_{s\bar{\alpha}}$.

We have $\Pi(\bar{\alpha}) = \alpha$, $\Pi(C \cap \bar{\alpha}) = C$ and $\Pi(p{\restriction}\bar{\alpha}) = p$ since $X^n_{s\bar{\alpha}} \cap \alpha = \alpha$. Set $\bar{s} = \Pi^{-1}(s)$, then by §2.5 Facts 2 and 4 $s \in \mathbf{S}_\alpha$ and $\mathfrak{A}^n_{s\bar{\alpha}} = \mathfrak{A}^n_{\bar{s}}$. Using the above and the isomorphism Π it easily follows that $|p{\restriction}\bar{\alpha}| = |\bar{s}|$ in $\mathbf{P}^{\bar{s}}_\tau$, $\Pi(\mathbf{P}^{\bar{s}}_\tau) = \mathbf{P}^s_\tau$ and $\mathfrak{A}^n_{\bar{s}} = \mathfrak{A}^n_{s\bar{\alpha}} \models \bar{\alpha}$ is regular. By Fact 3 we have $\Pi(\mathfrak{A}_{\bar{\xi}}) = \mathfrak{A}_\xi$ where $\Pi(\bar{\xi}) = \xi < |s|$.

Since $\Pi^{-1}(C) = C \cap \bar{\alpha}$ is cub in $\bar{\alpha}$ we have $\bar{\alpha} \in C$ and therefore $\dot{p}_{\bar{\alpha}} = \emptyset$ and $|p{\restriction}\bar{\alpha}| = |p_{\bar{\alpha}}|$ in $\mathbf{P}^{p_{\bar{\alpha}}}_\tau$. Hence it suffices to show $p_{\bar{\alpha}} = \bar{s}$ to complete the proof. We will show $p_{\bar{\alpha}}(\bar{\xi}) = \bar{s}(\bar{\xi})$ by induction on $\bar{\xi} \in [\bar{\alpha}, |\bar{s}|)$.

Suppose it holds for $\eta < \bar{\xi}$. We have $\Pi(\bar{s}{\restriction}\bar{\xi}) = s{\restriction}\xi$ where $\Pi(\bar{\xi}) = \xi$. By (iii) of the definition of $\mathbf{P}^s_\tau$ there is an $m < \omega$ s.t. for $i \ge m$ there are $D_i \in \mathfrak{A}_\xi$ s.t.

$D_i \subseteq \mathrm{card} \cap [\tau,\alpha)$ is cub in α and

$$* \ \gamma \in D_i \longrightarrow \widehat{p}(\rho^i_{\xi\gamma}) = s(\xi) \ .$$

Let D_i be the $\mathfrak{A}_\xi$ least as above for $i \geq m$. Then $D_i \in X^n_{s\bar{\alpha}}$ and $\Pi^{-1}(P_i) = D_i \cap \alpha \in \mathfrak{A}_{\bar{\xi}} = \Pi^{-1}(\mathfrak{A}_\xi)$. Applying Π^{-1} to * we get:

$$\gamma \in D_i \cap \bar{\alpha} \longrightarrow \widehat{p}(\Pi^{-1}(\rho^i_{\xi\pi(\gamma)})) = \bar{s}(\bar{\xi}) \ .$$

But $\rho^i_{\xi\gamma} \in (\gamma^+,\gamma^{++})$, $\gamma^+ < \bar{\alpha}$ and $\pi \restriction \bar{\alpha} = \mathrm{id} \restriction \bar{\alpha}$ so $\pi(\rho^i_{\bar{\xi}\gamma}) = \rho^i_{\xi\gamma}$ and $\rho^i_{\bar{\xi}\alpha} = \rho^i_{\xi\gamma}$. Moreover $\mathfrak{A}_{\bar{s}\restriction\bar{\xi}} = \mathfrak{A}_{p_{\bar{\alpha}}\restriction\bar{\xi}}$; hence $\rho^i_{p_{\bar{\alpha}}\restriction\bar{\xi},\gamma} = \rho^i_{\bar{s}\restriction\bar{\xi},\gamma} = \rho^i_{\xi\gamma}$ for $\gamma \in D_i \cap \bar{\alpha}$. Hence

$$\gamma \in D_i \cap \bar{\alpha} \longrightarrow \widehat{p}(\rho^i_{p_{\bar{\alpha}}\restriction\bar{\xi},\gamma}) = \bar{s}(\bar{\xi}) \ .$$

Since $D_i \cap \bar{\alpha}$ is cub in $\bar{\alpha}$ $(i \geq n)$ it follows that $p_{\bar{\alpha}}(\bar{\xi}) = \bar{s}(\bar{\xi})$ by (iii) of the definition of $\mathbf{P}^s_\tau$.

Q.E.D.

Lemma 2.13

(a) Let $D \in \mathbb{S}^+_{\alpha^+}$, $\tau \leq \alpha$. *Then* $\mathbf{P}^D_\tau$ *satisfies the* α^{++}-A.C. *in* $\mathfrak{A}_D$.

(b) If $s \in \mathbb{S}_{\alpha^+}$, *then* $\mathbf{P}^s_\alpha$ *is* α*-distributive in* $\mathfrak{A}_D$.

(c) If $D \in \mathbb{S}^+_{\alpha^+}$, *then* $\mathbf{P}^D_\alpha$ *is* α*-distributive in* $\mathfrak{A}_D$.

Proof (a) As in Theorem 1.1. (b),(c) by Lemmas 2.6 and 2.8 since $\mathbf{P}^s_\alpha \cong \mathbb{R}^s$ in $\mathfrak{A}^1_s$ and $\mathbb{P}^D_\alpha = \mathbb{R}^D$ in $\mathfrak{A}_D$.

Q.E.D.

2.7 EXTENSION OF CONDITIONS IN $\mathbb{P}^s_\tau$

Our aim is to prove that $\mathbf{P}^s_\tau$ is τ-distributive for $s \in \mathbb{S}_\alpha$ and $\tau \leq \alpha$. However, we must first prove a number of more elementary combinatorial lemmas. We recall in proving Theorem 1.5 the verification of the distributive laws was preceded by a non-trivial proof that any condition had extensions of 'arbitrary length'. We now prove a body of lemmas which say, collectively, that conditions may be extended arbitrarily.

Definition 1 $X \subseteq$ card if thin iff $X \cap \kappa$ is not stationary in κ for every regular κ. Let U be a transitive ZF^- model. $X \in U$ is thin in U if $U \models$ "X is thin". Note: If X is thin in U, then X is thin.

Lemma 2.14 The Extension Lemma *Let α be a limit cardinal, $s \in \mathbf{S}_\alpha$, $\tau < \alpha$ and $p \in \mathbb{P}^s_\tau$, then*

(a) There is a $q \leq p$ s.t. $|q| = |s|$.

(b) There is a $q \leq p$ s.t. *for all limit cardinals* $\gamma \in (\tau,\alpha)$ $|q/\gamma| = |q_\gamma|$.

(c) Let $\Omega^p_\alpha \leq |s|$, *there is a* $q \leq p$ s.t. $\Omega^{q_\gamma}_\gamma$ *exists and* $|q/\gamma| \geq \Omega_\gamma$ *for* $\gamma \in C_\alpha \cap (\tau,\alpha)$.

(d) Let $X \in \mathfrak{A}_s$ *be thin in* $\mathfrak{A}_s$ *where* $X \subseteq [\tau,\alpha)$. *Let* $\langle \xi_\gamma \mid \gamma \in X \rangle$ *be* s.t. $|p_\gamma| \leq \xi_\gamma < \gamma^+$. *Then there is a* $q \leq p$ s.t. $|q_\gamma| \geq \xi_\gamma$ *for* $\gamma \in X$.

Proof The proof is by induction on α. Suppose the lemma to hold below α. The proof will stretch out over several sublemmas.

Lemma 2.14.1 *Let* $\mathfrak{A}_s \models$ "α is regular" *and* $|p| = |s|$ *then (b) and (d) hold.*

Proof Pick an n, $1 \leq n < \omega$ s.t.

(i) $p, X, \langle \xi_\gamma \mid \gamma \in X \rangle \in \mathfrak{A}^n_s$.

(ii) There is a $C \in \mathfrak{M}^n_s$ s.t. $C \subseteq (\tau,\alpha)$ is a cub in α and $\gamma \in C$ γ is a limit cardinal, $\dot{p}_\gamma = \emptyset$, $|p\restriction\gamma| = |p_\gamma|$ and $C \cap X = \emptyset$.

Such an n can be chosen since X is thin in $\mathfrak{M}_s$, $\mathfrak{M}_s \models$ "α is regular" and hence $X \cap \alpha$ is not stationary in α. Set:

$$D = \{\bar{\alpha} < \alpha \mid \bar{\alpha} = X_{s\bar{\alpha}} \cap \alpha \text{ and } p, C, X, \langle \xi_\gamma \mid \gamma \in X\rangle \in X^n_s\}.$$

D is cub in α and every $\bar{\alpha} \in D$ satisfies the hypothesis of the collapsing lemma. Hence for $\bar{\alpha} \in D$ setting $\pi_{\bar{\alpha}} = \pi^n_{s\bar{\alpha}}$ we have: $\mathfrak{M}^n_{s\bar{\alpha}} = \mathfrak{M}^n_{p\bar{\alpha}}$, $p_{\bar{\alpha}} = \pi^{-1}_{\bar{\alpha}}(s)$, $\pi_{\bar{\alpha}}\restriction\bar{\alpha} = id\restriction\bar{\alpha}$, $\pi_{\bar{\alpha}}(\bar{\alpha}) = \alpha$, $\pi_{\bar{\alpha}}(C\cap\bar{\alpha}) = C$, $\pi_{\bar{\alpha}}(X\cap\bar{\alpha}) = X$, $\pi_{\bar{\alpha}}(\langle \xi_\gamma \mid \gamma \in X\cap\bar{\alpha}\rangle) = \langle \xi_\gamma \mid \gamma \in X\rangle$, $\pi_{\bar{\alpha}}(p\restriction\bar{\alpha}) = p$, $|p\restriction\bar{\alpha}| = |p_{\bar{\alpha}}|$, $\dot{p}_{\bar{\alpha}} = \emptyset$, $\mathfrak{M}^n_{p_{\bar{\alpha}}} \models$ "α is regular" and $\bar{\alpha} \in X$ (since $\bar{\alpha} \in C$ and $C \cap X = \emptyset$).

Let $\langle \alpha_i \mid i < \alpha\rangle$ enumerate $\{\tau\} \cup D$ monotonically. In constructing q from p we do not have to change P_{α_i} $i < \alpha$, since they already satisfy (b) and (d). The rest can be changed by the induction hypothesis as follows:

For $i < \alpha$ set:

$$q_i = \text{the } \mathfrak{M}^n_{p\alpha_{i+1}}\text{-least } q \leq p\restriction[\alpha_i,\alpha_{i+1}) \text{ in } \mathbb{P}^{p\alpha_{i+1}}_{\alpha_i} \text{ s.t.}$$

(i) $|q\restriction\gamma| = |q_\gamma|$ for limit cardinals $\gamma \in (\alpha_i,\alpha_{i+1})$.

(ii) $|q_\gamma| \geq \xi_\gamma$ for $\gamma \in X \cap (\alpha_i,\alpha_{i+1})$.

(iii) $q_{\alpha_i} = p_{\alpha_i}$ for $i > 0$.

Such a q exists by the induction hypothesis noting $X \cap (\alpha_i,\alpha_{i+1})$ is thin in $\mathfrak{M}^n_{p\alpha_{i+1}}$ since $C \cap \alpha_{i+1} \cap X = \emptyset$.

Set $q = \bigcup_{i<\alpha} q_i$; then, $q \in \mathbb{P}^s_\tau$, $q \leq p$, $|q\restriction\gamma| = |q_\gamma|$ for limit cardinals $\gamma \in (\tau,\alpha)$ and $|q_\gamma| \geq \xi_\gamma$ for $\gamma \in X$.

This is a standard construction we use often; fixing $p_{\bar{\alpha}}$ for $\bar{\alpha} \in C = \langle \alpha_i \mid i < \tau \rangle$ a cub subset in α and extending p only in intervals $\mathbb{P}_{\alpha_i}^{p_{\alpha_{i+1}}}$. The clauses of the definition of $\mathbb{P}_\tau^s$ are easily verified for q; in particular clause (i), $q/\gamma \in \mathfrak{A}_{q_\gamma}$ for limit cardinals γ, is satisfied since q/γ and q_i/γ differ only on an initial bounded segment for $\gamma \in \text{card} \cap (\alpha_i, \alpha_{i+1}]$. $q \in \mathfrak{A}_s$ since the whole definition can be carried out in $\mathfrak{A}_s$.

Q.E.D.

Lemma 2.14.2 *Let* $|p| \geq \Omega_\alpha$. *Then there is a* $q \leq p$ s.t. $\Omega_\gamma^{q_\gamma}$ *exists and* $|q/\gamma| \geq \Omega_\gamma$ *for* $\gamma \in C_\alpha \cap (\tau,\alpha)$.

Proof

Case 1 $\text{cf}(\alpha) = \omega$: Let $\langle \gamma_i \mid i < \omega \rangle \in \mathfrak{A}_s$ be a monotone sequence of cardinals converging to α where $\gamma_0 = \tau$.

Claim For $i < \omega$ there is a $q_i \leq p/[\gamma_i^+, \gamma_{i+1}^+)$ in $\mathbb{P}^{p_{\gamma_{i+1}^+}}$ s.t. Ω_γ is defined and $[q_i/\gamma] \geq \Omega_\gamma$ for $\gamma \in C_\alpha \cap [\gamma_i^+, \gamma_{i+1}^+]$ and $q_i(\gamma_i^+) = p(\gamma_i^+)$.

Proof Let $\delta = \max C_\alpha \cap [\gamma_i^+, \gamma_{i+1}^+)$ if such a δ exists (if not there is nothing to show). Take a $p'(\delta) \leq p(\delta)$ in $\mathbb{R}^{p_{\delta^+}}$ s.t. $\mathfrak{A}_{p'_\delta} \models$ "α is singular". Note that $p/[\gamma_i^+, \delta) \in \mathbb{P}_{\gamma_i^+}^{p_\delta} \subseteq \mathbb{P}_{\gamma_i^+}^{p'_\delta}$ and we can apply the induction hypothesis getting a $q \leq p/[\gamma_i^+, \delta)$, $q \in \mathbb{P}_{\gamma_i^+}^{p'_\delta}$, $|q| = |p'_\delta|$ satisfying the claim. Then

$$q_i = q \cup \langle \delta, p'(\delta) \rangle \cup p/[\delta^+, \gamma_{i+1}^+) \in \mathbb{P}^{p_{\gamma_{i+1}^+}}$$

satisfies the claim since we only changed a bounded section of $p/[\gamma_i^+, \gamma_{i+1}^+)$.

Q.E.D. (Claim)

Set q_i = the $\mathfrak{A}^1_{P_{\gamma^+_{i+1}}}$-least $q \le p \restriction [\gamma^+_i, \gamma^+_{i+1})$ s.t. the above holds.

Set $q = \bigcup_{i<\omega} q_i \cup \{\langle \tau, p(\tau)\rangle\}$, then $q \le p$ has the desired properties.

Q.E.D. (Case 1)

<u>Case 2</u> $cf(\alpha) > \omega$: By clause (iv) of the definition of $\mathbf{P}^s_\tau$ there is a $D \in \mathfrak{A}_{\Omega_\alpha}$ s.t. D is cub in α and

$$\gamma \in D \longrightarrow \Omega^{p_\gamma}_\gamma \text{ exists, } |p\restriction\gamma| \ge \Omega\gamma \text{ and } D \cap \gamma \in \mathfrak{A}_{\Omega\gamma} .$$

The above will still hold if we take $D' = D \cap C_\alpha$ because $C_\alpha \in \mathfrak{A}_{\Omega\gamma}$ and

$$\gamma \in D' = D \cap C_\alpha \longrightarrow D' \cap \gamma = D \cap C_\alpha \cap \gamma = D' \cap C_\gamma \in \mathfrak{A}_{\Omega\gamma} .$$

Let $\langle \alpha_i \mid i < \lambda \rangle$ be a monotone enumeration of $\{\tau\} \cup D$. By the induction hypothesis we can choose;

$$q_i = \text{the } \mathfrak{A}_{P_{\alpha_{i+1}}}\text{-least } q \le p\restriction[\alpha_i, \alpha_{i+1}) \text{ in } \mathbf{P}^{p_{\alpha_{i+1}}}_{\alpha_i} \text{ s.t. } \Omega\gamma$$

exists and

$$|q\restriction\gamma| \ge \Omega\gamma \text{ for } \gamma \in C_\alpha \cap (\alpha_i, \alpha_{i+1}) \text{ and } q(\alpha_i) = p(\alpha_i) .$$

Set $q = \bigcup_{i<\lambda} q_i$. Then $q \le p$ has the desired properties.

Q.E.D.

Before proceeding further we need some additional notation.

Definition 2 Let β be a singular cardinal, $s \in \mathbf{S}_\beta$ and $|s| \ge \Omega_\beta$. U_β = the set of $\delta = \langle\langle \rho, |C_\beta|\rangle, 3\rangle$ s.t. $\delta \in \bigcup_{i<\lambda_\beta} [\gamma^+_i, \gamma^{++}_i)$ where $\langle \gamma_i \mid i < \lambda_\beta \rangle$ are defined in §2.5 definition 4. Then $\tilde{\rho}_{\xi_i} \in U_\beta$ for $\Omega_\beta \le \xi \le |s|$ and $i < \lambda_\beta$. If $|C_{\beta'}| < |C_\beta|$ then $U_{\beta'} \cap U_\beta = \emptyset$ so;

(1) $\beta' \in C_\beta \longrightarrow U_{\beta'} \cap U_\beta = \emptyset$.

(2) If $\beta' < \beta$ then $U_{\beta'} \cap U_\beta$ is bounded in β' since if $|C_{\beta'}| = |C_\beta|$ then $\beta' \in C_\beta$.

Definition 3 Let $\tilde{\mathbb{P}}^s_\tau$ = the set of partial maps from $[\tau,\beta)$ to 2 s.t.

(a) $\mathrm{dom}(p) \subseteq U_\beta$ and $\mathrm{card}(p \restriction [\gamma_i^+, \gamma_i^{++})) \le \gamma_i^+$ for $i < \lambda_\beta$.

(b) $p \in \mathfrak{A}_s$. Set $|p| = |p|_{\tilde{\mathbf{P}}^s_\tau}$ = the least $\xi \le |s|$ s.t. $\xi \ge \Omega_\beta$ and $p \in \mathfrak{A}_\xi$.

(c) If $\Omega_\beta \le \xi < |p|$, then $\exists n < \lambda_\beta \ \forall i < \lambda_\beta \ (n \le i \longrightarrow p(\tilde{\rho}\xi_i) = s(\xi))$.

It follows exactly as in the proof of Theorem 1.5 that if $p \in \tilde{\mathbf{P}}^s_\tau$ and $|p| \le \xi \le |s|$, then there is a $q \le p$ in $\tilde{\mathbb{P}}^s_\tau$ s.t. $|q| = \xi$.

It is easy to see that if $p \in \mathbb{P}^s_\tau$ and $|p| \ge \Omega_\beta$ in $\mathbb{P}^s_\tau$ then $p \restriction U_\beta \in \tilde{\mathbf{P}}^s_\tau$ and $|p \restriction U_\beta|_{\tilde{\mathbf{P}}^s_\tau} \le |p|_{\mathbf{P}^s_\tau}$.

Making use of $\tilde{\mathbf{P}}^s_\tau$, we shall show that we can extend $|p| \ge \Omega_\alpha$ arbitrarily on a cub subset of α (except for the ω cofinal case) to a height of $|s|$.

Lemma 2.14.3 *Let* $p \in \mathbf{P}^s_\tau$, $|p| \ge \Omega_\alpha$ *and* $|p \restriction \gamma| \ge \Omega_\gamma$ *for* $\gamma \in C_\alpha \cap (\tau,\alpha)$. *Let* $\langle \beta_i \mid i < \rho \rangle$ *be the monotone enumeration of* $\{\gamma < \alpha \mid \gamma$ *is a limit point of* $(\tau,\alpha) \cap \{\gamma_i^\alpha \mid i < \lambda_\alpha\}\}$.
(Note that for $\mathrm{cf}(\alpha) = \omega$, ρ can be $\emptyset$ or a successor ordinal, while if $\mathrm{cf}(\alpha) > \omega$ then $\sup_{i<\rho} \beta_i = \alpha$.) In addition $\rho < \beta_0$ by definition.

Let $\langle r_i \mid i < \rho \rangle \in \mathfrak{A}_s$ s.t. $r_i \le p(\beta_i)$ in $\mathbf{R}^{p_{\beta_i^+}}$ for $i < \rho$. Then there is a $q \le p$ in $\mathbf{P}^s_\tau$ s.t. $|q| = |s|$, $|q \restriction \beta_i| = |q_{\beta_i}|$ and $q(\beta_i) \le r_i$ in $\mathbf{R}^{p_{\beta^+}}$ for $i < \rho$.

Proof Let $\tilde{q} \leq \overset{\frown}{p}{\restriction}U_\alpha$ in $\tilde{\mathbb{P}}^s_\tau$ s.t. $|\tilde{q}| = |s|$ in $\tilde{\mathbb{P}}^s_\tau$.

Let $r_i = \langle \dot{r}_i, r_i^! \rangle$ for $i < \rho$. We shall define

1. $r_{ij} \in \mathbf{S}_{\beta_i}$ for $i < \rho$ and $j < \omega$ s.t.

   $\langle \dot{r}_i, r_{ij} \rangle \leq \langle \dot{r}_i, r_{ik} \rangle \leq \langle \dot{r}_i, r_i^! \rangle$ in $\mathbf{R}^{P_{\beta_i^+}}$ for $k \leq j < \omega$.

2. $\tilde{q}_{ij} \in \tilde{\mathbb{P}}^{r_{ij}}$ for $i < \rho$; $j < \omega$ s.t.

   $\tilde{q}_{ij} \leq \tilde{q}_{ik} \leq \overset{\frown}{p}{\restriction}U_{\beta_i}$ in $\tilde{\mathbb{P}}^{r_{ij}}_\tau$ for $k \leq j < \omega$.

We define r_{ij}, q_{ij} by induction on j as follows:

$$r_{i_0} = r_i^!, \quad q_{i_0} = \overset{\frown}{p}{\restriction}U_{\beta_i}.$$

$r_{i,j+1}$ = the least r s.t. $|r|$ is the least η s.t.

$Z_4 \cap [|r_{ij}|, \eta)$ has order type β_i and

$r{\restriction}(Z_4 \cap [|r_{ij}|, |r|)$ codes $p{\restriction}\beta_i$,

$\langle q_{hk}{\restriction}\beta_i \mid h < \rho, k \leq j \rangle$, $\langle r_{hk} \quad h \leq i, k \leq j \rangle$ and $\tilde{q}{\restriction}\beta_i$ in a uniform manner.

We can do this in the interval $[|r_{ij}|, |r_{i,j+1}|)$ since $q_{hk}{\restriction}\beta_i$ is bounded in β_i for $h \neq i$ ($U_{\beta_h} \cap U_{\beta_i}$ is bounded for $h > i$) so $\mathrm{card}(\{q_{hk}{\restriction}\beta_i | h<\rho, k\leq j\}) \subseteq \mathrm{card}(p_{<\beta_i}(\beta_i)) = \beta_i$. Also note $\rho < \beta_i$.

$\tilde{q}_{i,j+1}$ = the $\mathfrak{A}_{r_{i,j+1}}$ - least $\tilde{q} \leq \tilde{q}_{ij}$ in $\tilde{\mathbb{P}}^{r_{i,j+1}}_\tau$ s.t.

$|\tilde{q}| = |r_{i,j+1}|$.

Set $r_i^* = \bigcup_{j<\omega} r_{ij}$, $\tilde{q}_i = \bigcup_{j<\omega} q_{ij}$.

<u>Claim (1)</u> $r_i^* \in \mathbf{S}_{\beta_i}$ and $\langle \dot{r}_i, r_i^* \rangle \leq \langle \dot{r}_i, r_i^! \rangle$ in $\mathbf{R}^{P_{\beta_i^+}}$.

Proof The only problem is $r_i^* \in \mathbf{S}_{\beta_i}$.

$|r_i^*|$ = the least η s.t. $[|r_i^!|, \eta) \cap Z_4$ has order type $\omega\beta_i$

and then $L_{|r_i^*|+\omega} \models \mathrm{card}(|r_i^*|) \le |r_i|$; hence if

$$L_\mu[A\cap\beta_i, r_i^!] \models \mathrm{card}(|r_i^!|) = \beta_i \text{ , where } \mathrm{card}(\mu)^L \le |r_i|$$

then $L_{\mu'}[A\cap\beta_i, r_i^*] \models |r_i^*| = \beta_i$ and

$$\mathrm{card}(\mu')^L \le |r_i^!| \le |r_i^*| \text{ , where } \mu' = \max(\mu, |r_i^*| + \omega) .$$

Q.E.D. (Claim 1)

Clearly p/β_i , $\langle \tilde{q}_{hk}/\beta_i \mid h < \rho, k < \omega\rangle$, $\tilde{q}/\beta_i$, $\langle r_{hk} \mid h < i, k < \omega\rangle \in \mathfrak{N}_{r_i^*}$. Moreover:

$$p, \langle \tilde{q}_{hk} \; h<\rho, k<\omega\rangle , \langle r_{hk} \; h<\rho, k<\omega\rangle , \tilde{q} \in \mathfrak{N}_s .$$

<u>Claim (2)</u> $\tilde{q}_i \in \tilde{\mathbb{P}}_\tau^{r_i^*}$ and $|\tilde{q}_i| = |r_i^*|$.

Proof Since $\tilde{q}_i = \bigcup_{j<\omega} q_{ij}$ no cardinality violations can occur in part (a) of the definition of $\tilde{\mathbb{P}}^s$ above. Part (b) holds since by the above comment $\langle \tilde{q}_{ik} \mid k < \omega\rangle \in \mathfrak{N}_{r_i^*}$ and $|q_{ij}| = |r_{i,j+1}|$ and $\sup_{j<\omega} |r_{i,j+1}| = |r_i^*|$. Part (c) holds because for each $|\tilde{q}_{ij}|$ and for $\xi < |r_i^*|$ there is a $j < \omega$ s.t. $|\tilde{q}_{ij}| \ge \xi$.

Q.E.D. (Claim 2)

Set $H = \{\gamma_i^+ \mid i < \lambda_\alpha \text{ and } \gamma_i \ge \tau\}$.

Define $q : \mathrm{card} \cap [\tau, \alpha) \longrightarrow V^2$ as follows:

1. $q(\beta) = \langle \dot{r}_i, r_i^*\rangle$ for $i < \rho$.
2. For $\delta \in H$ let $q_\delta^! = \bigcup_{i<\rho} \tilde{q}_i/[\delta,\delta^+) \cup \tilde{q}/[\delta,\delta^*)$.

   Note $\forall\, i < j < \rho$, $U_{\beta_i} \cap U_{\beta_j} = U_{\beta_i} \cap U_\alpha = \emptyset$ since $\beta_i \in C_{\beta_j} = \beta_j \cap C_\alpha$. So the $\tilde{q}_i$ - s and q have no domain in common and do not conflict.

   Let q_δ = the $L_{\delta^+}[A]$-least r s.t. $\langle \dot{p}_\delta, r\rangle \le p(\delta)$ in $\mathbb{R}^{p_{\delta^+}}$ and $r \ge q_\delta^!$ and $|r| = \sup(\mathrm{dom}(q_\delta^!))$. Such an r exists, since $\dot{p}_\delta$ places restrictions on $r(\xi)$ only for $\xi \in Z_0 \cup Z_1$, but $[\mathrm{dom}(q'_\delta) - \mathrm{dom}(p_\delta)] \subseteq Z_3$. $\bigcup_{n>3} Z_n$ leaves enough room to find an $r \supseteq q'_\delta$ s.t. $r \in \mathbf{S}_\delta$ and $|r| = \sup(\mathrm{dom}(q'_\delta))$. Finally for $\delta \in H$ set $q(\delta) = \langle \dot{p}_\delta, q_\delta \rangle$.

3. In all other cases set $q(\gamma) = p(\gamma)$.

Claim (3) $\forall i < \rho \ (q/\beta_i \in \mathfrak{A}_{q_{\beta_i}})$ and $q \in \mathfrak{A}_s$.

Proof Note that $\mathfrak{A}_{q_{\beta_i}} = \mathfrak{A}_{r_i^*}$ so the whole construction of q/β_i can be carried out in $\mathfrak{A}_{r_i^*}$. Clearly $q \in \mathfrak{A}_s$.

Q.E.D. (Claim 3)

Claim (4) $q \in \mathbf{P}_\tau^s$, $|q| = |s|$ and $|q/\beta_i| = |q_{\beta_i}|$ for $i < \rho$.

Proof We first show by induction on $\delta \in \mathrm{card} \cap [\tau,\alpha)$ that $q/\delta \in \mathbb{P}_\tau^{q_\delta}$ and if $\delta = \beta_i$ then $|q/\delta| = |q_\delta|$.

Case 1 $\delta = \tau$, trivial.

Case 2 $\delta = \gamma^+$, $\gamma \geq \tau$. It suffices to observe that $q(\gamma) \in \mathbf{R}^{q_{\gamma+}}$ and $q/\gamma \in \mathbf{P}_\tau^{q_\gamma}$ (by the induction hypothesis).

Case 3 δ is a limit cardinal and $\delta \notin \{\beta_i \mid i < \rho\}$. The conclusion follows because $q_\delta \supseteq p_\delta$ and q/δ differs from p/δ only on an initial segment. (We change $p(\gamma_i^+)$ for $i < \lambda_\alpha$ and β_i are limit points of the γ_i .)

Case 4 $\delta = \beta_i$. Since $\mathfrak{A}_{q_{\beta_i}} \models$ "α is singular" we must verify (i), (iii), (iv) and (v) of the definition of $\mathbf{P}_\tau^s$ for $s = q_\delta$ and $p = q/\delta$. Claim 2 establishes (i). (iii) is clear since $|p/\delta| \geq \Omega_\delta$ and $\overset{\frown}{q} \supseteq \overset{\frown}{p}$ so p already takes care of all cases where $\mathfrak{A}_{q_{\beta_i}/\xi} \models$ "α is regular".

(iv) is clear by taking $D = C_\delta = C_\alpha \cap \delta$. (v) follows with $|q{\restriction}\delta| = |q_\delta|$, since $\widehat{(q{\restriction}\delta)}{\restriction}U_\delta \supseteq \tilde{q}_i$, $\tilde{q}_i \in \tilde{\mathbb{P}}_\tau^{q_{\beta_i}} = \mathbf{P}_\tau^{|r_i^*|}$ and $|\tilde{q}_i| = |q_{\beta_i}|$. It remains to show $q \in \mathbf{P}_\tau^s$ and $|q| = |s|$ but this follows exactly as in Case 3.

Q.E.D. (Claim 4)

It easily follows that $q \le p$ has the desired properties.

Q.E.D. (Lemma)

Lemma 2.14.4 *Let* $|p| \ge \Omega_\alpha$. *Then Lemma 2.14 holds.*

Proof By Lemmas 2.14.2 and 2.14.3 we may assume that:

(a) $|p| = |s|$ and $|p{\restriction}\beta_i| = |p_{\beta_i}|$ where $\langle \beta_i \mid i < \rho \rangle$ is as in Lemma 2.14.3.

(b) Ω_γ exists and $|p{\restriction}\gamma| \ge \Omega_\gamma$ for $\gamma \in C_\alpha \cap (\tau,\alpha)$ (Lemma 2.14.2).

(c) $|p_{\beta_i}| \ge \xi_{\beta_i}$ for $i < \rho$, $\beta_i \in X$ (Lemma 2.14.3).

(d) $X \cap \beta_i$, $\langle \xi_\gamma \mid \gamma \in X \cap \beta_i \rangle \in \mathfrak{A}_{p_{\beta_i}}$ for $i < \rho$ (Lemma 2.14.3).

Case 1 cf(α) = ω . Let $\langle \delta_i \mid i < \omega \rangle$ be a monotone sequence of cardinals converging to α s.t. $\delta_0 = \tau$.

Set: q_i = the $\mathfrak{A}_{p_{\delta^+_{i+1}}}$ - least $q \le p{\restriction}[\gamma_i^+, \gamma_{i+1}^+)$ in $\mathbb{P}_{\delta_i^+}^{p_{\delta^+_{i+1}}}$ s.t. $|q{\restriction}\gamma| = |q_\gamma|$ for all γ a limit point of $(\text{card} \cap [\delta_i^+, \delta_{i+1}^+))$ and $|q_\gamma| \ge \xi_\gamma$ for $\gamma \in X \cap [\delta_i^+, \delta_{i+1}^+)$.
q_i exists by the induction hypothesis.
Let $r \le p{\restriction}\tau^+$ in $\mathbf{P}_\tau^{p_{\tau^+}}$ s.t. $|r_\tau| \ge \xi_\tau$ if $\tau \in X$.
Set: $q = r \cup \bigcup_{i<\omega} q_i$. Then $q \le p$ has the desired property.

Case 2 cf(α) > ω . Then $\sup_{i<\rho} \beta_i = \alpha$. For $i < \rho$ set:

q_i = the $\mathfrak{A}_{p_{\beta_{i+1}}}$ - least $q \le p/[\beta_i,\beta_{i+1}]$ in $\mathbf{P}_{\beta_i}^{p_{\beta_{i+1}}}$ s.t.

$|q/\gamma| = |q_\gamma|$ for γ a limit point of $(\mathrm{card} \cap (\beta_i,\beta_{i+1}))$,

$|q_\gamma| \ge \xi_\gamma$ for $\gamma \in X \cap (\beta_i,\beta_{i+1})$ and $q_{\beta_i} = p_{\beta_i}$.

q_i exists by the induction hypothesis as does:

q' = the $\mathfrak{A}_{p_{\beta_0}}$ - least $q \le p/[\tau,\beta_0)$ s.t. $|q/\gamma| = |q_\gamma|$

for γ a limit point of $(\mathrm{card} \cap (\tau,\beta_0))$ and $|q_\gamma| \ge \xi_\gamma$

for $\gamma \in X \cap [\tau,\beta_0)$.

Set $q = q' \cup \bigcup_{i<\rho} q_i$. Then $q \ge p$ has the desired property.

Q.E.D.

By Lemmas 2.14.1 and 2.14.4 it will suffice to prove Lemma 2.14(a) for the case: $|p| \not\ge \Omega_\alpha$ (including the case that Ω_α does not exist). But then it suffices to prove it for $|s| \not> \Omega_\alpha$, since for $|s| > \Omega_\alpha$, we can get the full theorem by applying Lemma 2.14(a) to s/Ω_α and then use Lemma 2.14.4. Thus it suffices to prove:

Lemma 2.14.5 *If* $|s| \not> \Omega_\alpha$ *, then there is a* $q \le p$ s.t. $|q| = |s|$.

Proof By induction on $|s|$ we show: If $|p| < |s|$ and $C \in \mathfrak{A}_{|p|}$ is cub in α , then there is a $q \le p$ s.t. $|q| = |s|$, $\dot{q}_\gamma = \dot{p}_\gamma$ everywhere and $q_\gamma = p_\gamma$ for $\gamma \not\in C \cup \{\delta^+ \mid \delta \in C\}$.

<u>Case 1</u> $|s| = \xi + 1$. We may assume w.l.o.g. that $|p| = \xi$. Let $C \in \mathfrak{A}_\xi$ be as given in the induction hypothesis. We may assume that $\gamma \in C \longrightarrow |p/\gamma| = |p_\gamma|$ and $\dot{p}_\gamma = \emptyset$ since we have such a $C \in \mathfrak{A}_\xi$ and could take the intersection of the two cub sets. Let $1 \le m < \omega$ be the least s.t. p , $C \in \mathfrak{A}_\xi^m$. For $m \le i < \omega$ set:

$$D_i = \{\bar\alpha < \alpha \mid \bar\alpha = \alpha \cap X^i_{\xi\bar\alpha} \text{ and } p, C \in X^i_{\xi\bar\alpha}\} .$$

Each $\bar\alpha \in D_i$ satisfies the hypothesis of the collapsing lemma, hence setting $\pi = \pi^i_{\bar\alpha} = \pi^i_{\xi\bar\alpha}$ we have for $\bar\alpha \in D_i$:

$$\pi\restriction\bar\alpha = id\restriction\bar\alpha , \ \pi(\bar\alpha) = \alpha , \ \pi(p\restriction\bar\alpha) = p , \ \pi(C \cap \bar\alpha) = C ,$$

$\pi(p_{\bar\alpha}) = s\restriction\xi$, $\mathfrak{A}^i_{p_{\bar\alpha}} = \mathfrak{A}^i_{\xi\bar\alpha}$ and $\mathfrak{A}_{p_{\bar\alpha}} \models \alpha$ is regular.

Clearly $C \subseteq D_m$.

Claim (1) $D_{i+1} \subseteq$ fixed points of D_i for $i \geq m$.

Proof Let $\bar\alpha \in D_{i+1}$, $X^i_{\xi\bar\alpha} \in X^{i+1}_{\xi\bar\alpha}$ so we can define $D_i \cap \bar\alpha$ in $X^{i+1}_{\xi\bar\alpha}$ by,

$$\{\beta < \alpha \mid X^i_{\xi\beta} \cap \alpha = \beta \text{ and } p, C \in X^i_{\xi\beta}\}$$

since $\alpha \cap X^{i+1}_{\xi\bar\alpha} = \bar\alpha$. $D_i \cap \alpha$ must have order type $\bar\alpha$ (since D_i has order type α and $\pi^{i+1}_{\xi\bar\alpha}(D_i \cap \bar\alpha) = D_i$) . Hence $\bar\alpha$ is a fixed point of D_i .

Q.E.D. (Claim 1)

Note that in the above proof we showed that

$$\pi^i_{\bar\alpha}(D_k \cap \bar\alpha) = D_k \quad \text{for} \quad m \leq k < i .$$

Set $D = \bigcap_i D_i$ and $\pi_{\bar\alpha} = \bigcup_i \pi^i_{\bar\alpha}$ for $\bar\alpha \in D$. $\pi_{\bar\alpha}$ is well defined since by §2.5 FACT 5

$$\pi^{i+1}_{s\bar\alpha} \supseteq \pi^i_{s\bar\alpha} \ (\bar\alpha = \alpha \cap X^i_{\xi\bar\alpha} \text{ for all } i)$$

$\pi_{\bar\alpha} : \mathfrak{A}_{p_{\bar\alpha}} \xrightarrow[\Sigma_1]{} \mathfrak{A}_\xi$ cofinally since $\mathrm{dom}(\pi_{\bar\alpha}) = \bigcup_i \mathrm{dom}(\pi^i_{\bar\alpha}) = \bigcup_i \mathfrak{A}^i_{p_{\bar\alpha}} = \mathfrak{A}_{p_{\bar\alpha}}$ and $\mu^i_\xi \in X^{i+1}_{\xi\bar\alpha}$ and $\mu^\omega_\xi = \mu_\xi = \sup_{i<\omega} \mu^i_\xi = \sup_i (\mathrm{rng}(\pi^i_{\bar\alpha}) = \sup (\mathrm{rng}(\pi_{\bar\alpha}))$, hence is cofinal and Σ_1 .

We wish to extend p to a height of $\xi+1$. By the definition of $\mathbf{P}_\tau^s$ (iii) if $p' \le p$, $|p'| = \xi+1$ there must be a m_0 s.t. for $i \ge m_0$ there is a cub $C_i \in \mathfrak{M}_\xi$ s.t. $\gamma \in C_i \longrightarrow \frown{p}'(\rho^i_{\xi\gamma}) = s(\xi)$. As a first step we will define $q_i \le p$, $m \le i < \omega$ s.t. $|q_i| = |p| = \xi$ (in fact $q_i \in \mathfrak{M}^{i+1}_\xi$) and $\gamma \in D_i \longrightarrow q_i(\rho^i_{\xi\gamma}) = s(\xi)$.

We define q_i by induction on $m \le i < \omega$ as follows: Set $q_{m-1} = p$ and define q_i $(i \ge m)$ by:

1. If $\gamma \not\in \{\bar{\alpha}^+ \mid \bar{\alpha} \in D_i\}$ then $q_i(\gamma) = q_{i-1}(\gamma)$.
2. If $\gamma = \bar{\alpha}^+$ for $\bar{\alpha} \in D_i$, set $\dot{q}_{i\gamma} = \dot{q}_{i-1,\gamma} = \dot{p}_\gamma$ and $q_{i\gamma}$ = the $L_{\gamma^+}[A]$-least $r \in \mathbf{S}_\gamma$ s.t. $|r| = \rho^i_{\xi\bar{\alpha}}+1$, $\langle \dot{p}_\gamma, r\rangle \le q_{i-1}(\gamma)$ in $\mathbf{R}^{p_{\gamma^+}}$ and $r(\rho^i_{\xi\bar{\alpha}}) = s(\xi)$.

Such an r exists since q_{i-1} is $\mathfrak{M}^1_\xi$-definable from p and C; so $q_{i-1} \in X^i_{\xi\bar{\alpha}} \subseteq X^i_{\xi\bar{\alpha}^+}$ and $X^i_{\xi\bar{\alpha}^+} \cap \bar{\alpha}^{++}$ is transitive. As before $\mathrm{dom}(\frown{q}_{i-1}) \subseteq \mathfrak{M}^i_{\xi\bar{\alpha}^+}$ and $\rho^i_{\xi\bar{\alpha}} \not\in \mathrm{dom}(\frown{q}_{i-1})$. In addition $\rho^i_{\xi\bar{\alpha}} \in Z_2$ and $\dot{p}_\gamma$ restricts only $Z_0 \cup Z_1$.

We will show that $q_i \le p \wedge |q_i| = |p|$.

<u>Claim (2)</u> $q_i{\restriction}\gamma \in \mathbf{P}_\tau^{q_{i\gamma}}$ for $\gamma \in \mathrm{card} \cap [\tau,\alpha)$ and $|q_i{\restriction}\gamma| = |q_\gamma|$ for $\gamma \in D_i$.

Proof By induction on γ.

(a) For $\gamma = \tau$, trivial.

(b) For $\gamma = \beta^+ > \tau$, it suffices to observe that $q_i(\beta) \in \mathbf{R}^{q_{i\beta^+}}$ and $q_i{\restriction}\beta \in \mathbf{P}_\tau^{q_{i\beta}}$ by the induction hypothesis.

(c) For γ a limit cardinal and $\gamma \not\in D_i$. This follows since $q_i{\restriction}\gamma$ differs from $q_{i-1}{\restriction}\gamma$ only on an initial segment; hence

$$q_i{\restriction}\gamma \in \mathbf{P}_\tau^{q_{i-1,\gamma}} \subseteq \mathbf{P}_\tau^{q_{i\gamma}}.$$

(d) For $\gamma = \alpha \in D_i$; we must verify (i)-(v) in the definition of $\mathbf{P}^s_\tau$ with $s = q_{i\bar\alpha}$ and $p = q_i{\restriction}\bar\alpha$. We have $q_{i\bar\alpha} = p_{\bar\alpha} = (\pi^i_{\bar\alpha})^{-1}(s{\restriction}\xi)$ and $\mathfrak{A}^i_{p\bar\alpha} = \mathfrak{A}^i_{\xi\bar\alpha}$. Using $\pi^i_{\bar\alpha}$ we see that $D_i \cap \bar\alpha$ can be defined from $\mathfrak{A}^i_{p\bar\alpha}$, $p{\restriction}\bar\alpha$ and $C \cap \bar\alpha$ as D_i was defined from $\mathfrak{A}^i_\xi$, p , C . Since $q_i{\restriction}\bar\alpha$ is definable in $\mathfrak{A}^i_{p\bar\alpha}$ from $D_i \cap \bar\alpha$ we have $q_i{\restriction}\bar\alpha \in \mathfrak{A}^{i+1}_{p\bar\alpha} \subseteq \mathfrak{A}_{p\bar\alpha}$. This establishes (i). Since $\mathfrak{A}^{i+1}_{p\bar\alpha} \models$ "α is regular", we only have to show (ii)-(iv). (ii) follows by the fact that $C \cap \bar\alpha \in \mathfrak{A}_{p\bar\alpha}$ and $(q_i{\restriction}\gamma) \leq (p{\restriction}\gamma)$ in $\mathbf{P}^{p_\gamma}_\tau$, $|q_i{\restriction}\gamma| = |p{\restriction}\gamma| = |p_\gamma| = |q_\gamma|$ for $\gamma \in C \cap \bar\alpha$. (iii) and (iv) hold by the fact that $|q_i{\restriction}\bar\alpha| = |p{\restriction}\bar\alpha|$, since $q_i{\restriction}\bar\alpha \in \mathfrak{A}_{p\bar\alpha}$ and q_i extends p .

Q.E.D. (Claim 2)

But applying the same argument as (d) to q_i at α shows that (i)-(v) hold with $p = q_i$, $s = s{\restriction}|p|$ and $|q_i| = |p|$. Hence $q_i \in \mathbf{P}^s_\tau$, $q_i \leq p$ and $|q_i| = \xi$.

Now define $q\colon \mathrm{card} \cap [\tau,\alpha) \longrightarrow V^2$ as follows:

$\dot q_\gamma = \dot q_{i\gamma} = \dot p_\gamma$ everywhere.

$q_\gamma = \bigcup_i q_{i\gamma}$ for $\gamma \notin D = \bigcap_{i<\omega} D_i$.

$q_\gamma = p_\gamma \cup \{\langle \bar\xi, s(\xi)\rangle\}$ for $\gamma \in D$, where $\pi_\gamma(\bar\xi) = \xi$.

We shall show that $q \leq p$ and $|q| = |s|$ completing Case 1.

<u>Claim (3)</u> $q{\restriction}\gamma \in \mathbf{P}^{q_\gamma}_\tau$ for $\gamma \in \mathrm{card} \cap [\tau,\alpha)$ and $|q{\restriction}\gamma| = |q_\gamma|$ if $\gamma \in D$.

Proof By induction on γ .

(a) For $\gamma = \tau$ trivial as before.

<u>FACT</u> If $\alpha_i < \delta_i$, $\alpha_i > \omega_2$ and $\mathrm{card}(\alpha_i)^L = \mathrm{card}(\delta_i)^L$ for $i < \omega$ then $\mathrm{card}(\sup_i \alpha_i)^L = \mathrm{card}(\sup_i \delta_i)^L$.

<u>Proof</u> We use $\neg 0^\#$ and "Marginalia" [DJ]. Let $\alpha = \sup_i \alpha_i$ and $\delta = \sup_i \delta_i$. If $\mathrm{card}(\alpha)^L < \mathrm{card}(\delta)^L$

then $\mathrm{card}(\delta)^L$ must be the successor of $\mathrm{card}(\alpha)^L$ in L and hence regular in L. But then $\mathrm{cf}(\delta) = \mathrm{card}(\delta) \geq \omega_2 > \omega$ which contradicts $\mathrm{cf}(\delta) = \omega$.

(b) Let γ be a successor cardinal. If $\gamma = \alpha^+$ and $\alpha \notin D$ then the claim is trivial. If $\gamma = \alpha^+$ and $\alpha \in D$ then $q_\gamma = \bigcup_{i<\omega} q_{i\gamma}$. The only non-trivial property we have to show is $q_\gamma \in \mathbf{S}_\gamma$.
$L[A \cap \gamma, q_\gamma] \models$ "$\mathrm{card}(|q_\gamma|) = \gamma$" since $q_\gamma = \bigcup_{i<\omega} q_{i\gamma}$ and $\langle q_{i\gamma} \mid i < \omega \rangle$ is $L[A \cap \gamma, q_\gamma]$ definable from q_γ and $\langle \rho^i_{\xi\bar{\alpha}} \mid i < \omega \rangle$, and $\langle \rho^i_{\xi\bar{\alpha}} \mid i < \omega \rangle$ is $L[A \cap \gamma, q_\gamma]$ definable. In addition if μ_i is s.t. $L_{\mu_i}[A \cap \gamma, q_{i\gamma}] \models$ "$\mathrm{card}(|q_{i\gamma}|) = \gamma$" and $\mathrm{card}(\mu_i)^L = \mathrm{card}(|q_{i\gamma}|)^L$, then by the above fact $L_\mu[A \cap \gamma, q_\gamma] \models$ "$\mathrm{card}(|q_\gamma|) = \gamma$" and $\mathrm{card}(\mu)^L = \mathrm{card}(|q_\gamma|)^L$ where $\mu = \sup_{i<\omega}(\mu_i) + \omega$. Hence $q_\gamma \in \mathbf{S}_\gamma$.

(c) For γ a limit cardinal, $\gamma \notin D$. There is an $i_0 < \omega$ s.t. $\gamma \notin D_i$ for $i \geq i_0$, hence $D_i \cap \gamma$ is bounded in γ for $i \geq i_0$ and $q \restriction \gamma$ differs from $q_{i_0} \restriction \gamma$ only on an initial segment.

(d) Now let $\gamma = \bar{\alpha} \in D$. We have $\pi^{-1}_{\bar{\alpha}}(s \restriction \xi) = p_{\bar{\alpha}}$ and by the definition of $q_{\bar{\alpha}}$, $p_{\bar{\alpha}} = q_{\bar{\alpha}} \restriction \bar{\xi}$, where $\bar{\xi} = \pi^{-1}_{\bar{\alpha}}(\xi)$. Hence $q_{\bar{\alpha}} = \pi^{-1}_{\bar{\alpha}}(s)$, since $|s| = \xi + 1$ and $\langle \bar{\xi}, s(\xi) \rangle \in q_{\bar{\alpha}}$. Since $\mathfrak{A}_{q_{\bar{\alpha}} \restriction \bar{\xi}} = \mathfrak{A}_{p_{\bar{\alpha}}}$, we have $\pi_{\bar{\alpha}}: \mathfrak{A}_{\bar{\xi}} \xrightarrow[\Sigma_1]{} \mathfrak{A}_\xi$ cofinally.
In particular, $\pi_{\bar{\alpha}}(p \restriction \bar{\alpha}) = p$, $\pi_{\bar{\alpha}}(C \cap \bar{\alpha}) = C$, $\pi_{\bar{\alpha}}(D_i \cap \bar{\alpha}) = D_i$ and $\pi_{\bar{\alpha}}(q_i \restriction \bar{\alpha}) = q_i$. But then we can define $q \restriction \bar{\alpha}$ from $\mathfrak{A}_{\bar{\xi}}$, $p \restriction \bar{\alpha}$, $C \cap \bar{\alpha}$ in $\mathfrak{A}_{q_{\bar{\alpha}}}$ as we defined q from $\mathfrak{A}_\xi$, p, C in $\mathfrak{A}_s$; hence $q \restriction \bar{\alpha} \in \mathfrak{A}_{q_{\bar{\alpha}}}$ establishing (i). $q \restriction \bar{\alpha} \notin \mathfrak{A}_{p_{\bar{\alpha}}}$, since otherwise $q \restriction \bar{\alpha} \in \mathfrak{A}^n_{p_{\bar{\alpha}}}$ for some $n \geq m$. But then $\rho^n_{p_{\bar{\alpha}},\gamma} \in \mathrm{rng}(\widehat{q \restriction \bar{\alpha}})$ for sufficiently large $\gamma \in D_m \cap \bar{\alpha}$, contradicting the fact that $q \restriction \bar{\alpha}$ extends $q_n \restriction \bar{\alpha}$; so $|q \restriction \bar{\alpha}| = |q_{\bar{\alpha}}|$. If

$\mathfrak{A}_{q_{\bar\alpha}} \models$ "$\bar\alpha$ is regular" then $D \cap \bar\alpha = \bigcap_i (D_i \cap \bar\alpha)$ is cub in $\bar\alpha$, $D \cap \bar\alpha \in \mathfrak{A}_{q_{\bar\alpha}}$ and $\delta \in D \cap \bar\alpha \longrightarrow \dot q_\delta = \dot p_\delta = \emptyset$ and $|q/\delta| = |q_\delta|$. ($|q/\delta| = |q_\delta|$ by the induction hypothesis.) This proves (ii). (iii) holds for $\gamma < \bar\xi$, since then $\gamma < |p_{\bar\alpha}|$. But (iii) holds for $\bar\xi$ since $D_i \cap \bar\alpha \in \mathfrak{A}_{\bar\xi}$ and for $\delta \in D_i \cap \bar\alpha$, $\widehat{q}(\rho^i_{\bar\xi\delta}) = \widehat{q}_i(\rho^i_{\bar\xi\delta}) = s(\xi)$.
$\pi_{\bar\alpha}(\mathfrak{A}_{q_{\bar\alpha}/\bar\xi,\delta}) = \mathfrak{A}_{q/\xi,\delta}$ so $\pi_{\bar\alpha}(\rho^i_{\bar\xi\delta}) = \rho^i_{\xi\delta}$ but $\rho^i_{\bar\xi\delta} \in (\delta^+, \delta^{++})$ and $\pi_{\bar\alpha}/\bar\alpha = \mathrm{id}/\bar\alpha$ so $\rho^i_{\bar\xi\delta} = \rho^i_{\xi\delta}$.
We defined $q_{\bar\alpha}(\bar\xi) = s(\xi)$ so we have:
$\delta \in D_i \cap \bar\alpha \longrightarrow \widehat{q}(\rho^i_{\xi\delta}) = q_{\bar\alpha}(\bar\xi)$ for $i \geq m$.
This completes (iii).

It remains to show that if $\mathfrak{A}_{q_{\bar\alpha}} \models$ "$\bar\alpha$ is singular" then (iv) holds. In this case $|q_{\bar\alpha}| = \Omega_{\bar\alpha}$ because $|q_{\bar\alpha}| = \xi + 1$ and $\mathfrak{A}_{q_{\bar\alpha}/\xi} \models$ "$\bar\alpha$ is regular". Note that we need not show (v) since $[\Omega_{\bar\alpha}, |q_{\bar\alpha}|)$ is empty. Assume $\mathrm{cf}(\bar\alpha) > \omega$ since otherwise there is nothing to show. We are going to use Lemma 2.4. $q_{\bar\alpha} \in \mathbb{S}^*_{\bar\alpha}$ since $\mathfrak{A}_{q_{\bar\alpha}} \models$ "$\bar\alpha$ is singular" and $\mathfrak{A}^0_{q_{\bar\alpha}} = \mathfrak{A}_{p_{\bar\alpha}} \models$ "$\bar\alpha$ is regular" and Lemma 2.4 applies. Let $D_{q_{\bar\alpha}}$, $\langle s_\delta \mid \delta \in D_{q_{\bar\alpha}} \rangle$ and $\langle \pi^{q_{\bar\alpha}}_\delta \mid \delta \in D_{q_{\bar\alpha}} \rangle$ be given by Lemma 2.4. Set:

$$W = \{\delta \in D_{q_{\bar\alpha}} \mid p/\bar\alpha, C \cap \bar\alpha \in \mathrm{rng}(\pi^{q_{\bar\alpha}}_\delta) \text{ and } \sup(\mathrm{rng}(\pi^{q_{\bar\alpha}}_\delta)) = \mu^0_{q_{\bar\alpha}}\}$$

By the remarks after Lemma 2.4 W is cub in $\bar\alpha$ since $|q_{\bar\alpha}| = \bar\xi + 1$ is a successor and $\mathrm{cf}(\bar\alpha) > \omega$. $W \in \mathfrak{A}_{q_{\bar\alpha}}$ since by Lemma 2.4 D_q, $\langle p_\delta \mid \delta \in D_{q_{\bar\alpha}} \rangle$ and $\langle \pi^{q_{\bar\alpha}}_\delta \mid \delta \in D_{q_{\bar\alpha}} \rangle \in \mathfrak{A}_{q_{\bar\alpha}}$. $W \in \mathfrak{A}_{q_{\bar\alpha}} = \mathfrak{A}_{\Omega_{\bar\alpha}}$ so in order to show (iv) it is sufficient to show

$$\alpha' \in W \longrightarrow \Omega_{\alpha'} \text{ exists}, \quad |q/\alpha'| \geq \Omega_{\alpha'} \text{ and } W \cap \alpha' \in \mathfrak{A}_{\Omega_{\alpha'}}.$$

Let $\alpha' \in W$ and set $\pi' = \pi^{q_{\bar\alpha}}_{\alpha'}$. By Lemma 2.4

$\pi' : \langle \mathfrak{A}^0_{s_{\alpha'}}, C_{s_{\alpha'}} \rangle \xrightarrow[\Sigma_1]{} \langle \mathfrak{A}^0_{q_{\bar\alpha}}, C_{q_{\bar\alpha}} \rangle$ cofinally

$(\alpha' \in W \longrightarrow \sup(\text{rng}(\pi')) = \mu^0_{q_{\overline{\alpha}}})$. By Lemma 2.3 (v)

$s_{\alpha'} = (\pi')^{-1} q_{\overline{\alpha}}$. $\mathfrak{A}^0_{q_{\overline{\alpha}}} = \mathfrak{A}^0_{q_{\overline{\alpha}} / \overline{\xi}}$ and $\mathfrak{A}^0_{s_{\alpha'}} = \mathfrak{A}_{s' / \xi'}$ where $\xi' = (\pi')^{-1}(\overline{\xi})$. By Lemma 2.4 $s_{\alpha'} \in \mathbf{S}^*_{\alpha'}$ hence $|s_{\alpha'}| = \xi' + 1 = \Omega^{s_{\alpha'}}_{\alpha'}$. So if we can show that $s_{\alpha'} = q_{\alpha'}$ and $|q/\alpha'| = |q_{\alpha'}|$ we will have shown $\Omega^{q_{\alpha'}}_{\alpha'}$ exists and $|q/\alpha'| \geq \Omega_{\alpha'}$.

We now apply §2.5 Fact 1 to π' getting $\pi'(\mathfrak{A}^1_{\xi'}) = \mathfrak{A}^1_{\overline{\xi}}$, $\mathfrak{A}^i_{\xi'} = \mathfrak{A}^i_{\overline{\xi}_{\alpha'}}$, $\pi' / \mathfrak{A}^i_{\xi'} = \Pi^i_{\overline{\xi}_{\alpha'}}$ and $(\pi')''\mathfrak{A}^i_{\xi'} = X^i_{\overline{\xi}_{\alpha'}}$.

We can now apply the collapsing lemma to $p_{\overline{\alpha}} = q_{\overline{\alpha}} / \overline{\xi}$ since $|p/\overline{\alpha}| = |p_{\overline{\alpha}}|$, $C \cap \overline{\alpha}$, $p \cap \overline{\alpha} \in \mathfrak{A}^m_{\overline{\xi}}$, $\alpha' < \overline{\alpha}$, $\alpha' = \overline{\alpha} \cap X^m_{\overline{\xi}_{\alpha'}}$. We get $q_{\alpha'} / \xi' = p_{\alpha'} = (\pi')^{-1}(p_{\overline{\alpha}}) = (\pi')^{-1}(q_{\overline{\alpha}} / \overline{\xi}) = s_{\alpha'} / \xi'$ and $|p/\alpha'| = |p_{\alpha'}|$ (since $\alpha' \in C$).

$D_i \cap \overline{\alpha}$, $i \geq m$ are definable in $\mathfrak{A}^0_{q_{\overline{\alpha}}}$ from $C \cap \overline{\alpha}$ and $p/\overline{\alpha}$ since $C \cap \overline{\alpha}$, $p/\overline{\alpha} \in \text{rng}(\pi')$ we have $D_i \cap \overline{\alpha} \in \text{rng}(\pi')$ for $i \geq m$. Hence $(\pi')^{-1}(D_i \cap \overline{\alpha}) = D_i \cap \alpha'$ for $i \geq m$ and $\alpha' \in \bigcap_i D_i$. But then by the definition of q , $q_{\alpha'}(\xi') = s(\xi)$ and we also have $s_{\alpha'}(\xi') = q_{\overline{\alpha}}(\overline{\xi}) = s(\xi)$ hence $s_{\alpha'}(\xi') = q_{\alpha'}(\xi')$ which together with $q_{\alpha'} / \xi' = s_{\alpha'} / \xi'$ gives $s_{\alpha'} = q_{\alpha'}$. $\alpha' \in D \longrightarrow |q/\alpha'| = |q_{\alpha'}|$ by the induction hypothesis of Claim 3. It remains to prove $W \cap \alpha' \in \mathfrak{A}_{q_{\alpha'}}$

$$W \cap \alpha' = \{\delta \in D_{s_{\alpha'}} \mid p/\alpha', C \cap \alpha' \in \text{rng}(\Pi^{s_{\alpha'}}_{\delta}) \text{ and } \sup(\text{rng } \Pi^{s_{\alpha'}}_{\delta}) = \mu^0_{s_{\alpha'}}\}$$

The above can be seen by observing the following facts, most of which come from Lemma 2.4:

$$D_{s_{\alpha'}} = D_{q_{\overline{\alpha}}} \cap \alpha' \in \mathfrak{A}_{q_{\alpha'}} = \mathfrak{A}_{s_{\alpha'}} \ , \quad \Pi^{q_{\overline{\alpha}}}_{\delta} \cdot \Pi^{s_{\alpha'}}_{\delta} = \Pi^{q_{\overline{\alpha}}}_{\delta}$$

for $\delta \in D_{s_{\alpha'}}$, $\pi'(p/\alpha') = \Pi^{q_{\overline{\alpha}}}_{\alpha'}(p/\alpha') = p/\overline{\alpha}$,

$\Pi^{q_{\overline{\alpha}}}_{\alpha'}(C \cap \alpha') = C \cap \overline{\alpha}$ and $\sup \text{rng}(\Pi^{q_{\overline{\alpha}}}_{\alpha'}(\mu^0_{s_{\alpha'}})) = \mu^0_{q_{\overline{\alpha}}}$.

Clearly then we have $W \cap \alpha' \in \mathfrak{A}_{q_{\alpha'}}$.

Q.E.D. (Claim 3)

Applying to q , s the same argument we applied to $q/\bar{\alpha}$, $q_{\bar{\alpha}}$ for $\bar{\alpha} \in D$ we then get $q \in \mathbb{P}^s_\tau$, $|q| = |s|$ and $q \le p$.

Q.E.D. (Case 1)

<u>Case 2</u> $\lim(|s|)$.

Let $B \in \mathfrak{A}_{|p|}$ be cub in α . We wish to construct a $q \le p$ s.t. $|q| = |s|$, $\dot{q}_\gamma = \dot{p}_\gamma$ for all $\gamma \in \text{card} \cap [\tau,\alpha)$ and $q(\gamma)$ differs from $p(\gamma)$ only for $\gamma \in B \cup \{\beta^+ \mid \beta \in B\}$.
Let $C = C_s$ be as in Lemma 2.3.
Define $\langle \eta_i \mid i \le \theta \rangle$ as follows:

$$\eta_0 = \min(C \setminus \mu^0_{|p|}) .$$

$$\eta_{i+1} = \text{the least } \eta \in C \text{ s.t. } \exists\xi(\eta_i < \mu^0_\xi \le \eta) \text{ if } \eta_i < \mu^0_s ,$$
$$\text{otherwise } \eta_{i+1} \text{ is undefined and } i = \theta .$$

$$\eta_\lambda = \sup_{i<\lambda}(\eta_i) \quad \text{for } \lim(\lambda) .$$

Note that $\theta \le \text{ot}(C)$.

<u>Claim</u> For $\lim(\lambda)$, η_λ is a limit point of $C \setminus \mu^0_{|p|}$; moreover every $\mu^0_\xi \in C \setminus \mu^0_{|p|}$ and μ^0_s is an η_i for some $i < \theta$.

Proof By Lemma 2.3, if γ is a limit point of C then $\gamma = \mu^0_\xi$ where $C_{s/\xi} = \gamma \cap C_s$. So for $\lim(\lambda)$ $\exists\xi(\eta_\lambda = \mu^0_\xi)$ and μ^0_ξ is a limit point. In addition if $\mu^0_\xi \in C \setminus \mu^0_{|p|}$ then η_{i+1} cannot pass μ^0_ξ if $\eta_i < \mu^0_\xi$; hence $\exists i(\eta_i = \mu^0_\xi)$.

Q.E.D. (Claim)

Set: ξ_i = the least $\xi < |s|$ s.t. $\eta_i \le \mu^0_\xi$.
Then $\langle \xi_i \mid i < \theta \rangle$ is monotone, $\sup_{i<\theta} \xi_i = |s|$ and $\mu^0_{\xi_\lambda} = \eta_\lambda$ for $\lim(\lambda)$. $C \cap \mu^0_{\xi_i} \in \mathfrak{A}^i_{\xi_i}$ if $\mu^0_{\xi_i}$ is a limit point of C,

because then $C \cap \mu^0_{\xi_i} = C_s \upharpoonright \xi_i \in \mathfrak{A}^1_{\xi_i}$ by Lemma 2.3 or if $\mu^0_{\xi_i}$ is not a limit point in C, then $C \cap \mu^0_{\xi_i}$ is bounded in $\mu^0_{\xi_i}$. Hence $C \cap \mu^0_{\xi_i} \in \mathfrak{A}^1_{\xi_i}$.

Define $q_i \leq p$ for $i \leq \rho \leq \theta$ s.t. $|q_i| = \xi_i$ and $D_i \in \mathfrak{A}_{\xi_i}$ for $i < \rho$ s.t. $\bigcap_{j<i} D_j \in \mathfrak{A}_{\xi_i}$ by induction as follows:

q_0 = the $\mathfrak{A}_{\xi_0}$-least $q \leq p$ s.t. $|q| = \xi_0$ and $q(\gamma)$ differs from $p(\gamma)$ only for $\gamma \in B \cup \{\beta^+ \mid \beta \in B\}$ and $\dot{q}_\gamma = \dot{p}_\gamma$ for all γ.

Let $m = m_i$ = the least $m \geq 1$ s.t. $q_i, B, \bigcap_{j<i} D_j \in \mathfrak{A}^m_{\xi_i}$.

Set $D_i = \{\bar{\alpha} < \alpha \mid \bar{\alpha} = \alpha \cap X^m_{\xi_i \bar{\alpha}}$ and $C \cap \mu^0_{\xi_i}, q_i, B, \bigcap_{j<i} D_j \in X^m_{\xi_i \bar{\alpha}}\}$.

q_{i+1} = the $\mathfrak{A}_{\xi_{i+1}}$-least $q \leq q_i$ s.t. $|q| = \xi_{i+1}$ and $q(\gamma)$ differs from $p(\gamma)$ only for $\gamma \in D_i \cup \{\beta^+ \mid \beta \in D_i\}$ and $\dot{p}_\gamma = \dot{q}_\gamma$ for all $\gamma \in \text{card} \cap [\tau, \alpha)$.

For $\lim(\lambda)$ set $q = \bigcup_{i<\lambda} q_i$. If $q \in \mathbb{P}^s_\tau$, $|q| = \xi_\lambda$ set: $q_\lambda = q$; otherwise q_λ is undefined.

We claim that q_i is defined for $i \leq \theta$. From this it follows that $q_\theta \leq p$ and $|q_\theta| = |s|$, which will prove Lemma 2.14.5 and hence Lemma 2.14. We prove the claim by induction on $i \leq \theta$. For $i = 0$ or i a successor it follows as in Case 1, Claim 3. In particular if $\gamma = \alpha^+$, $\alpha \in D = \bigcap_{i<\lambda} D_i$ we use $\neg 0^\#$ as in (b).

Now consider $\lim(\lambda)$, $q = \bigcup_{i<\lambda} q_i$. We must show $q \in \mathbb{P}^s_\tau$ and $|q| = \xi_\lambda$. (Note: we set $\xi_\theta = |s|$.)

<u>Case 2.1</u> $\lambda < \alpha$: Set $D = \bigcap_{i<\lambda} D_i$.

<u>Claim 1</u> $q \upharpoonright \gamma \in \mathbb{P}^{q_\gamma}_\tau$ for $\gamma \in \text{card} \cap [\tau, \alpha)$ and $|q \upharpoonright \gamma| = |q_\gamma|$ for $\gamma \in D$.

Proof By induction on γ. As before the only non-trivial case is $\gamma = \bar{\alpha} \in D$.

Set: $\bar{\mathfrak{A}} = \bigcup_{i<\lambda} \mathfrak{A}^{m_i}_{\xi_i\bar{\alpha}}$ and $\pi = \bigcup_{i<\lambda} \Pi^{m_i}_{\xi_i\bar{\alpha}}$.

Since $\bar{\alpha} \in D \subseteq D_i \subseteq \{\beta \mid \beta = \alpha \cap X^{m_i}_{\xi_i\beta}\}$ for $i < \lambda$ we can use §2.5 Fact 5 to show that π is well defined.

$\eta_i = \sup(C \cap \mu^0_{\xi_i}) \in X^{m_i}_{\xi_i\bar{\alpha}}$; $(C \cap \mu^0_{\xi_i} \in X^{m_i}_{\xi_i\bar{\alpha}}$ by definition of D_i), hence $\eta_i \in \mathrm{rng}(\Pi^{m_i}_{\xi_i\bar{\alpha}})$ for $\bar{\alpha} \in D$.

$\sup_{i<\lambda} \eta_i = \eta_\lambda = \mu^0_{\xi_\lambda}$ so $\pi\colon \bar{\mathfrak{A}} \xrightarrow[\Sigma_1]{} \mathfrak{A}^0_{\xi_\lambda}$ cofinally.

Note: It can be shown (similarly to Lemma 6.10), that if $\pi\colon L_{\bar{\alpha}}[B] \xrightarrow[\Sigma_0]{} L_\alpha[B]$, $A \subseteq L_\alpha[B]$, $\langle L_\alpha[B], A\rangle$ is amenable and there is a sequence $\langle \delta_i \rangle$ s.t. $\sup \delta_i = \alpha$ and $A \cap L_{\delta_i}[B] \in \mathrm{rng}(\pi)$, then there is a unique $\bar{A} \subseteq L_{\bar{\alpha}}[B]$ s.t. $\pi\colon \langle L_{\bar{\alpha}}[B], \bar{A}\rangle \xrightarrow[\Sigma_0]{} \langle L_\alpha[B], A\rangle$ and $\langle L_{\bar{\alpha}}[B], \bar{A}\rangle$ is amenable, where $\bar{A} = \bigcup_i \pi^{-1}(A \cap L_{\delta_i}[B])$.

We use the above, noting that $\langle \mathfrak{A}^0_{\xi_\lambda}, C_{\xi_\lambda}\rangle$ is amenable in defining $\bar{C} = (\pi^{-1})''C = \bigcup_{i<\lambda} (\Pi^{m_i}_{\xi_i\bar{\alpha}})^{-1}(C \cap \mu^0_{\xi_i})$ getting

$$\pi\colon \langle \bar{\mathfrak{A}}, \bar{C}\rangle \xrightarrow[\Sigma_1]{} \langle \mathfrak{A}^0_{\xi_\lambda}, C_{\xi_\lambda}\rangle \text{ cofinally.}$$

Set $\bar{s}_i = \pi^{-1}(s \upharpoonright \xi_i)$ for $i < \lambda$. Then by §2.5 Fact 1 $\pi(\mathfrak{A}_{\bar{s}_i}) = \mathfrak{A}_{\xi_i}$, $\mathfrak{A}_{\bar{s}_i} = \mathfrak{A}_{\xi_i\bar{\alpha}}$ and $\pi \upharpoonright \mathfrak{A}_{\bar{s}_i} = \pi_{\xi_i\bar{\alpha}}$. We can use the collapsing lemma with respect to $s \upharpoonright \xi_i$, q_i since $\bar{\alpha} = X_{\xi_i\bar{\alpha}} \cap \alpha$, $\pi(q_i \upharpoonright \bar{\alpha}) = q_i \in X_{\xi_i\bar{\alpha}}$ and $|q_i| = |s \upharpoonright \xi_i|$ (note we are using $n = \omega$). This gives $\bar{s}_i = q_{i\bar{\alpha}}$, $\mathfrak{A}_{q_i\bar{\alpha}} = \mathfrak{A}_{\xi_i\bar{\alpha}}$, $\dot{q}_{i\bar{\alpha}} = \emptyset$ and $|q_i \upharpoonright \bar{\alpha}| = |q_{i\bar{\alpha}}| = |\bar{s}_i|$.

Set: $\bar{s} = \bigcup_{i<\lambda} s_i = (\pi^{-1})(s \upharpoonright \xi_\lambda)$.

By the above we have $\bar{s} = \bigcup_{i<\lambda} q_i = q_{\bar{\alpha}}$.

$$\pi\colon \langle \bar{\mathfrak{A}}, \bar{C}\rangle \xrightarrow[\Sigma_1]{} \langle \mathfrak{A}^0_{\xi_\lambda}, C_{s \upharpoonright \xi_\lambda}\rangle \text{ cofinally,}$$

where $\pi\restriction\bar{\alpha} = \mathrm{id}\restriction\bar{\alpha}$ and $\pi(\bar{\alpha}) = \alpha$, hence by Lemma 2.3:

$$\bar{\mathfrak{A}} = \mathfrak{A}^0_{q_{\bar{\alpha}}} \quad \text{and} \quad \bar{C} = C_{q_{\bar{\alpha}}} .$$

$\langle q_i \mid i < \lambda \rangle$ and $\langle D_i \mid i < \lambda \rangle$ are $\Sigma_1(\langle \mathfrak{A}^0_{\xi_\lambda}, C_{s\restriction\xi_\lambda}\rangle)$ in p, B . Hence $\langle q_i\restriction\bar{\alpha} \mid i < \lambda \rangle$ and $\langle D_i \cap \bar{\alpha} \mid i < \lambda \rangle$ are $\Sigma_1(\mathfrak{A}^0_{q_{\bar{\alpha}}}, C_{q_{\bar{\alpha}}})$ in $p\restriction\bar{\alpha}$, $B \cap \bar{\alpha}$ by the same definition. This gives us

$$q\restriction\bar{\alpha} = \bigcup_{i<\lambda} q_i\restriction\bar{\alpha} \in \mathfrak{A}_{q_{\bar{\alpha}}} \quad \text{and} \quad D \cap \bar{\alpha} = \bigcap_i D_i \cap \bar{\alpha} \in \mathfrak{A}_{q_{\bar{\alpha}}}$$

(and in fact $q, D \in \mathfrak{A}_{\xi_\lambda}$).

Thus (i) of the definition of $\mathbf{P}^s_\tau$ holds. To see that (ii) holds, let $\mathfrak{A}_{q_{\bar{\alpha}}} \models$ "$\bar{\alpha}$ is regular". Then $\gamma \in D \cap \bar{\alpha} \longrightarrow \dot{q}_\gamma = \emptyset$ and $|q\restriction\gamma| = |q_\gamma|$ by the induction hypothesis (of Claim 1). In addition $|q\restriction\bar{\alpha}| = |q_{\bar{\alpha}}|$ since

$$|q\restriction\bar{\alpha}| \geq \sup_{i<\lambda} |q_i\restriction\alpha| = \sup |s_i| = \xi_\lambda = |q_{\bar{\alpha}}| .$$

(iii) holds since, if $\xi < |q_{\bar{\alpha}}|$, then $\xi < |q_i\restriction\bar{\alpha}|$ for some $i < \lambda$.

It remains to show that (iv) holds if $\Omega_{\bar{\alpha}} = |q_{\bar{\alpha}}|$. As before we do not have to show (v) since $[\Omega_{\bar{\alpha}}, |q_{\bar{\alpha}}|) = \emptyset$. Then $s \in \mathbb{S}^*_\alpha$ (note that in this case $\lambda = \theta$) and we can use Lemma 2.4. Let $D_{q_{\bar{\alpha}}}$, $\langle s_\delta \mid \delta \in D_{q_{\bar{\alpha}}} \rangle$ and $\langle \Pi^{q_{\bar{\alpha}}}_\delta \mid \delta \in D_{q_{\bar{\alpha}}} \rangle$ be as in Lemma 2.4.

Set: $W = \{\delta \in D_{q_{\bar{\alpha}}} \mid p\restriction\bar{\alpha}, B \cap \bar{\alpha} \in \mathrm{rng}(\Pi^{q_{\bar{\alpha}}}_\delta)\}$.

We can assume $\mathrm{cf}(\bar{\alpha}) > \omega$ since otherwise there is nothing to prove; hence $D_{q_{\bar{\alpha}}}$ is cub in α which implies W is cub in $\bar{\alpha}$. $W \in \mathfrak{A}_{q_{\bar{\alpha}}}$ and as in Case 1 it is sufficient to show that $\alpha' \in W \longrightarrow \Omega_{\alpha'}$ exists, $|q\restriction\alpha'| \geq \Omega_{\alpha'}$ and $W \cap \alpha' \in \mathfrak{A}_{\Omega_{\alpha'}}$.

Set: $\pi' = \Pi^{q_{\bar{\alpha}}}_{\alpha'}$. By the remarks after Lemma 2.4 we have $\sup(\pi')''(\mu^0_{s_{\alpha'}}) = \mu^0_{q_{\bar{\alpha}}\restriction\bar{\xi}}$ for $\bar{\xi} \leq |q_{\bar{\alpha}}|$ and $\mu^0_{q_{\bar{\alpha}}\restriction\bar{\xi}}$ is a limit point of $C_{q_{\bar{\alpha}}}$.

Hence;

$$\pi': \langle \mathfrak{A}^0_{s_{\alpha'}}, C_{s_{\alpha'}} \rangle \xrightarrow[\Sigma_1]{} \langle \mathfrak{A}^0_{\bar\xi}, C_{\bar\xi} \rangle \text{ cofinally.}$$

The full diagram being:

$$\langle \mathfrak{A}^0_{s_{\alpha'}}, C_{s_{\alpha'}} \rangle \xrightarrow[\text{cofinally}]{\Sigma_1} \langle \mathfrak{A}^0_{\bar\xi}, C_{\bar\xi} \rangle \subseteq \langle \mathfrak{A}_{q_{\bar\alpha}}, C_{q_{\bar\alpha}} \rangle \xrightarrow[\text{cofinally}]{\pi} \langle \mathfrak{A}^0_{\xi_\lambda}, C_{\xi_\lambda} \rangle .$$

Since $\mu^0_{q_{\bar\alpha}} \wedge \bar\xi$ is a limit point of $C_{q_{\bar\alpha}}$ and by the definition of $\langle \xi_i \mid i \le \theta \rangle$ $\exists \rho \le \lambda$ s.t. $\bar\xi = \bar\xi_\rho = \pi^{-1}(\xi_\rho)$. This gives us:

$$\bigcup_{i<\rho} \Pi^{m_i}_{\xi_i \bar\alpha}: \langle \mathfrak{A}^0_{\bar\xi_\rho}, C_{\bar\xi_\rho} \rangle \xrightarrow[\Sigma_1]{} \langle \mathfrak{A}^0_{\xi_\rho} . C_{\xi_\rho} \rangle \text{ cofinally.}$$

The final diagram being, setting $\pi^* = \bigcup_{i<\rho} \pi^{m_i}_{\xi_i \bar\alpha}$:

$$\langle \mathfrak{A}^0_{s_{\alpha'}}, C_{s_{\alpha'}} \rangle \xrightarrow[\Sigma_1]{\pi'} \langle \mathfrak{A}^0_{\bar\xi_\rho}, C_{\bar\xi_\rho} \rangle \xrightarrow[\Sigma_1]{\pi^*} \langle \mathfrak{A}^0_{\xi_\rho}, C_{\xi_\rho} \rangle ,$$

where both π' and π^* are cofinal. $\langle q_i \mid i < \rho \rangle$ and $\langle D_i \mid i < \rho \rangle$ are uniformly $\Sigma_1(\langle \mathfrak{A}^0_{\xi_\rho}, C_{\xi_\rho} \rangle)$ in p and B since by Claim 2.2 $\langle \mathfrak{A}^i_\xi \mid \xi < \xi_\rho , i < \omega \rangle$ is uniformly $\Sigma_1(\mathfrak{A}^0_{\xi_\rho})$ in α . $\pi'(p \wedge \alpha') = p \wedge \bar\alpha$, $\pi'(B \cap \alpha') = B \cap \bar\alpha$, $\pi^*(p \wedge \bar\alpha) = p$, $\pi^*(B \cap \bar\alpha) = B$ so $\langle q_i \wedge \alpha' \mid i < \rho \rangle$ and $\langle D_i \cap \alpha' \mid i < \rho \rangle$ are $\Sigma_1(\langle \mathfrak{A}^0_{s_{\alpha'}}, C_{s_{\alpha'}} \rangle)$ in $p \wedge \alpha'$ and $B \cap \alpha'$ by the same definition as $\langle q_i \mid i < \rho \rangle$ and $\langle D_i \mid i < \rho \rangle$ are $\Sigma_1(\langle \mathfrak{A}^0_{\xi_\rho}, C_{\xi_\rho} \rangle)$ in p and B . Hence $\pi'(D_i \cap \alpha') = D_i \cap \alpha$ for $i < \rho$ and $\alpha' \in \bigcap_{i<\rho} D_i$. In addition

$$\bar\xi_i \in \operatorname{rng}(\pi') \text{ for } i < \rho \text{ where } \pi^*(\bar\xi_i) = \xi_i .$$

We apply §2.5 Fact 1 to π' getting:

$$\pi'(\mathfrak{A}_{\xi_i'}) = \mathfrak{A}_{\bar\xi_i}, \ \mathfrak{A}_{\xi_i'} = \mathfrak{A}_{\bar\xi_i \alpha'} \text{ and } \pi' \wedge \mathfrak{A}_{\xi_i'} = \pi_{\bar\xi_i \alpha'} \text{ for } i < \rho$$

where $\pi'(\xi_i') = \bar\xi_i$.

We can apply the collapsing lemma to $q_{i\bar{\alpha}}$, $q_i/\bar{\alpha}$ for $i < \rho$ getting: $s_{\alpha'}/\xi_i = q_{i\alpha'}$. Hence $s_{\alpha'} = \bigcup_{i<\rho} q_{i\alpha'} = q_{\rho_{\alpha'}}$ (writing q_ρ for q if $\rho = \lambda$). Since $s_{\alpha'} \in \mathbf{S}^*_{\alpha'}$ and $s_{\alpha'} = q_{\rho_{\alpha'}}$, $\Omega_{\alpha'}^{q_{\rho_{\alpha'}}}$ exists. $\alpha' \in \bigcap_{i<\rho} D_i$ implies by the induction hypothesis that $|q_\rho/\alpha'| = |q_{\rho_{\alpha'}}|$ hence:

$$|q/\alpha'| \geq |q_\rho/\alpha'| = |q_{\rho_{\alpha'}}| = \Omega_{\alpha'}^{q_{\rho_{\alpha'}}} = \Omega_{\alpha'}^{q_{\alpha'}}$$

The last thing to show is,

$$W \cap \alpha' = \{\delta \in D_{s_{\alpha'}} \mid p/\alpha', B\cap\alpha' \in \mathrm{rng}(\Pi_\delta^{s_{\alpha'}})\} \in \mathfrak{N}_{s_{\alpha'}} \text{ and } s_{\alpha'} = q_{\rho_{\alpha'}}$$

so $W \cap \alpha' \in \mathfrak{N}_{q_{\alpha'}}$.

Q.E.D. (Claim 1)

But we can apply to q and s/ξ_λ the same proof given for $q/\bar{\alpha}$ and $q_{\bar{\alpha}}$ for $\bar{\alpha} \in D$ to get $q \in \mathbf{P}_\tau^s$ and $q \leq p$ and $|q| = \xi_\lambda$.

Q.E.D. (Case 2.1)

<u>Case 2.2</u> $\lambda = \alpha$ (hence $\lambda = \theta$).

Set $D = \{\bar{\alpha} \mid \bar{\alpha} \in \bigcap_{i<\bar{\alpha}} D_i\}$ the diagonal intersection. D is cub in α.

<u>Claim 1</u> $\bar{\alpha} \in D \longrightarrow \bar{\alpha} = \min \bigcap_{i<\bar{\alpha}} D_i$ and $\bar{\alpha} \notin D_{\bar{\alpha}+1}$.

Proof $D_{i+1} \subseteq$ fixed points of D_i can be shown as we did in Case 1. If $\beta = \min \bigcap_{i<\gamma} D_i$; $\bigcap_{i<\gamma} D_i \supseteq D_\gamma$ so $\beta \notin D_{\gamma+1}$. We can prove the claim by induction on $\bar{\alpha} \in D$.

Q.E.D. (Claim 1)

<u>Claim 2</u> $q/\alpha \in \mathbf{P}_\tau^{q_\gamma}$ for $\gamma \in \mathrm{card} \cap [\tau,\alpha)$. Moreover $|q\ \bar{\alpha}| = |q_{\bar{\alpha}}|$ for $\bar{\alpha} \in D$.

Proof Since $\sup_{i<\theta}(\min(D_i)) = \alpha$, there is an $i < \theta$ s.t. $q \upharpoonright \gamma = q_i \upharpoonright \gamma$ and $q_\gamma = q_{i\gamma}$ since q_{i+1} differs from q_i only for $D_i \cap \{\beta^+ \mid \beta \in D_i\}$ and we can choose an i s.t. $\min(D_i) > \gamma$. Hence $q \upharpoonright \gamma \in \mathbb{P}_\tau^{q_\gamma}$. Now let $\bar{\alpha} \in D$. Then $q \upharpoonright \bar{\alpha} = q_{\bar{\alpha}} \upharpoonright \bar{\alpha}$ and $q_{\bar{\alpha}} = (q_{\bar{\alpha}})_{\bar{\alpha}}$ since by Claim 1 $\bar{\alpha} \in \bigcap_{i<\alpha} D_i$ and $(\alpha+1) \cap D_j = \emptyset$ for $i > \bar{\alpha}$, but then $|q \upharpoonright \bar{\alpha}| = |q_{\bar{\alpha}}|$ by the claim in Case 2.1.

Q.E.D. (Claim 2)

The argument of Case 2.1 then yields $q \in \mathbb{P}_\tau^s$, $q \leq p$ and $|q| = |s|$.

3.1 INTRODUCTION

The aim of this chapter is to show that $\mathbb{P}_\tau$ is τ-distributive. The proof is fairly long and it will help to present an outline of the idea behind it. First note that it is sufficient to deal with a class dense in $\mathbb{P}_\tau$.

Definition $\mathbb{P}^*_\tau = \{p \in \mathbb{P}_\tau \mid \mathrm{dom}(p) = \mathrm{card} \cap [\tau,\alpha] \text{ for some } \alpha\}$. For $p \in \mathbb{P}^*_\tau$ set $\alpha_p = \max(\mathrm{dom}(p))$. $\mathbb{P}^*_\tau$ is dense in $\mathbb{P}_\tau$ since $p \in \mathbb{P}^s_\tau$ implies $p \cup \{\langle\alpha,\langle 0,s\rangle\rangle\} \in \mathbb{P}^*_\tau$ (by Lemma 2.11). Let $\langle\Delta_i\ i<\tau\rangle$ be a class of dense classes in $\mathbb{P}^*_\tau$, i.e. $\{\langle i,p\rangle \mid p \in \Delta_i\}$ is definable in $\langle M,A\rangle$. We will inductively construct a sequence $p_i \in \mathbb{P}^*_\tau$ for $i \le \tau$ s.t. $p_{i+1} \in \Delta_i$. They must be constructed s.t. $\bigcup_{i<\lambda} p_i = p_\lambda \in \mathbb{P}^*_\tau$ for $\lim(\lambda)$. We must show that $p_\lambda \restriction \gamma \in \mathbb{P}^{p_{\lambda,\gamma}}$ for $\gamma \in \mathrm{card} \cap [\tau,\alpha_{p_\lambda}]$, dealing simultaneously with all cardinals in $\mathrm{card} \cap [\tau,\alpha_{p_\lambda}]$. A difficult step is showing $p_\lambda \restriction \gamma \in \mathfrak{A}_{p_{\lambda,\gamma}}$. This is done by making $p_{\lambda,\gamma} = \bigcup_{i<\lambda} p_{i\gamma}$ generic over some collapsed model, b , that believes "$|p_\gamma| = \gamma^+$" . The model b is obtained by collapsing a $X \prec_{\Sigma_n} \langle M,A\rangle$ s.t. $\gamma \subseteq X$. $\langle p_i \mid i<\lambda\rangle$ is Σ_n definable in $\langle M,A\rangle$ and p_i collapses to p'_i s.t. $p_i \restriction \gamma = p'_i \restriction \gamma$ and $p'_{i\gamma} = p_{i\gamma}$.

X is defined s.t. $b \models$ "$p'_{i\gamma}$ is $\mathbb{P}^*_\tau$-generic" and $p'_{i\gamma} = p_{i\gamma}$ so we may use the coding ability that $p_{i\gamma}$ has via genericity to code all of b . From b we define $\langle p'_i \mid i<\lambda\rangle$ and with some help from $\neg 0^{\#}$ get $p_\lambda \restriction \gamma = p'_\lambda \restriction \gamma = \bigcup_{i<\lambda} p'_i \restriction \gamma \in \mathfrak{A}_{p_{\lambda\gamma}}$.

To carry out the above we must first show coding and preservation properties of $\mathbb{P}^s_\tau$ for $s \in \mathbb{S}_{\alpha^+}$.

Theorem 3.1 *Let* $s \in \mathbb{S}_{\alpha+}$ *and* $\tau \leq \alpha$ *, then* $\mathbb{P}^s_\tau$ *is* τ-*distributive in* $\mathfrak{A}^1_s$.

Theorem 3.2 *Let* s , α *and* τ *be as in Theorem 3.1, where* α *is inaccessible. Let* X , $\langle \Delta_\nu \mid \nu \in X \rangle \in \mathfrak{A}^1_s$ *be s.t.* $X \subseteq \text{card} \cap [\tau,\alpha)$ *is thin in* $\mathfrak{A}^1_s$ *and* Δ_ν *is dense in* $\mathbb{P}^s_\nu$. *Then* Δ *is dense in* $\mathbb{P}^s_\tau$ *, where:*

$$\Delta = \{p \in \mathbb{P}^s_\tau \mid \forall \nu \in X \quad p/\!\!\restriction[\nu,\alpha) \in \Delta_\nu\}.$$

Theorems 3.1 and 3.2 will be proved by simultaneous induction on α . Then by a corollary to the proof we will get:

Theorem 3.3 $\mathbb{P}_\tau$ *is* τ-*distributive.*

3.2 CONSEQUENCES OF THEOREMS 3.1 and 3.2

First we will give some definitions and notation.

Definition 1 Let $p \in \mathbb{P}$, then $(p)_\tau = p/\!\!\restriction(\text{card}\setminus\tau)$ and $(p)^\tau = p/\!\!\restriction\tau$ for $\tau \in \text{card}$.

Clearly for $\gamma \leq \tau$ we have:

$$\mathbb{P}^s_\tau = \{(p)_\tau \mid p \in \mathbb{P}^s_\gamma\} \quad \text{and} \quad \mathbb{P}_\tau = \{(p)_\tau \mid p \in \mathbb{P}_\gamma\} .$$

Definition 2 If $G \subseteq \mathbb{P}^s_\tau$ is $\mathbb{P}^s_\tau$-generic over $\mathfrak{A}^1_s$ for $s \in \mathbb{S}_\alpha$, define $D_\gamma = \bigcup_{p \in G} \tilde{p}_\gamma$ for $\gamma \in \text{card} \cap [\tau,\alpha)$ (where as before $\tilde{p}_\gamma = \{\xi \mid p_\gamma(\xi) = 1\}$) and $D = D_G = \bigcup_{\gamma \in \text{card} \cap [\tau,\alpha)} D_\gamma$. We call such a D $\mathbb{P}^s_\tau$-generic over $\mathfrak{A}^1_s$.

Clearly $G \in \mathfrak{A}^1_s[D]$ and $D \in \mathfrak{A}^1_s[G]$. If D is as above then denote by G_D the $\mathbb{P}^s_\tau$-generic G s.t. $D = D_G$. The notion $D \subseteq [\tau,\infty)$ is $\mathbb{P}_\tau$-generic over $\langle M,A \rangle$ is defined similarly.

Definition 3 Let $\tau \le \gamma < \alpha$ and $s \in \mathbb{S}_\alpha$. If D is $\mathbb{P}^s_\gamma$-generic over $\mathfrak{A}^1_s$, set $\mathbb{P}^D_\tau = \mathbb{P}^{D_\gamma}_\tau$, noting that $D_\gamma \in \mathbb{S}^+_\gamma$.

Definition 4 Let $\tau \le \gamma < \alpha$ and $s \in \mathbb{S}_{\alpha+}$. Let Δ be dense in $\mathbb{P}^s_\tau$. For $p \in \mathbb{P}^s_{\gamma+}$ set:

$$\Delta^p = \{q \in \mathbb{P}^{p_{\gamma+}}_\tau \mid q \cup p \in \Delta\} .$$

Lemma 3.4 *Let* $\tau \le \gamma < \alpha$ *and* $s \in \mathbb{S}_{\alpha+}$. *Let* $\mathbb{P}^s_{\gamma+}$ *be* (γ^+,∞)*-distributive in* $\mathfrak{A}^1_s$. *Let* $\Delta_\nu \in \mathfrak{A}^1_s$ *be dense in* $\mathbb{P}^s_\tau$ *for* $\nu < \gamma^+$. *Then* Δ *is dense in* $\mathbb{P}^s_{\gamma+}$ *where,*

$$\Delta = \{p \in \mathbb{P}^s_{\gamma+} \mid \forall \nu < \gamma^+\ \Delta^p_\nu \textit{ is dense in } \mathbb{P}^{p_{\gamma+}}_\tau\} .$$

Proof Let $p_0 \in \mathbb{P}^s_{\gamma+}$. We must find a $q \le p_0$ in $\mathbb{P}^s_{\gamma+}$ s.t. $q \in \Delta$. Let $D \subseteq [\gamma^+,\alpha^+)$ be $\mathbb{P}^s_{\gamma+}$-generic over $\mathfrak{A}^1_s$ s.t. $p_0 \in G_D$. Set

$$\Delta^D_\nu = \{q \in \mathbb{P}^D_\tau \mid \exists p \in G_D(q \cup p \in \Delta_\nu)\} .$$

<u>Claim</u> Δ^D_ν is dense in $\mathbb{P}^D_\tau$.

Proof Let $q_0 \in \mathbb{P}^D_\tau$, then

$$H = \{p \in \mathbb{P}^s_{\gamma+} \mid p/p_0 \vee \exists q \le q_0(q \cup p \in \Delta_\nu)\} \text{ is dense in } \mathbb{P}^s_{\gamma+}\} .$$

So $H \cap G_D \ne 0$ giving $p \in G_D$ s.t. $\exists q \le q_0(q \cup p \in \Delta_\nu)$. Note $p \in G_D \longrightarrow p$ is compatible with p_0.

Q.E.D (Claim)

Choose a maximal set $X_\nu \subseteq \Delta^D_\nu$, of mutually incompatible conditions. Since $\mathbb{P}^D_\tau$ satisfies γ^{++}-A.C., we have $\text{card}(X_\nu) \le \gamma^+$, $X_\nu \in \mathfrak{A}^1_s$ and $\langle X_\nu \mid \nu < \gamma^+\rangle \in \mathfrak{A}^1_s$, since $\gamma < \alpha$ and $\alpha^+ < \mu^1_s$. For each $r \in X_\nu$ choose $p_{r\nu} \in G_D$

s.t. $r \cup p_{r\nu} \in \Delta_\nu$. The sets $\Delta_{r\nu} = \{p \in \mathbb{P}^s_{\gamma+} \mid p/p_{r\nu} \vee p \le p_{r\nu}\}$ are dense in $\mathbb{P}^s_{\gamma+}$ and $\langle \Delta_{r\nu} \mid r \in X_\nu, \nu < \gamma^+ \rangle \in \mathfrak{A}^1_s$; hence by (γ^+,∞)-distributivity there is a $q \in G_D$ s.t. $q \le p_0$ and $q \le p_{r\nu}$ for $\nu < \gamma^+$ and $r \in X_\nu$. By Lemma 2.14 we may assume w.l.o.g. that $X_\nu \in \mathfrak{A}^1_{q_{\gamma+}}$ for $\nu < \gamma^+$.

$\chi_{D \cap |q_{\gamma+}|} \wedge |q_{\gamma+}| = q_{\gamma+}$ so $X_\nu \subseteq \mathbb{P}^{q_{\gamma+}}_\tau$ for $\nu < \gamma^+$.

$r \in X_\nu \longrightarrow r \cup q \le r \cup p_{r\nu} \in \Delta_\nu$ so $X_\nu \subseteq \Delta^q_\nu$ for $\nu < \gamma^+$.

X_ν is predense in $\mathbb{P}^{q_{\gamma+}}_\tau$ since X_ν is predense in $\mathbb{P}^D_\tau$.

Hence $\forall \nu < \gamma^+$ (Δ^q_ν is dense in $\mathbb{P}^{q_{\gamma+}}_\tau$).

Q.E.D.

Lemma 3.5 *Let* $s \in \mathbb{S}_\alpha$ *and let* D *be* $\mathbb{P}^s_\tau$*-generic over* $\mathfrak{A}^1_s$.

(a) If $\tau \le \delta < \alpha$ *, then* $D \cap [\delta,\alpha)$ *is* $\mathbb{P}^s_\delta$*-generic over* $\mathfrak{A}^1_s$.

(b) If $\delta^+ < \alpha$ *then, setting* $D^1 = D \cap [\delta^+,\alpha)$ *and* $D^0 = D \cap \delta^+$ *we have that* D^0 *is* $\mathbb{P}^{D^1}_\tau$*-generic over* $\mathfrak{A}_{D^1 \cap \delta^{++}}$*, where* $\mathfrak{A}_{D^1 \cap \delta^{++}} = \langle L_{\delta^{++}}[D^1], D^1 \rangle$.

Proof (a) follows directly from the definitions. So we prove (b). Let $\Delta \subseteq L_{\delta^{++}}[D] \cap P^{D^1}$ be definable in $\mathfrak{A}_{D^1 \cap \delta^{++}}$ and dense in $P^{D^1}_\tau$. Let $\Delta' \subseteq \Delta$ be a maximal set of mutually incompatible conditions. Then by δ^{++}-A.C. $\mathrm{card}(\Delta') < \delta^{++}$ and $\Delta' \in \mathfrak{A}^1_{(D^1 \cap \xi) \wedge \xi}$ for $\xi < \delta^+$, $\chi_{D^1 \cap \xi} \in \mathbb{S}_{\delta^+}$. Δ' is predense in $\mathfrak{A}^1_{(D^1 \cap \xi) \wedge \xi}$. Let,

$$\Delta^* = \{q \in \mathbb{P}^s_\tau \mid (q)^{\delta^+} \not\in \mathbb{P}^{D^1 \cap \xi}_\tau \vee (q)^{\delta^+} \text{ meets } \Delta'\}.$$

$\Delta^* \in \mathfrak{A}^1_s$ and is predense in $\mathbb{P}^s_\tau$. So there is a $q \in G_D$ that meets Δ^*, but then $(q)^{\delta^+}$ meets Δ' so $(q)^{\delta^+} \in \Delta \cap G_{D^0}$.

Q.E.D.

Corollary 3.5.1 D^0 *is* $\mathbb{P}_\tau^{p_{\delta+}}$*-generic over* $\mathfrak{A}^1_{p_{\gamma+}}$ *for* $p \in G_{D^1}$.

Proof By Corollary 2.11.1

Lemma 3.6 *Let Theorems 3.1 and 3.2 hold for* $\gamma \leq \alpha$. *Let* $s \in \mathbf{S}_{\alpha+}$ *and* $\tau \leq \alpha$. *Let* $D \subseteq [\tau, \alpha^+)$ *be* $\mathbb{P}^s_\tau$*-generic over* $\mathfrak{A}^1_s$. *Then cardinals and cofinalities are preserved in* $\mathfrak{A}^1_s[D]$.

Proof α^+ is the largest cardinal in $\mathfrak{A}^1_s$, so it suffices to show that $cf(\kappa)$ is preserved for cardinals $\kappa \leq \alpha^+$.

<u>Claim</u> If $\gamma \in card$, $\gamma \leq \alpha$ and $f \in \mathfrak{A}^1_s[D]$ s.t. $f\colon \gamma \longrightarrow On$, then there is an $X \in \mathfrak{A}^1_s$ s.t. $card(X) = \gamma$ and $f''\gamma \subseteq X$.

The claim clearly implies the lemma. We prove the claim by induction on α .

<u>Case 1</u> $\gamma^+ \leq \alpha$. Let $\gamma \leq \delta$, $\tau \leq \delta^+ < \alpha$.

Let $\mathring{f}$ be a term for f and let $p_0 \in \mathbb{P}^s_\tau$ s.t. $p_0 \Vdash \mathring{f}\colon \check{\gamma} \longrightarrow \check{On}$. For $\nu < \gamma$ set

$$\Delta_\nu = \{p \in \mathbb{P}^s_\tau \mid p/p_0 \vee \exists \xi (p \Vdash \mathring{f}(\check{r}) = \check{\xi})\} .$$

Set $D^1 = D \cap [\delta^+, \alpha)$ and $D^0 = D \cap \delta^+$. By Lemma 3.5 and its corollary D^1 is $\mathbb{P}^s_{\delta^+}$-generic over $\mathfrak{A}^1_s$ and D^0 is $\mathbb{P}_\tau^{p_\delta}$ generic over $\mathfrak{A}^1_{p_{\delta+}}$ for $p \in G_{D^1}$. By Lemma 3.4 we can choose a $p \in G_{D^1}$ s.t. Δ^p_ν is dense in $\mathbb{P}_\tau^{p_{\delta+}}$ for $\nu < \gamma$. Let $q \leq p$, $q \in G_{D^1}$ s.t. $\Delta^p_\nu \in \mathfrak{A}^1_{q_{\delta+}}$ for $\nu < \gamma$. $\Delta^p_\nu \subseteq \Delta^q_\nu$ and by Lemma 2.11 Δ^p_ν is predense in $\mathbb{P}_\tau^{q_{\delta+}}$.

Setting $(\Delta^p_\nu)^* = \{r \in \mathbb{P}_\tau^{q_{\delta+}} \mid r \text{ meets } \Delta^p_\nu\}$ we have

$(\Delta_\nu^p)^* \subseteq \Delta_\nu^q$, $(\Delta_\nu^p)^* \in \mathfrak{N}^1_{q_{\delta+}}$ and $(\Delta_\nu^p)^*$ is dense in $\mathbb{P}_\tau^{q_{\delta+}}$. D^0 is $\mathbf{P}_\tau^{q_{\delta+}}$-generic over $\mathfrak{N}^1_q$. For $\nu < \gamma$ define

$$g(\nu) = \text{the } \mathfrak{N}^1_{q_{\delta+}}\text{-least } r \in (\Delta_\nu^p)^* \cap G_{D^0} .$$

By the induction hypothesis there is a $Y \subseteq \mathbf{P}_\tau^{q_{\delta+}}$ s.t. card(Y) $\leq \gamma$ and $g''\gamma \subseteq Y$. For $\nu < \gamma$ and $r \in Y$ define

$$\xi_{\nu r} = \begin{cases} \xi \text{ where } \xi \text{ is the unique } \xi \text{ s.t.} \\ \quad r \cup q \Vdash \mathring{f}(\check{\nu}) = \check{\xi} \text{ if such a } \xi \text{ exists} \\ 0 \text{ otherwise.} \end{cases}$$

Clearly $f''\gamma \subseteq \{\xi_{\nu r} \mid r \in Y \text{ and } \nu < \gamma\}$.

Q.E.D. (Case 1)

<u>Case 2</u> $\gamma = \alpha$ and $\alpha = \delta^+$.
Similar to Case 1 except that we can take $Y = \mathbf{P}_\tau^{p_{\delta+}}$ where $\text{card}(\mathbb{P}_\tau^{p_{\delta+}}) = \delta^+ = \gamma$.

<u>Case 3</u> $\gamma = \alpha$ and α is singular. The claim must hold for α , since otherwise $\mathfrak{N}^1_s[D] \models \text{cf}(\alpha^+) < \alpha$ and a cardinal less than α would be collapsed contradicting the induction hypothesis.

<u>Case 4</u> $\gamma = \alpha$ and α is inaccessible.

Define Δ_ν for $\nu < \alpha$ as in Case 1 and set:

$$\Delta_\nu^* = \{q \in \mathbf{P}^s_{\omega_{\tau+\nu+1}} \mid \Delta_\nu^q \text{ is dense in } \mathbb{P}_\tau^{q_{\omega_{\tau+\nu+1}}}\} .$$

By Lemma 3.4 each Δ_ν^* is dense in $\mathbf{P}^s_{\omega_{\tau+\nu+1}}$. Note that the set of successor cardinals less than α is thin in $\mathfrak{N}^1_s$; hence by Theorem 3.2 for α we get

$$\Delta = \{p \in \mathbb{P}_\tau^s \mid \forall \nu < \alpha (p)_{\omega_{\tau+\nu+1}} \in \Delta_\nu^*\} \text{ is dense in } \mathbf{P}_\tau^s .$$

Let $p \in G_D \cap \Delta$. For $\nu < \alpha$ and $r \in \mathbb{P}_\tau^{P\omega_{\tau+\nu+1}}$ set:

$$\xi_{r\nu} = \begin{cases} \xi & \text{where } \xi \text{ is the unique } \xi \text{ s.t. } r \cup (p)_{\omega_{\tau+\nu+1}} \Vdash \mathring{f}(\check{\nu}) = \check{\xi} \text{ if such a } \xi \text{ exists} \\ 0 & \text{otherwise.} \end{cases}$$

Let $X = \{\xi_{r\nu} \mid \nu < \alpha \text{ and } r \in \mathbb{P}_\tau^{P\omega_{\tau+\nu+1}}\}$, then $\mathrm{card}(X) \le \alpha$ and $p \Vdash \mathring{f}''\check{\alpha} \subseteq \check{X}$.

Q.E.D.

Lemma 3.7 *Let Theorems 3.1 and 3.2 hold for* $\gamma \le \alpha$. *Let* $s \in \mathbb{S}_{\alpha^+}$, $\tau \le \alpha$ *and let* $D \subseteq [\tau,\alpha^+)$ *be* $\mathbf{P}_\tau^s$*-generic over* $\mathfrak{A}_s^1$. *Then* $A \cap \alpha^+, s, D \in L_{\mu_s^1}[A \cap \tau, D_\tau]$.

Proof Let $\langle \gamma_i \mid i \le \delta \le \alpha \rangle$ enumerate $\mathrm{card} \cap [\tau,\alpha^+)$. We will work in $L_{\mu_s^1}[A \cap \tau, D_\tau]$ and define by induction on i, γ_i, $a_i = A \cap \gamma_i$ and $d_i = D\gamma_i$ as follows:

(1) $\gamma_0 = \tau$, $a_0 = A \cap \tau$, $d_0 = D_\tau$.

(2) Assume we have already defined γ_i, a_i and d_i.
$\gamma_{i+1} = \sup(d_i)$ and

$$a_{i+1} = \{\nu < \gamma_{i+1} \mid d_i \cap V_\nu \text{ is bounded in } \gamma_{i+1}\}.$$

We cannot define $\mathbb{S}_{\gamma_{i+1}}$ yet since we may not recognize γ_{i+1}^+ $(= \gamma_{i+2})$ in $L[a_{i+1}]$. Instead we define $\mathbb{S}'_{\gamma_{i+1}} \subseteq \mathbb{S}_{\gamma_{i+1}}$ as follows

$$r \in \mathbb{S}'_{\gamma_{i+1}} \text{ iff } r\colon [\gamma_{i+1}, |r|] \longrightarrow 2 \text{ s.t.}$$

1. $|r| \ge \gamma_{i+1}$.
2. $\forall \xi \le |r|\ L_{\mu_s^1}[a_{i+1}, r \upharpoonright \xi] \models \mathrm{card}(\xi) = \gamma_{i+1}$.
3. If $L_{\mu_s^1}[a_{i+1}] \models \gamma_{i+1} \ge \omega_2$ then

$$\forall \xi \le |r| \; L_\delta[a_{i+1}, r/\xi] \models \mathrm{card}(\xi) = \gamma_{i+1} \,,$$

for a δ s.t. $L_{\mu_s^1} \models \mathrm{card}(\delta) \le \xi$.

We are going to define d_{i+1} so that initial segments of d_{i+1} are in $\mathbf{S}'_{\gamma_{i+1}}$. We work in $L_{\mu_s^1}[A \cap \tau, D_\tau]$ and using the coding we will be able to choose a path all the way up to the "real" γ_{i+2} .

We define a function f on an initial segment of $[\gamma_{i+1}, \infty)$ as follows: if f is already defined on $[\gamma_{i+1}, \xi)$ and $f \upharpoonright \xi \notin \mathbf{S}'_{\gamma_{i+1}}$, then $f(\xi)$ is undefined and we have completed f . Otherwise let $r = f \upharpoonright \xi \in \mathbf{S}'_{\gamma_{i+1}}$. By Lemma 2.5,

$$b_r \in L_{\mu_r^2}[a_{i+1}, r] \subseteq L_{\mu_s^1}[a_{i+1}, d_i] \subseteq L \subseteq [a_0, d_0] \,.$$

Set

$$f(\xi) = \begin{cases} 1 & \text{if } S(b_r) \cap d_i \text{ is bounded in } \gamma_{i+1} \\ 0 & \text{if not.} \end{cases}$$

Finally set $d_{i+1} = \{\xi \mid f(\xi) = 1\}$.

3. Assume γ_i, a_i, d_i are defined for $i < \lambda$, for $\lim(\lambda)$. Set $\gamma_\lambda = \sup_{i<\lambda} \gamma_i$, $a_\lambda = \bigcup_{i<\lambda} a_i$ and $d = \bigcup_{i<\lambda} d_i$. Define $\mathbf{S}'_{\gamma_\lambda}$ as in 2. and in order to define d_λ we define a function, f , on an initial segment of $[\gamma_\lambda, \infty)$. Let f be defined on $[\gamma_\lambda, \xi)$. If $f \upharpoonright \xi \notin \mathbf{S}'_{\gamma_\lambda}$ then $f(\xi)$ is undefined and we have completed f . Otherwise set $r = f \upharpoonright \xi \in \mathbf{S}'_{\gamma_\lambda}$.

Define $\langle \mathfrak{A}_r^i \mid i < \omega \rangle$ as in Chapter 2. If $\mathfrak{A}_r \models$ "γ is regular", define $\langle \rho_{r\eta}^i \mid \eta \in (\mathrm{card} \cap \gamma_\lambda)_{\mathfrak{A}_r}$ and $i < \omega \rangle$ as in Chapter 2 and set $f(\xi) = 1$ iff there is a $m < \omega$ s.t. for all $i \ge m$ a cub $C \subseteq \gamma_\lambda$ exists s.t. $C \in \mathfrak{A}_r$ and

$\eta \in C \longrightarrow \rho^i_{r\eta} \in d$;

otherwise set $f(\xi) = 0$.

If $\mathfrak{A}_r \models$ "γ_λ is singular" then define $\langle \tilde{\rho}_{r_i} \mid i < \lambda_{\gamma_\lambda} \rangle$ as in Chapter 2. Set $f(\xi) = 1$ iff there is a $m < \lambda_{\gamma_\lambda}$ s.t. for $i \geq m$, $\tilde{\rho}_{r_i} \in d$. Otherwise set $f(\xi) = 0$.

It is clear that the above can be carried out in $L_{\mu^1_s}[a_0, d_0]$. Hence $\langle\langle \gamma_i, a_i, d_i \rangle \mid i \leq \alpha \rangle \in L_{\mu^1_s}[a_0, d_0]$. But then $A \cap \alpha^+$ can be defined from d_α and $A \cap \alpha^+$ the way f was defined from d_i and a_{i+1} in 2.

Thus $A \cap \alpha^+, s \in L_{\mu^1_s}[a_0, d_0]$.

Q.E.D. (Lemma 3.7)

Lemma 3.8 *Let Theorems 3.1 and 3.2 hold for* $\gamma \leq \alpha$. *Let* $s \in \mathbf{S}_{\alpha^+}$. *Let* $\Delta_\nu \in \mathfrak{A}^1_s$ *s.t.* Δ_ν *is dense in* $\mathbb{P}^s_\tau$ *for* $\nu < \alpha$. *Let* $\Delta = \{p \in \mathbb{P}^s_\tau \mid \forall \nu < \alpha (p \in \Delta_\nu \quad \exists\, \gamma \in \mathrm{card} \cap [\tau, \alpha)$ *s.t.* $\Delta_\nu^{(p)\gamma+}$ *is dense in* $\mathbb{P}_\tau^{p_{\gamma^+}})\}$. *Then* Δ *is dense in* $\mathbb{P}^s_\tau$.

Proof

Case 1 α is a successor cardinal. Let $\alpha = \gamma^+$. $\mathbb{P}^s_\alpha \cong \mathbb{R}^s$ and $\mathbb{R}^s$ is α-distributive by Lemma 2.6. By Lemma 3.4, $\Delta^* = \{q \in \mathbb{P}^s_\alpha \mid \forall \nu < \alpha \; \Delta^q_\nu$ is dense in $\mathbb{P}^{p_\alpha}_\tau\}$ is dense in $\mathbb{P}^s_\alpha$. Set $\Delta' = \{p \in \mathbb{P}^s_\tau \mid (p)_\alpha \in \Delta^*\}$. Then Δ' is dense in $\mathbb{P}^s_\tau$ and $\Delta' \subseteq \Delta$, hence Δ is dense in $\mathbb{P}^s_\tau$.

Case 2 α is singular. Let $\kappa = \mathrm{cf}(\alpha) < \alpha$ and let $\langle \alpha_i \mid i < \kappa \rangle$ be a monotone sequence of cardinals greater than τ and κ converging to α . Let γ be the least s.t. $\tau, \kappa \leq \gamma < \alpha$. For $i < \kappa$ set

$$\Delta^*_i = \{q \in \mathbb{P}^s_\gamma \mid \forall \nu < \alpha^*_i \; (\Delta^{(q)\alpha^+_i} \text{ is dense in } \mathbb{P}_\gamma^{q_{\alpha^+_i}})\} .$$

By assumption $\mathbb{P}^s_{\alpha^+_i}$ is α^+_i-distributive, so by Lemma 3.4 Δ^*_i is dense in $\mathbb{P}^s_\gamma$. Hence, $\Delta^* = \bigcap_{i<\kappa} \Delta^*_i$ is dense in $\mathbb{P}^s_\gamma$

$(\kappa \leq \gamma)$. Finally this implies, $\Delta' = \{p \in \mathbb{P}^s_\tau \mid (p)_\gamma \in \Delta^*\}$ is dense in $\mathbb{P}^s_\tau$ and $\Delta' \subseteq \Delta$.

Case 3 α is inaccessible. Let $\langle \alpha_\nu \mid \nu < \alpha \rangle$ be a monotone enumeration of $\mathrm{card} \cap [\tau,\alpha)$. Set:

$$\Delta^*_\nu = \{q \in \mathbb{P}^s_{\alpha^+_\nu} \mid \Delta^q_\nu \text{ is dense in } \mathbb{P}^{q_{\alpha^+_\nu}}_\tau\} .$$

By Lemma 3.4 Δ^*_ν is dense in $\mathbb{P}^s_{\alpha^+_\nu}$ and $\{\alpha^+_\nu \mid \nu < \alpha\}$ is thin in $\mathfrak{N}^1_s$. So $\Delta^* = \{p \in \mathbb{P}^s_\tau \mid \forall \nu ((p)_{\alpha^+_\nu} \in \Delta^*_\nu)\}$ is dense in $\mathbb{P}^s_\tau$ by Theorem 3.2. But $\Delta^* \subseteq \Delta$.

Q.E.D. (Lemma 3.8)

We will need a somewhat stronger form of Lemma 3.8, for whose formulation a few definitions are necessary.

Definition 5 Let $\tau \leq \alpha$, $s \in \mathbf{S}_{\alpha+}$ and $p \in \mathbb{P}^s_\tau$. $\mathbb{F}(p)$ is the set of maps f s.t.

1. $\mathrm{dom}(f) \subseteq \mathrm{card} \cap [\tau,\alpha]$ is thin in $\mathfrak{N}^1_s$.
2. For $\nu \in \alpha \cap \mathrm{dom}(f)$, $f(\nu) \in \mathfrak{N}^1_{p_{\nu+}}$.
3. $f(\alpha) \subseteq \mathfrak{N}^1_s$ if $\alpha \in \mathrm{dom}(f)$.
4. $\mathrm{card}(f(\nu)) \leq \nu$ for $\nu \in \mathrm{dom}(f)$.

It follows from 4. that $f(\nu) \in \mathfrak{N}^1_{p_{\nu+}}$ for $\nu \in \alpha \cap \mathrm{dom}(d)$ and $f(\alpha) \in \mathfrak{N}^1_s$ if $\alpha \in \mathrm{dom}(f)$.

Remark We will essentially be interested in the $\Delta \in f(\nu)$ s.t. Δ is predense in $\mathbb{P}^{p_{\nu+}}_\tau$.

Definition 6 Let $\tau \leq \alpha$, $s \in \mathbf{S}_{\alpha+}$, $p \in \mathbb{P}^s_\tau$ and $f \in \mathbb{F}(p)$.

$\Sigma_f = \Sigma^{s,p}_f$ = the set of $q \in \mathbb{P}^s_\tau$ s.t. q/p or $q \leq p$ and

1. If $\alpha \in \mathrm{dom}(f)$ and $\Delta \in f(\alpha)$ is predense in $\mathbb{P}^s_\tau$, then

either q meets Δ or there is a γ s.t. $\tau \le \gamma < \alpha$ and $\Delta^{(q)_{\gamma+}}$ is dense in $\mathbb{P}_\tau^{q_{\gamma^+}}$.

2. If $\nu \in \alpha \cap \mathrm{dom}(f)$ and $\Delta \in f(\nu)$ is predense in $\mathbb{P}_\tau^{p_{\nu+}}$, then either $(q)^{\nu+}$ meets Δ in $\mathbb{P}_\tau^{q_{\nu+}}$ or there is a γ s.t. $\tau \le \gamma < \nu$ and $\Delta_\nu^{(q)^{\nu+}_{\gamma^+}}$ is dense in $\mathbb{P}_\tau^{(q)_{\gamma+}}$ where $(q)^\nu_\gamma = ((q)^\nu)_\gamma = q/\!\!\upharpoonright[\gamma,\nu)$.

For the following facts let $s \in \mathbb{S}_{\alpha+}$, $p \in \mathbb{P}_\tau^s$ and $\tau \le \gamma \le \beta \le \alpha$.

Fact 3.9.1 If $f \in \mathbb{F}((p)^{\beta+}_\gamma)$ then $f \in \mathbb{F}((p)_\gamma)$. Moreover if $\Sigma_f^{p_{\beta+},(p)^{\beta+}_\gamma}$ is dense in $\mathbb{P}_\gamma^{p_{\beta+}}$ then $\Sigma_f^{s,(p)_\gamma}$ is dense in $\mathbb{P}_\gamma^s$.

Proof Directly from the definitions.

Definition 7 Let $f \in \mathbb{F}((p)^{\beta^+})$. For $\gamma \le \beta$ we define $(f/\!\!\upharpoonright[\gamma^+,\beta])^* = f^*$ s.t. $\mathrm{dom}(f^*) = \mathrm{dom}(f/\!\!\upharpoonright[\gamma^+,\beta])$. For each $\nu \in \mathrm{dom}(f^*)$ and $\Delta \in f(\nu)$ s.t. Δ is predense in $\mathbb{P}_\tau^{p_{\nu+}}$ set

$$\Delta^* = \{q \in \mathbb{P}_{\gamma+}^{p_{\nu+}} \mid \Delta^q \text{ is dense in } \mathbb{P}_\tau^{q_{\gamma^+}}\}.$$

Finally define:

$$f^*(\nu) \coloneqq \{\Delta^* \mid \Delta \in f(\nu)\,,\ \Delta \text{ is predense in } \mathbb{P}_\tau^{p_{\nu+}}\}.$$

Fact 3.9.2 Let $f \in \mathbb{F}((p)^{\beta^+})$ and let $q \in \Sigma_{(f/\!\upharpoonright[\gamma^+,\beta])^*}^{p_{\beta+},(p)^{\beta^+}_{\gamma^+}}$. Then for any $r \in \mathbb{P}_\tau^{p_{\beta+}}$ s.t. $(r)_{\gamma^+} \le q$, $r \in \Sigma_{f/\!\upharpoonright[\gamma^+,\beta]}^{p_{\beta+},(p)^{\beta^+}_{\gamma^+}}$.

Proof We can assume that $q \le (p)^{\beta^+}_{\gamma+}$ and $r \le (p)^{\beta^+}$. Let $\nu \in \mathrm{dom}(f/\!\!\upharpoonright[\gamma^+,\beta])$ and let Δ be predense in $\mathbb{P}_\tau^{p_{\nu+}}$.

By Lemma 3.4 $\Delta^* = \{h \in \mathbb{P}_{\gamma^+}^{p_{\nu^+}} \mid \Delta^h$ is dense in $\mathbb{P}_\tau^{h_{\nu^+}}\}$ is dense in $\mathbb{P}_{\gamma^+}^{p_{\nu^+}}$.

Since $(r)_{\gamma^+} \le q$ we have $(r)_{\gamma^+} \in \Sigma_{(f\upharpoonright[\gamma^+,\beta])^*}^{p_{\beta^+},(p)_{\gamma^+}^{\beta^+}}$ and by definition of Σ either,

(a) $(r)_{\gamma^+}^{\nu^+}$ meets Δ^* and then $\Delta^{(r)_{\gamma^+}^{\nu^+}}$ is dense in $\mathbb{P}_\tau^{r_{\gamma^+}}$

or (b) $\exists \eta(\gamma^+ \le \eta < \nu)$ s.t. $\Delta^{*(r)_{\eta^+}^{\nu^+}}$ is dense in $\mathbb{P}_\tau^{r_{\eta^+}}$.

$h \in \Delta^{*(r)_{\eta^+}^{\nu^+}} \longrightarrow \Delta^{h \cup (r)_{\eta^+}^{\nu^+}}$ is dense in $\mathbb{P}_\tau^{r_{\eta^+}}$.

Claim $\Delta^{(r)_{\eta^+}^{\nu^+}}$ is dense in $\mathbb{P}_\tau^{r_{\eta^+}}$.

Proof Let $h_0 \in \mathbb{P}_\tau^{r_{\eta^+}}$. Take $h_1 \le h_0$ in $\mathbb{P}_\tau^{r_{\eta^+}}$ s.t. $(h_1)_{\gamma^+} \in \Delta^{*(r)_{\eta^+}^{\nu^+}}$. Then setting $x = (h_1)_{\gamma^+} \cup (r)_{\eta^+}^{\nu^+}$, we have Δ^x is dense in $\mathbb{P}_\tau^{(h_1)_{\gamma^+}}$. Choose $u \in \Delta^x$, $u \le (h_1)^{\gamma^+}$, then $u \cup (h_1)_{\gamma^+} \le h_1$ and $(u \cup (h_1)_{\gamma^+}) \cup (r)_{\eta^+}^{\nu^+} \in \Delta$.

Q.E.D. (Claim)

Since (a) or (b) is true for every $\Delta \in f(\nu)$, we have $r \in \Sigma_{f\upharpoonright[\gamma^+,\beta]}^{p_{\beta^+},(p)^{\beta^+}}$.

Q.E.D. (Fact)

Lemma 3.10 *Let Theorems 3.1 and 3.2 hold for* $\gamma \le \alpha$. *Let* $s \in \mathbf{S}_{\alpha^+}$, $\tau < \alpha$, $p \in \mathbb{P}_\tau^s$ *and* $f \in \mathbf{F}(p)$. *Then* $\Sigma_f^{s,p}$ *is dense in* $\mathbf{P}_\tau^s$.

Proof By induction on α. If $\alpha \in \mathrm{dom}(f)$ set:

$H = \{q \in \mathbb{P}_\tau^s \mid \forall \Delta \in f(\alpha)$ (Δ is predense in $\mathbf{P}_\tau^s \longrightarrow q$ meets Δ or $\exists \gamma \in \mathrm{card} \in [\tau,\alpha)$ s.t. $\Delta^{(q)_{\gamma^+}}$ is dense in $\mathbb{P}_\tau^{q_{\gamma^+}}\}$.

Noting that $\mathrm{card}(f(\alpha)) \le \alpha$, Lemma 3.8 implies H is dense in $\mathbb{P}^s_\tau$. It is clear by definition that $\Sigma_f = \Sigma_{f/\alpha} \cap H$, hence we only have to show that $\Sigma_{f/\alpha}$ is dense and we may assume $\alpha \not< \mathrm{dom}(f)$. If $\alpha = \gamma^+$ then since $\alpha \not< \mathrm{dom}(f)$, $\sum_f^{s,p} = \sum_f^{p_\gamma, p/\gamma}$ and the induction hypothesis implies that $\sum_f^{p_\gamma, p/\gamma}$ is dense. So we may assume that α is a limit cardinal.

Case 1 α is inaccessible. Let $\langle \alpha_i \mid i < \alpha \rangle$ be a normal sequence of limit cardinals s.t. $\sup_{i<\alpha} \alpha_i = \alpha$, $\alpha_i > \tau$ and $\{\alpha_i \mid i < \alpha\} \cap \mathrm{dom}(f) = 0$ ($\mathrm{dom}(f)$ is thin in $\mathfrak{N}^1_s$).

For $i < \alpha$ define $f^*_i : \mathrm{dom}(f) \cap (\alpha_i, \alpha_{i+1}) \longrightarrow V$ by replacing each $\Delta \in f(\nu)$ s.t. Δ is predense in $\mathbb{P}^{p_{\nu+}}_\tau$ by

$$\Delta^* = \{q \in \mathbb{P}^{p_{\nu+}}_{\alpha_i^+} \mid \Delta^q \text{ is dense in } \mathbb{P}^{q_{\alpha_i^+}}_\tau\} .$$

Then $f^*_i \in \mathbb{F}((p)^{\alpha^+_{i+1}}_{\alpha^+_i})$ and by the induction hypothesis $\sum_{f^*_i}^{p_{\alpha^+_{i+1}}, (p)^{\alpha^+_{i+1}}_{\alpha^+_i}}$ is dense in $\mathbb{P}^{p_{\alpha^+_{i+1}}}_{\alpha^+_i}$. By Fact 3.9.1 we have $\sum_{f^*_i}^{s,(p)_{\alpha^+_i}}$ is dense in $\mathbb{P}^s_{\alpha^+_i}$. Now using Theorem 3.2 we have that $\Delta = \{q \in \mathbb{P}^s_\tau \mid \forall i < \alpha (q)_{\alpha^+_i} \in \sum_{f^*_i}^{s,(p)_{\alpha^+_i}}\}$ is dense in $\mathbb{P}^s_\tau$. By the induction hypothesis $\sum_{f/\alpha_0}^{p_{\alpha_0^+}, (p)^{\alpha_0^+}}$ is dense in $\mathbb{P}^{p_{\alpha^+_0}}_\tau$ and by Fact 3.9.1 $\sum^{s,p}_{f/\alpha_0}$ is dense in $\mathbb{P}^s_\tau$. Let $p' \le p$ and pick a $q \le p'$ s.t. $q \in \Delta \cap \Sigma_{f/\alpha_0}$.

Claim $q \in \Sigma_f$.

Proof For $\nu \in \mathrm{dom}(f)$ and Δ predense in $\mathbb{P}^{p_{\nu+}}_\tau$ we must show that q satisfies clause 2 of the definition of Σ_f . If $\nu < \alpha_0$ then $q \in \sum^{s,p}_{f/\alpha_0}$ so q easily satisfies clause (b). If $\nu \in (\alpha_i, \alpha_{i+1})$ it is sufficient to show $q \in \Sigma^{s,p}_{f/\![\alpha_i^+,\alpha_{i+1})}$. By Fact 3.9.2 this is true since

$(q)_{\alpha_i^+} \in \Sigma^{s,(p)_{\alpha_i^+}}_{f_i^*}$.

Q.E.D. (Case 1)

Case 2 α is singular. Let $\kappa = \mathrm{cf}(\alpha)$ and $\langle \alpha_i \mid i < \kappa \rangle$ be a normal sequence of cardinals converging to α s.t. $\alpha_0 \geq \tau, \kappa$. Define $f_i^*: \mathrm{dom}(f) \cap (\alpha_0, \alpha_i) \longrightarrow V$ by replacing each $\Delta \in f(\nu)$ s.t. Δ is predense in $\mathbb{P}_\tau^{p_{\nu^+}}$ by $\Delta^* = \{q \in \mathbb{P}^{p_{\nu^+}}_{\alpha_0^+} \mid \Delta^q \text{ is dense in } \mathbb{P}_\tau^{q_{\alpha_0^+}}\}$.

Then each $\Sigma_{f_i^*}$ is dense in $\mathbb{P}^s_{\alpha_0^+}$ by the induction hypothesis. Hence so is $\Sigma_{f^*} = \bigcap_{i<\kappa} \Sigma_{f_i^*}$ $(\alpha_0 \geq \kappa)$, where $f^* = \bigcup_i f_i^*$. As in Case 1 $\Sigma_{f/\![\tau,\alpha_0]}$ is dense in $\mathbb{P}^s_\tau$ and for a $p' \leq p$ we can take a $q \leq p'$ s.t. $(q)_{\alpha_0^+} \in \Sigma_f$ and $q \in \Sigma_{f/\![\tau,\alpha_0]}$. We can show $q \in \Sigma_f$ as in Case 1 (using Fact 3.9.2).

Q.E.D. (Lemma)

3.3 PRELIMINARIES OF THE PROOF OF THEOREMS 3.1 AND 3.2

We now turn to the proof of Theorems 3.1 and 3.2. Assume they both hold for $\gamma < \alpha$. Let $s \in \mathbf{S}_{\alpha^+}$ and $\tau \leq \alpha$. The initial case $\tau = \alpha = \omega$ is trivial, indeed if $\tau = \alpha$ Theorem 3.1 holds since $\mathbf{P}^s_\alpha \cong \mathbf{R}^s$ and Theorem 3.2 is vacuous. So we may assume $\omega \leq \tau < \alpha$.

Lemma 3.11 *Let $\mathbf{P}^s_{\gamma^+}$ be γ^+-distributive, where $\tau \leq \gamma < \alpha$. Then $\mathbf{P}^s_\tau$ is τ-distributive.*

Proof Let $\Delta_\nu \in \mathfrak{A}^1_s$ be dense in $\mathbf{P}^s_\tau$ for $\nu < \tau$. Let $p \in \mathbf{P}^s_\tau$. We must find a $q \leq p$, $q \in \bigcap_{\nu<\tau} \Delta_\nu$. By

Lemma 3.4 there is a $q' \leq (p)_{\gamma^+}$ in $\mathbb{P}^s_{\gamma^+}$ s.t. $\Delta^{q'}_\nu$ is dense in $\mathbb{P}^{q'_{\gamma+}}_\tau$ for $\nu < \tau$. As in Lemma 3.6, Case 1, we can choose a $q \leq q'$ in $\mathbb{P}^s_{\gamma^+}$ s.t. $\Delta^{q'}_\nu \in \mathfrak{N}^1_{q_{\gamma+}}$ for $\nu < \tau$ and get $\Delta^*_\nu = \{u \in \mathbb{P}^{q_{\gamma+}}_\tau | u \text{ meets } \Delta^{q'}_\nu\} \in \mathfrak{N}^1_{q_{\gamma+}}$, $\Delta^*_\nu \subseteq \Delta^q_\nu$ and Δ^*_ν is dense in $\mathbb{P}^{q_{\gamma+}}_\tau$.

$\mathbb{P}^{q_{\gamma+}}_\tau$ is τ-distributive (by the induction hypothesis if $\tau < \gamma$ or by $\mathbb{P}^{q_{\gamma+}}_\gamma \cong \mathbb{R}^{q_{\gamma+}}$ if $\tau = \gamma$). Hence there is a $r \leq (p)^{\gamma^+}$ in $\mathbb{P}^{q_{\gamma+}}_\tau$ s.t. $r \in \bigcap_{\nu<\tau} \Delta^*_\nu$. But $\bigcap_{\nu<\tau} \Delta^*_\nu \subseteq \bigcap_{\nu<\tau} \Delta^q_\nu$, hence $r \cup q \leq p$ and $r \cup q \in \bigcap_{\nu<\tau} \Delta_\nu$.

Q.E.D.

Corollary 3.11.1 *Let* α *be a successor cardinal. Then Theorem 3.1 holds.*

Proof Let $\alpha = \gamma^+$. Then $\mathbb{P}^s_{\gamma^+} \cong \mathbb{R}^s$ and is γ^+-distributive; hence Lemma 3.4 implies the corollary.

From now on we assume α to be a limit cardinal. We consider the inaccessible case first.

Lemma 3.12 *Let* α *be inaccessible. Let* $p \in \mathbb{P}^s_\tau$ *and* $f \in \mathbb{F}(p)$ *s.t.* $\text{dom}(f) \subseteq \alpha$. *Then* $\Sigma^{s,p}_f$ *is dense in* $\mathbb{P}^s_\tau$.

Proof Let $p' \leq p$. We must find a $q \leq p'$ s.t. $q \in \Sigma_f$. By the extension Lemma 2.14 we may assume w.l.o.g. that, $|p' \wedge \gamma| = |p'_\gamma|$ for limit cardinals $\gamma \in (\tau,\alpha]$, $f \in \mathfrak{N}_{p'_\alpha}$ and $\text{dom}(f)$ is thin in $\mathfrak{N}_{p'_\alpha}$. We shall construct a $q \leq p' \wedge \alpha$ in $\mathbb{P}^{p'_\alpha}_\tau$ s.t. $q \cup \{\langle\alpha,p'(\alpha)\rangle\} \in \Sigma_f$. Set $X = \text{dom}(f)$. Let $i \leq n < \omega$ be s.t. $f,q \in \mathfrak{N}^n_{p'_\alpha}$, $\mathfrak{N}^n_{p'_\alpha} \models$ "X is thin", there is a cub $C \in \mathfrak{N}^n_{p'_\alpha}$ s.t.

$C \cap X = 0$, $C \subseteq \text{card} \cap (\tau,\alpha)$ and $\gamma \in C \longrightarrow \dot{p}'_\nu = 0$.

Set $D = \{\bar\alpha \mid \bar\alpha = \alpha \cap X^n_{p'_\alpha,\bar\alpha} \cap p'/\alpha , f , C \in X^n_{p'_\alpha,\bar\alpha}\}$.

Set $\pi = \pi_{p'_\alpha,\bar\alpha}$, then for $\bar\alpha \in D$, $\pi(C \cap \bar\alpha) = C$, $\pi(f/\bar\alpha) = f$ and $\pi(p'/\bar\alpha) = p'/\alpha$.

Using the collapsing Lemma (2.12) we get for $\bar\alpha \in D$, $p'_{\bar\alpha} = \pi^{-1}(p'_\alpha)$ and $\mathfrak{N}^n_{p'_{\bar\alpha}} = \mathfrak{N}^n_{p'_\alpha,\bar\alpha}$ and $\dot{p}'_{\bar\alpha} = 0$. In addition we have $\bar\alpha \in C$ and $\bar\alpha \notin X$ for $\bar\alpha \in D$. Let $\langle \alpha_i \mid i < \alpha \rangle$ be a monotone enumeration of D . For $i < \alpha$ define $f^*_i : X \cap (\alpha_i, \alpha_{i+1}) \longrightarrow V$ by replacing each $\Delta \in f(\nu)$ s.t. Δ is predense in $\mathbb{P}_\tau^{p'_{\nu+}}$ by

$$\Delta^* = \{p \in \mathbb{P}_{\alpha_i^+}^{p'_{\nu+}} \mid \Delta^p \text{ is dense in } \mathbb{P}_\tau^{p'_{\alpha_i^+}}\} .$$

Define

q_i = the $\mathfrak{N}^1_{p'_{\alpha^+_{i+1}}}$ -least $q \le (p')^{\alpha^+_{i+1}}_{\alpha^+_i}$ in $\mathbb{P}^{p'_{\alpha^+_{i+1}}}_{\alpha^+_i}$ s.t. $q \in \Sigma_{f^*_i}$ for $i < \alpha$.

q^* = the $\mathfrak{N}^1_{p'_{\alpha^+_0}}$ -least $q \le (p')^{\alpha^+_0}$ in $\mathbb{P}_\tau^{p'_{\alpha^+_0}}$ s.t. $q \in \Sigma_{f/\alpha_0}$.

Let D^* = the limit points of D .

Finally set

$$q = q^* \cup \bigcup_{i<\alpha} q_i \cup p'/D^* .$$

Claim $q \in \mathbb{P}_\tau^{p'_\alpha}$ and $|q| = |p'_\alpha|$.

Proof We must show for every $\gamma \in \text{card} \cap [\tau,\alpha)$ that $q/\gamma \in \mathbb{P}_\tau^{q_\gamma}$ and if $\gamma \in D^*$ then $|q/\gamma| = |q_\gamma|$ and $\dot{q}_\gamma = 0$. If $\gamma \notin D^*$ then there is an i s.t. q/γ differs from q_i/γ (or q^*/γ) only on an initial segment.

We can define D , $\langle q_i \mid i < \alpha \rangle$, q^* and hence q in $\mathfrak{N}_{p'_\alpha}$

from $p'\restriction\alpha$, f and C . We noted that for $\gamma \in D^* \subseteq D$; $\pi(C \cap \gamma) = C$, $\pi(f\restriction\gamma) = f$, $\pi(p'\restriction\gamma) = p'\restriction\alpha$, $\pi(p'_\gamma) = p'_\alpha$ and $\mathfrak{A}^n_{p'_\gamma} = \mathfrak{A}^n_{p'_\alpha,\gamma}$ where $\pi = \pi_{p'_\alpha,\gamma}$. Hence we can define $D \cap \gamma$, $\langle q_i\restriction\gamma \mid i < \gamma\rangle$, q^* and $q\restriction\gamma$ in $\mathfrak{A}_{p'_\gamma}$ by the same definition from $C \cap \gamma$, $f\restriction\gamma$ and $p'\restriction\gamma$. But $\gamma \in D^*$, so by definition $\mathfrak{A}_{p'_\gamma} = \mathfrak{A}_{q_\gamma}$ and we get $q\restriction\gamma \in \mathfrak{A}_{q_\gamma}$, so (i) holds. For $\gamma \in D^*$, $|q\restriction\gamma| \geq |p'\restriction\gamma| = |p'_\gamma| = |q_\gamma|$ and $\dot{q}_\gamma = \dot{p}'_\gamma = 0$. α is inaccessible so for $\gamma \in D^*$, $\mathfrak{A}_{q_\gamma} \models$ " γ is regular". $D^* \cap \gamma \in \mathfrak{A}_{q_\gamma}$ and must be cub in γ since otherwise $D \cap \gamma \setminus (\max(D^* \cap \gamma))$ would make γ singular. Hence (ii) using $D^* \cap \gamma$. (iii) holds since $q\restriction\gamma \leq p'\restriction\gamma$, $|q_\gamma| = |p'_\gamma|$ and (iii) holds for $p'\restriction\gamma$.

Q.E.D. (Claim)

The same argument at α for $q\restriction\alpha$ yields $q\restriction\alpha \in \mathbf{P}^{p'_\alpha}_\tau$ and $|q\restriction\alpha| = |p'_\alpha|$. Hence $q' = q\restriction\alpha \cup \{\langle\alpha,p'(\alpha)\rangle\} \in \mathbf{P}^s_\tau$ and $q' \leq p'$. We can show that $q' \in \Sigma_f$ by methods similar to that of Lemma 3.10 (using Fact 3.9.2).

Q.E.D.

3.4 THE LEMMA

We now come to a central lemma, whose method of proof is the essence of the chapter (and perhaps the book). We shall refer to the proof of the lemma many times.

Lemma 3.13 *Let α be inaccessible. Then $\mathbf{P}^s_\tau$ is τ-distributive in $\mathfrak{A}^1_s$.*

Proof Let $\Delta_\nu \in \mathfrak{A}^1_s$ be dense in $\mathbf{P}^s_\tau$ for $\nu < \tau$. Let $p_0 \in \mathbf{P}^s_\tau$. We must find a $p \leq p_0$, $p \in \bigcap_{\nu<\tau} \Delta_\nu$. We may assume w.l.o.g. that $|p_0\restriction\gamma| = |p_{0\gamma}|$ for all limit cardinals $\gamma \in (\tau,\alpha]$.

We construct $p_i \in \mathbb{P}^s_\tau$, $Y_i \prec \mathfrak{A}^1_s$ for $\gamma \in \mathrm{card} \cap [\tau,\alpha]$ and $f_i \in \mathbb{F}(p_i)$ by induction on $i < \rho \leq \tau + 1$ as follows.

p_o is given

Y^γ_i = the smallest $Y \prec \mathfrak{A}^1_s$ s.t. $\gamma \subseteq Y$ and p_j, Y^γ_j, $\langle \Delta_\nu \mid \nu < \tau \rangle \in Y$ for $j < i$.

Hence $\lim(\lambda) \longrightarrow Y^\gamma_\lambda = \bigcup\limits_{i<\lambda} Y^\gamma_i$.

$$f_i(\gamma) = \begin{cases} Y_i \cap \mathfrak{A}^1_{p_{i\gamma^+}} & \text{if } \gamma^+ \in Y^\gamma_i \\ \\ \text{undefined otherwise} \end{cases}$$

for $\gamma \in \mathrm{card} \cap [\tau,\alpha)$.

Claim $\mathrm{dom}(f_i)$ is thin in $\mathfrak{A}^1_s$.

Proof We have to show that $\mathrm{dom}(f_i) \cap \kappa$ is not stationary for κ regular. We only have to worry about inaccessibles ($\mathrm{dom}(f_i) \subseteq \mathrm{card}$) and the set $H = \{\gamma \mid \gamma = Y^\gamma_i \cap \delta\}$ is cub in δ. Hence $\gamma \in H \longrightarrow \gamma \not\in Y^\gamma_i \longrightarrow \gamma^+ \not\in Y_i \longrightarrow \gamma \not\in \mathrm{dom}(f_i)$.

Q.E.D. (Claim)

p_{i+1} = the $\mathfrak{A}^1_s$-least $p \leq p_i$ s.t.

(a) $p \in \Delta_i$.

(b) $p \in \sum^{s,p_i}_{f_i}$.

(c) $p(\alpha) \in \bigcap \{\Delta \in Y^\alpha_i \mid \Delta$ is dense in $\mathbf{R}^s\}$.

(d) $|p{\restriction}\gamma| = |p_\gamma|$ for limit cardinals $\gamma \in (\tau,\alpha]$.

For $\lim(\lambda)$ set $p\colon \mathrm{card} \cap [\tau,\alpha] \longrightarrow V^2$ s.t. $p(\gamma) = \langle \bigcup\limits_{i<\lambda} \dot{p}_{i\gamma}, \bigcup\limits_{i<\lambda} p_{i\gamma} \rangle$.

$$p_\lambda = \begin{cases} p & \text{if } p \in \mathbb{P}^s_\tau \\ \text{undefined otherwise.} \end{cases}$$

We will prove that p_i is defined for $i \le \tau$ by induction on i. The lemma will follow from this since then $p_\tau \in \bigcap_{i<\tau} \Delta_i$.

The above definition, for p_{i+1}, is a canonical way of defining $p_i \in \mathbf{P}^s_\tau$ s.t. for $\lim(\lambda)$, $\bigcup_{i<\lambda} p_i \in \mathbf{P}^s_\tau$. The clauses that put $p \in \mathbb{P}^s_\tau$ are (b), (c) and (d) ((c) only if $s \in \mathbf{S}_{\alpha^+}$). (a) is for a specific purpose and will be altered in later applications of the construction. Note that Lemma 3.12 guarantees that $\sum^{s,p_i}_{f_i}$ is dense in $\mathbf{P}^s_\tau$. For $i = 0$ or i is a successor $p_i \in \mathbf{P}^s_\tau$ trivially. Assume $i = \lambda$ and $\lim(\lambda)$. We will prove $p = \bigcup_{i<\lambda} p_i \in \mathbf{P}^s_\tau$. The proof will stretch out over the following lemmas.

Lemma 3.14 *Let* $\tau \le \delta \le \gamma \le \beta < \alpha$. *Let* $\Delta \in Y^\delta_\lambda$ *be pre-dense in* $\mathbf{P}^{p_i\beta^+}_\gamma$ *for some* $i < \lambda$. *Then there is a* $j < \lambda$ *s.t.* $(p_j)^{\beta^+}_\gamma$ *meets* Δ.

Proof Suppose that for every $j < \lambda$, $(p_j)^{\beta^+}_\gamma$ does *not* meet Δ. Let $i \le i_o < \lambda$ be s.t. $\Delta \in Y^\delta_{i_o}$. $\gamma, \beta \in Y^\delta_{i_o}$ since $\Delta \ne 0$ and $\gamma = \min(\mathrm{dom}(r))$ and $\beta = \max(\mathrm{dom}(r))$ for $r \in \Delta$. Since $p_{i_o+1} \in \sum^{s,p_{i_o}}_{f_{i_o}}$ it follows

Claim $(p_{i_o+1}) \in \sum^{p_{i_o}\beta^+,(p_{i_o})^{\beta^+}_\gamma}_{f_{i_o}\restriction[\gamma,\beta]}$.

Proof This can only be shown because of the way f_i was defined.

$\sum_{f_{i_o}}^{s,p_{i_o}}$ only deals with $\Delta' \in f_{i_o}(\beta)$ s.t. Δ' is predense in $\mathbb{P}^{p_{i_o}\beta^+}$. If $\Delta \in f_{i_o}(\beta)$ is predense in $\mathbb{P}_\gamma^{p_{i_o}\beta^+}$ we can define

$$\Delta' = \{h \in \mathbb{P}_\tau^{p_{i_o}\beta^+} \mid (h)_\gamma \in \Delta\} \in Y_{i_o}^\delta$$

and hence $\Delta' \in f_{i_o}(\beta)$ and Δ' is predense in $\mathbb{P}_\tau^{p_{i_o}\beta^+}$. So either $(p_{i_o+1})^{\beta^+}$ meets Δ', in which case $(p_{i_o+1})_\gamma^{\beta^+}$ meets Δ, or $\exists \eta(\tau \leq \eta < \beta^+)$ s.t. $(\Delta')^{(p_{i_o+1})_{\eta^+}^{\beta+}}$ is dense in $\mathbb{P}_\tau^{(p_{i_o+1})_{\eta^+}}$. If $\eta^+ \leq \gamma$ then $(p_{i_o+1})_\gamma^{\beta+}$ meets Δ. If $\gamma < \eta^+$ then $\Delta^{(p_{i_o+1})_{\eta^+}^{\beta+}}$ is dense in $\mathbb{P}_\gamma^{(p_{i_o+1})_{\eta^+}}$. Either way we get

$$(p_{i_o+1})_\gamma^{\beta+} \in \sum_{f_{i_o}\restriction[\gamma,\beta]}^{p_{i_o}\beta^+,(p_{i_o})_\gamma^{\beta^+}} .$$

Q.E.D. (Claim)

We have $\Delta \in f_{i_o}(\beta)$ and that $(p_{i_o+1})_\gamma^{\beta+}$ does not meet Δ, hence by the definition of $\Sigma_{f_{i_o}}$ there is a least $\beta_o \in [\gamma,\beta)$ s.t. $\Delta_o = \Delta^{(p_{i_o+1})_{\beta_o^+}^{\beta^+}}$ is dense in $\mathbb{P}_\gamma^{p_{i_o+1},\beta_o^+}$. We have just defined Δ_o in $Y_{i_o+2}^\delta$ from Δ and $(p_{i_o+1})_\gamma^{\beta+}$, hence $\Delta_o \in f_{i_o+2}(\beta_o)$. But $(p_{i_o+3})_\gamma^{\beta_o^+}$ does not meet Δ_o, since otherwise $(p_{i_o+3})_\gamma^{\beta+}$ would meet Δ $((p_{i_o+3})_{\beta_o^+}^{\beta+} \leq (p_{i_o+1})_{\beta_o^+}^{\beta+}$ and if $(p_{i_o+3})_\gamma^{\beta_o^+}$ met Δ_o then $(p_{i_o+3})_\gamma^{\beta_o^+} \leq r$ for some r s.t. $r \cup (p_{i_o+1})_{\beta_o^+}^{\beta^+}$ meets Δ). But then by the same argument, there is a $\beta_1 \in [\gamma,\beta^o)$ s.t.

$$\Delta_1 = \Delta_o^{(p_{i_o+3})_{\beta_1^+}^{\beta_o^+}} \text{ is dense in } \mathbb{P}_\gamma^{p_{i_o+3},\beta_1^+} .$$

Continuing in this fashion we get $\beta_o > \beta_1 > \beta_2 > \ldots$ contradiction.

Q.E.D. (Lemma)

Lemma 3.15 *Let* $\Delta \in Y^\alpha_\lambda$ *be dense in* $\mathbf{R}^s$. *Then there is an* $i < \lambda$ *s.t.* $p_i(\alpha) \in \Delta$.

Proof $\exists j < \lambda\ \Delta \in Y^\gamma_j \longrightarrow p_{j+1}(\alpha) \in \Delta$ by the definition of p_{i+1} , clause (c).

Definition 1 For $\gamma \in \text{card} \cap [\tau,\alpha]$ define $\sigma = \sigma^\gamma_\lambda$ and $b = b^\gamma_\lambda$ by $\alpha\colon b \cong Y^\gamma_\lambda$ and b is transitive. We further define

(a) $s' = s^\gamma_\lambda = \sigma^{-1}(s)$.

(b) $A' = A^\gamma_\lambda = \sigma^{-1}(A \cap \alpha^+)$.

(c) $\alpha' = \alpha^\gamma_\lambda = \sigma^{-1}(\alpha)$.

(d) $\alpha^* = \alpha^{*\gamma}_\lambda = \sigma^{-1}(\alpha^+)$.

(e) $\mu' = \mu^\gamma_\gamma = \text{On} \cap b$.

(f) $p'_i = p^\gamma_{i\lambda} = \sigma^{-1}(p_i)$ for $i < \lambda$.

(g) For $\delta \in [\text{card}]_d \cap [\gamma,\alpha']$, $D'_\delta = D^\gamma_{\lambda\delta} = \bigcup_{i<\lambda} \tilde{p}'_{i\delta}$ (where $\tilde{p}'_{i\delta} = \{\xi \mid p'_{i\delta}(\xi) = 1\})$.

And $D' = D^\gamma_\lambda = \bigcup_{\delta\in[\text{card}]_b} D'_\delta$.

(h) For $\delta \in [\text{card}]_b \cap [\gamma,\alpha']$, $\mathbf{S}_\delta = \mathbf{S}^b_\delta = \sigma^{-1}(\mathbf{S}_{\sigma(\delta)})$ and if $r \in \mathbf{S}^b_{(\delta+)_b}$ then $\mathbf{R}^r = \mathbf{R}^r_b = \sigma^{-1}(\mathbf{R}^{\sigma(r)})$.

(i) For $\eta \in \delta \cap [\text{card}]_b$ and $r \in \mathbf{S}^b_\delta$,

$$\mathbf{P}^r_\eta = \mathbf{P}^r_{\eta b} = \sigma^{-1}(\mathbf{P}^{\sigma(r)}_{\sigma(\eta)}) .$$

Finally note that $b = \langle L_\mu, [A',s'], A', s'\rangle$ and that the above structures have the same definitions relativised to b .

Lemma 3.16 *Let* $\gamma \le \eta \le \delta < \alpha'$ *s.t.* $\eta, \delta \in (\mathrm{card})_b$. *Let* $\sigma((\delta^+)_b) = \beta^+$ *and* $r = \sigma^{-1}(p_{i\beta^+})$ *for some* $i < \lambda$. *Then* $D' \cap [\eta, (\delta^+)_b)$ *is* $\mathbb{P}^r_\eta$*-generic over* $(\mathfrak{A}^1_r)_b$.

Proof Let $\Delta' \in (\mathfrak{A}^1_r)_b$ be predense in $\mathbb{P}^r_\eta$. Applying σ we get $\sigma(\Delta') = \Delta$, $\Delta \in Y^\gamma_\lambda$ and Δ is predense in $\mathbb{P}^{p_{i\beta^+}}_{\sigma(\eta)}$. By Lemma 3.14 $\exists j < \lambda$ s.t. $(p_j)^{\beta^+}_{\sigma(\eta)}$ meets Δ . Applying σ^{-1} we get $(p'_j)^{(\delta^+)_b}_\eta$ meets Δ' .
$D' \cap [\eta, (\delta^+)_b)$ corresponds to $\bigcup_{i<\lambda} (p'_i)^{(\delta^+)_b}_\eta$, hence $(p'_j)^{(\delta^+)_b}_\eta \in G_{D' \cap [\eta, (\delta^+)_b)}$ and meets Δ' .

Q.E.D.

Corollary 3.16.1 $D' \cap (\delta^+)_b$ is $\mathbb{P}^{(\delta^+)_b}_\gamma$-generic over $(\mathfrak{A}^1_{(\delta^+)_b})_b$.

Proof Lemma 3.16 essentially says that if $r \subseteq \chi_{D'_{(\delta^+)_b}}$, then $D' \cap (\delta^+)_b$ is $\mathbb{P}^r_\gamma$-generic over $(\mathfrak{A}^1_r)_b$. By Lemma 2.11 this implies that $D' \cap (\delta^+)_b$ is $\mathbb{P}^{(\delta^+)_b}_\gamma$-generic over $(\mathfrak{A}^1_{(\delta^+)_b})_b$.

Lemma 3.17 $D'_{\alpha'}$ *is* $\mathbf{R}^{s'}$*-generic over* b .

Proof A direct application of Lemma 3.15.

Lemma 3.18 *Let* $\delta < \alpha'$, $\delta \in (\mathrm{card})_b$, *then* $L_{\alpha'}[A' \cap \alpha', D' \cap \alpha'] \models$ " δ is a cardinal".

Proof It is clear that Lemma 3.6 relativises to b . Let $\delta < \theta < \alpha'$, $\theta \in (\mathrm{card})_b$. Then by Corollary 3.16.1, $D' \cap (\theta^+)_b$ is $\mathbb{P}^{(\theta^+)_b}_\gamma$-generic over $\mathfrak{A}^1_{(\theta+)_b}$. By Lemma 3.6,

δ is preserved in $(\mathfrak{A}^1_{(\theta^+)_b})_b$. In particular

$$L_{(\theta+)_b}[A' \cap (\theta^+)_b, D' \cap (\theta^+)_b] \models \text{" } \delta \text{ is a cardinal" } .$$

But this holds for arbitrarily large $\theta < \alpha'$.

Q.E.D.

Lemma 3.19 $p_{i\gamma} = p'_{i\gamma} = (p^{\gamma}_{i\lambda})_{\gamma}$ *for* $i < \lambda$ *and* $\gamma \in \text{card} \cap [\tau, \alpha]$ *hence* $D'_{\gamma} = (D^{\gamma}_{\lambda\gamma}) = \tilde{p}_{\gamma}$) .

Proof If $\gamma \in Y^{\gamma}_{\lambda}$ then $\sigma \restriction (\gamma+)_b = \text{id} \restriction (\gamma+)_b$ since $Y^{\gamma}_{\lambda} \cap \gamma+$ is transitive and $(\gamma+)_b = Y^{\gamma}_{\lambda} \cap \gamma+$. Hence $p_{i\gamma} = \sigma(p_{i\gamma}) = p'_{i\gamma}$. So we assume $\delta \not\in Y^{\gamma}_{\lambda}$. Then γ is a limit point of $\text{card} \cap [\tau, \alpha]$. Let $\sigma(\gamma) = \gamma^*$. We have that $\gamma < \gamma^*$ and γ^* is also a limit cardinal. $\gamma^*, p_i \in Y^{\gamma}_{\lambda}$ so $p_{i\gamma^*} \in Y^{\gamma}_{\lambda}$, $p'_{i\gamma} = \sigma^{-1}(p_{i\gamma^*})$ and $\sigma(\mathfrak{A}_{p'_{i\gamma}}) = \mathfrak{A}_{p_{i\gamma^*}}$. By the argument of §2.5 Fact 1 (b) we can show that $\sigma''\mathfrak{A}_{p'_{i\gamma}} = X_{p_{i\gamma^*},\gamma}$ and hence $\mathfrak{A}_{p'_{i\gamma}} = \mathfrak{A}_{p_{i\gamma^*,\gamma}}$ and $\sigma \restriction \mathfrak{A}_{p'_{i\gamma}} = \pi_{p_{i\gamma^*,\gamma}}$.

Since $p_{i \restriction \gamma^*} \in \mathbf{P}^{p_{i\gamma^*}}_{\tau}$ and $|p_i \restriction \gamma^*| = |p_{i\gamma^*}|$, we can apply the collapsing lemma (2.12) to get $p'_{i\gamma} = p_{i\gamma}$.

Q.E.D.

Lemma 3.20 $A', s' \in L_{\mu'}[A \cap \gamma, p_{\gamma}]$.

Proof Remember $b = \langle L_{\mu'}[A', s'], A', s' \rangle$.

Claim 1 $A' \cap \alpha', D' \cap \alpha' \in L_{\mu'}[A \cap \gamma, p_{\gamma}]$.

Proof For $\gamma = \alpha'$ there is nothing to prove. So assume $\gamma < \alpha'$. Let $\langle \gamma_i \mid i < \alpha' \rangle$ enumerate $(\text{card})_b \cap [\gamma, \alpha')$. We will use the proof of Lemma 3.7 relativised to b . By Corollary 3.16.1 $\forall \theta < \alpha'$ $D' \cap (\theta^+)_b$ is $\mathbf{P}^{(\theta+)_b}_{\gamma}$-generic over

$\mathfrak{N}^1_{(\theta+)_b}$, so we will be able to continue the construction up to α' .

Define γ_i , $a_i = A' \cap \gamma'$ and $d_i = D'_{\gamma_i}$ by induction on $i < \alpha'$ (starting with $\gamma_o = \gamma$, $a_o = A' \cap \gamma$ and $d_o = \tilde{\mathfrak{p}}_\gamma$) exactly as in Lemma 3.7 except that we define $\mathbf{S}'_{\gamma_i}$ by

$$r \in \mathbf{S}'_{\gamma_i} \quad \text{iff} \quad r\colon [\gamma_i, |r|) \longrightarrow 2 \quad \text{s.t.}$$

1. $|r| \geq \gamma_i$.

2. $\forall \xi \leq |r| \; L_{\alpha'}[a_i, r{\restriction}\xi] \models \text{card}(\xi) \leq \gamma_i$.

3. If $L_{\alpha'}[a_i] \models \gamma_i \geq \omega_2$ then

   $\forall \xi \leq |r| \; L_\delta[a_i, r{\restriction}\xi] \models$ "$\text{card}(\xi) \leq \gamma_i$" for δ s.t.

   $L_{\alpha'} \models$ "$\text{card}(\delta) = \text{card}(\delta)$" .

By Lemma 3.18 $L_{\alpha'}[A' \cap \alpha', D' \cap \alpha']$ preserves cardinals of b so during the construction cardinals of b are not collapsed. The above construction can be carried out in $L_{\mu'}[A \cap \gamma, p_\gamma]$ since $\mu' > \alpha'$.

Q.E.D. (Claim 1)

Claim 2 $D'_{\alpha'} \in L_{\mu'}[A \cap \gamma, p_\gamma]$.

Proof $b \models$ "$p'_i{\restriction}\alpha \in \mathbf{P}_\tau^{p'_{i\alpha'}}$ and $|p'_i{\restriction}\alpha'| = |p'_{i\alpha'}| \;\; \forall i < \lambda$" . Hence $p'_i{\restriction}\alpha'$ correctly codes $p'_{i\alpha'}$, $\forall i < \lambda$. $p'_\alpha = \bigcup_{i<\lambda} p'_{i\alpha'}$ and $D'_{\alpha'} = \tilde{p}'_\alpha$. Hence $D' \cap A'$ contains the coding information of $\bigcup_{i<\lambda} p'_i{\restriction}\alpha$ for $D'_{\alpha'}$ and we may employ the argument of Lemma 3.7 (3) to define $D'_{\alpha'}$ from $A \cap \alpha', D' \cap \alpha'$. Note that it is not necessary for $\bigcup_{i<\lambda} p'_i{\restriction}\alpha'$ to be a condition for it to correctly code $D'_{\alpha'}$. While carrying out the argument of Lemma 3.7 (3) we define $\mathbf{S}'_{\alpha'}$ as follows

$r \in \mathbb{S}'_{\alpha'}$ iff $r: [\alpha', |r|) \longrightarrow 2$ s.t.

1. $|r| \geq \alpha'$.

2. $\forall \xi \leq |r| \; L_\delta[A \cap \alpha', r \upharpoonright \xi] \models$ "$\mathrm{card}(\xi) \leq \alpha'$" for a δ s.t.

   $L_{\mu'} \models$ "$\mathrm{card}(\delta) = \mathrm{card}(\xi)$" .

The definition of $D'_{\alpha'}$ can be carried out in $L_{\mu'}[A \cap \gamma, p_\gamma]$ since $A \cap \alpha'$, $D' \cap \alpha' \in L_{\mu'}[A \cap \gamma, p_\gamma]$ by Claim 1.

Q.E.D. (Claim 2)

Claim $A', s' \in L_{\mu'}[A \cap \gamma, p_\gamma]$.

Proof By Lemma 3.17 $D'_{\alpha'}$ is $\mathbb{R}^{s'}$-generic over b , hence we may employ the argument of Lemma 3.7 (2) to define A' and s' from $D'_{\alpha'}$.

Q.E.D. (Lemma 3.20)

Lemma 3.21 $b, \langle p'_i \mid i < \lambda \rangle \in L_{\mu'+\omega}[A \cap \gamma, p_\gamma]$.

Proof By Lemma 3.20 $b \in L_{\mu'+\omega}[A \cap \gamma, p_\gamma]$ and $\langle p'_i \mid i < \lambda \rangle$ can be defined from b the way $\langle p_i \mid i < \lambda \rangle$ was defined from $\mathfrak{A}^1_s$. Hence

$$\langle p_i \mid i < \lambda \rangle \in L_{\mu'+\omega}[A \cap \gamma, p_\gamma] .$$

Q.E.D.

Lemma 3.22 *(a)* $p_\gamma \in \mathbb{S}_\gamma$ *and (b)* $\gamma \geq \omega_2 \longrightarrow b \in \mathfrak{A}^1_{p_\gamma}$.

Proof

(a) We first show:

$$\forall \xi \leq |p_\gamma| \longrightarrow L[A \cap \gamma, p \upharpoonright \xi] \models \text{"card}(\xi) \leq \gamma\text{"} .$$

For $\xi < |p|$, it follows from $p_\gamma = \bigcup_{i<\lambda} p_{i\gamma}$ and

$\xi < |p_{i\gamma}|$ for some $i < \lambda$. Now let $\xi = |p_\gamma|$. Then by Lemma 3.19 $p_{i\gamma} = p'_{i\gamma}$ for $i < \lambda$. $\lambda \le \tau \le \gamma$ so by Lemma 3.21

$$L_{\mu'+\omega}[A \cap \gamma, p_\gamma] \models \text{"card}(\xi) = \text{card}(\bigcup_{i<\lambda} |p'_{i\gamma}|) \le \lambda.\gamma = \gamma\text{"} .$$

If $\gamma \ge \omega_2$ we must also show:

$$\xi \le |p_\gamma| \longrightarrow L_\delta[A \cap \gamma, p/\xi] \models \text{"card}(\xi) \le \gamma\text{"} \text{ for a } \delta \text{ s.t. }$$

$$L \models \text{card}(\delta) \le \xi .$$

For $\xi < |p_\gamma|$ this is again trivial. Let $\xi = |p_\gamma|$ and take $\delta = \mu' + \omega$. By negation assume $L \models \xi < \text{card}(\mu')$. Then there is a $\eta \in (\xi, \mu']$ s.t. $L \models$ "η is regular" by the marginalia [DJ], since $\gamma \ge \omega_2$, $\text{cf}(\eta) = \gamma > \omega$ (in V).

Set $\sigma' = \sigma/L_\eta$ and let $\eta' = \sup(\text{rng}(\sigma'))$, then $\sigma' : L_\eta \xrightarrow[\Sigma_1]{} L_{\eta'}$ cofinally where η is regular in L.
By [DJ] this implies $0^\#$ exists if $\sigma' \ne \text{id}/L_\eta$.
$\xi = |p_\gamma| = (\gamma+)_b$ (since $\tilde{p}_\gamma = \bigcup_{i<\lambda} \tilde{p}'_{i\gamma} = \tilde{D}'_\gamma$ and $\sup(\tilde{D}'_\gamma) = (\gamma+)_b$). But $\xi < \eta$ and hence $\sigma'(\xi) = \gamma+ > \xi$ and $\sigma' \ne \text{id}/L_\eta$. This is a contradiction and (a) is proved.

(b) Recall that $\mathfrak{A}^1_{p_\gamma}$ has the form $\langle L_\mu[A \cap \gamma, p_\gamma], A \cap \gamma, p_\gamma \rangle$ where $\exists \eta < \mu$ s.t. $L_\eta[A \cap \gamma, p_\gamma] \models \text{card}(|p_\gamma|) = \gamma$ and $L_\mu \models$ "$\text{card}(\eta) = |p_\gamma|$". Also recall $b = \langle L_{\mu'}[A', s'], A', s' \rangle$.

It will be sufficient to show $\mu' < \mu$ since by Lemma 3.21,

$$b \in L_{\mu'+\omega}[A \cap \gamma, p_\gamma] \subseteq L_\mu[A \cap \gamma, p_\gamma] .$$

Set $\xi = |p_\gamma|$. $b \models$ "$\xi = \gamma+$" and by Lemma 3.18 $L_{\alpha'}[A \cap \gamma, p_\gamma] \models$ "ξ is a cardinal". So $\alpha' < \eta$ since $L_\eta[A \cap \gamma, p_\gamma] \models$ "$\text{card}(\xi) = \gamma$". $L_{\mu'} \models$ "$\xi < \text{card}((\xi^+)_b)$", while $L_\mu \models$ "$\text{card}(\eta) = \xi$" and $(\xi^+)_b < \alpha' < \eta$. Hence $\mu' < \mu$.

Q.E.D. (Lemma 3.22)

Lemma 3.23 $p(\alpha) \in \mathbb{R}^s$ *and* $p(\gamma) \in \mathbb{R}^{p_{\gamma+}}$ *for* $\gamma \in \text{card} \cap [\tau,\alpha)$.

Proof By Lemma 3.22 $p_\gamma \in \mathbb{S}_\gamma$ for $\gamma \in \text{card} \cap [\tau,\alpha]$. The constraints $\dot{p}_\gamma = \bigcup_{i<\lambda} \dot{p}_{i\gamma}$ and $\dot{p}_\alpha = \bigcup_{i<\lambda} \dot{p}_{i\alpha}$ cause no problem at unions.

Q.E.D.

All that remains in order to prove $p \in \mathbb{P}^s_\tau$ is

Lemma 3.24 $p{\restriction}\gamma \in \mathbb{P}^{p_\gamma}_\tau$ *for* $\gamma \in \text{card} \cap [\tau,\alpha]$.

Proof By induction on γ . The cases $\gamma = \tau$ and γ is a successor are trivial. Now let γ be a limit cardinal. We must prove (i)-(v) of the definition of $\mathbb{P}^s_\tau$ with p_γ for s and $p{\restriction}\gamma$ for p . By Lemmas 3.21 and 3.22 we have $\langle p_i' \mid i<\lambda \rangle \in \mathfrak{A}^1_{p_\gamma}$, but $p_i'{\restriction}\gamma = p_i{\restriction}\gamma$. Hence $p{\restriction}\gamma = \bigcup_{i<\lambda} p_i{\restriction}\gamma \in \mathfrak{A}^1_{p_\gamma}$. This proves (i).

(iii) and (v) hold along with $|p{\restriction}\gamma| = |p_\gamma|$ since

$$\xi < |p_\gamma| \longrightarrow \exists i<\lambda(\xi < |p_i{\restriction}\gamma|) .$$

We turn to (ii). Let $\mathfrak{A}_{p_\gamma} \models$ "γ is regular" . Let D_i = the $\mathfrak{A}_{p_{i\gamma}}$-least $D \subseteq \gamma$ s.t. D is cub in γ and $\delta \in D \longrightarrow \delta$ is a limit cardinal, $|p_i{\restriction}\delta| = |p_{i\delta}|$ and $\dot{p}_{i\delta} = 0$. Set $D = \bigcap_{i<\lambda} D_i$. Then $D \in \mathfrak{A}_{p_\gamma}$ since $\langle D_i \mid i<\lambda \rangle$ is definable from $\langle p_i{\restriction}\gamma \mid i<\lambda \rangle$ and D is cub in γ since $\lambda<\gamma$ and $\mathfrak{A}_{p_\gamma} \models$ "γ is regular" . For $\delta \in D$, $\dot{p}_\delta = \bigcup_i \dot{p}_i = 0$. Moreover, for all limit cardinals $\delta \in (\tau,\gamma)$ we have

$$|p{\restriction}\delta| \geq \sup_{i<\lambda} |p_i{\restriction}\delta| = \sup_{i<\lambda} |p_{i\delta}| = |p_\delta| .$$

This proves (ii).

All that remains to show is (iv) for the case that $\Omega_\gamma \leq |p_\gamma|$. If $\Omega_\gamma < |p_\gamma|$ there is an $i < \lambda$ s.t. $\Omega_\gamma \leq |p_{i\gamma}|$ and (iv) follows from the fact that it holds for p_i/γ . Now let $\Omega_\gamma = |p_\gamma|$. We assume that $cf(\gamma) > \omega$ since otherwise there is nothing to prove.

We must find a $D \subseteq \gamma$, $D \in \mathfrak{M}_{\Omega_\gamma} = \mathfrak{M}_{p_\gamma}$ s.t. D is cub and $\delta \in D \longrightarrow \Omega_\delta^{p_\delta}$ exists, $|p/\delta| \geq \Omega_\delta$ and $D \cap \delta \in \mathfrak{M}_{\Omega_\delta}$. $p_\gamma \in \mathbb{S}^*_\gamma$ so we may use Lemma 2.4 to get

$$D_{p_\gamma} , \quad \langle r_\delta \mid \delta \in D_{p_\gamma} \rangle \quad \text{and} \quad \langle \Pi_\delta^{p_\gamma} \mid \delta \in D_{p_\gamma} \rangle .$$

For $\delta \in D_{p_\gamma}$ set

$$I_\delta = \{ i < \lambda \mid (p_i/\gamma) \in \operatorname{rng}(\Pi_\delta^{p_\gamma}) \} .$$

Case 1 $\quad \exists \delta \in D_{p_\gamma} \ (\sup(I_\delta) = \lambda)$.

Let δ_o be the least such and set $I_{\delta_o} = I$ and $D = D_{p_\gamma} \setminus \delta_o \in \mathfrak{M}_{p_\gamma}$. We will show that D has the desired properties. Let $\delta \in D$; by Lemma 2.4 $D \cap \delta = D_{r_\delta} \setminus \delta_o \in \mathfrak{M}_{r_\delta}$ and $r_\delta \in \mathbb{S}^*_\delta$, which implies that $\Omega_\delta^{r_\delta}$ exists and $\Omega_\delta^{r_\delta} = |r_\delta|$. Therefore it is sufficient to show that $r_\delta = p_\delta$ (since we already have $|p/\delta| = |p_\delta|$ for limit cardinals).

Set $\pi = \Pi_\delta^{p_\gamma}$. By Lemma 2.4 we have

$$\pi : \langle \mathfrak{M}^o_{r_\delta}, C_{r_\delta} \rangle \xrightarrow[\Sigma_o]{} \langle \mathfrak{M}^o_{p_\gamma}, C_{p_\gamma} \rangle .$$

For $i < \lambda$ we have $\pi(p_i/\delta) = p_i/\gamma$ and $p_{i\gamma_o} = p_\gamma / |p_i/\gamma| \in \operatorname{rng}(\pi)$. Set $r_i = \pi^{-1}(p_{i\gamma})$, then

$$\pi(\mu^o_{r_i}) = \mu^o_{p_{i\gamma}} , \quad |p_\gamma| = \bigcup_{i<\lambda} |p_{i\gamma}| \quad \text{and} \quad \mu^o_{p_\gamma} = \sup_{i \in I}(\mu^o_{p_{i\gamma}}) .$$

Hence π is cofinal and Σ_1 . By Lemma 2.3,

$r_\delta = (\pi^{-1})"p_\gamma$. Then $r_\delta = \bigcup_{i\in I} \pi^{-1}(p_{i\gamma})$. We now apply §2.5 Fact 1 and get $\mathfrak{A}_{r_i} = \mathfrak{A}_{p_{i\gamma,\delta}}$ and $\pi/\mathfrak{A}_{r_o} = \pi_{p_{i\gamma},\delta}$. Hence $p_i/\gamma \in \mathrm{rng}(\pi_{p_{i\gamma},\delta})$ and we have $|p_i/\gamma| = |p_{i\gamma}|$. So we can apply the collapsing lemma (2.12) getting $r_i = p_{i\delta}$. Thus $r_\delta = \bigcup_{i\in I} r_i = \bigcup_{i\in I} p_{i\delta} = p_\delta$.

Q.E.D. (Case 1)

Case 2 $\forall \delta \in D_{p_\gamma} (\sup(I_\delta) < \lambda$ and $\sup(\Pi_\delta^{p_\gamma}"|r_\delta|) < |p_\gamma|)$.

By Lemma 2.4 $\sup(\Pi_\delta^{p_\gamma}"|r_\delta|) \leq \sup(\Pi_\eta^{p_\gamma}"|r_\eta|)$ for $\delta \leq \eta \in D_{p_\gamma}$ and $\sup_{\delta\in D_{p_\gamma}} (\sup(\Pi_\delta^{p_\gamma}"(|r_\delta|)) = |p_\gamma|$.

Set $\eta_\delta = \sup(\Pi_\delta^{p_\gamma}"|r_\delta|)$ and

$\rho_\delta = \sup\{|p_{i\gamma}| < \eta_\delta \mid p_i/\gamma \in \mathrm{rng}(\Pi_\delta^{p_\gamma})\}$ for $\delta \in D_{p_\gamma}$.

$\langle \eta_\delta \mid \delta \in D_{p_\gamma}\rangle$ and $\langle \rho_\delta \mid \delta \in D_{p_\gamma}\rangle$ are continuous non-decreasing functions s.t. $\sup_\delta(\eta_\delta) = \sup_\delta(\rho_\delta) = |p_\gamma|$. $\mathrm{cf}(\gamma) > \omega$ so if we set $D = \{\delta \in D_{p_\gamma} \mid \eta_\delta = \rho_\delta\}$ then D is cub in γ . $\langle p_i/\gamma \mid i < \lambda\rangle \in \mathfrak{A}_{p_\gamma}$ and by Lemma 2.4

$$D_{p_\gamma} , \quad \langle r_\delta \mid \delta \in D_{p_\gamma}\rangle \text{ and } \langle \Pi_\delta^{p_\gamma} \mid \delta \in D_{p_\gamma}\rangle \in \mathfrak{A}_{p_\gamma} .$$

Thus $\langle \Pi_\delta^{p_\gamma}"|r_\delta| \mid \delta \in D_{p_\gamma}\rangle \in \mathfrak{A}_{p_\gamma}$ and $D \in \mathfrak{A}_{p_\gamma}$. We shall show that D has the desired properties. Let $\delta \in D$, $\eta = \eta_\delta = \rho_\delta$. There is a $\theta < \lambda$ s.t. $\sup_{i<\theta}|p_{i\gamma}| = \eta$ and $p_{\theta,\gamma} = p_\gamma/\eta$ (since $\eta_\delta = \rho_\delta$).

Set $I = \{i < \theta \mid p_i/\gamma \in \mathrm{rng}(\pi)\}$ (where $\pi = \Pi_\delta^{p_\gamma}$)

$$\pi \colon \langle \mathfrak{A}^o_{r_\delta}, C_{r_\delta}\rangle \xrightarrow[\Sigma_1]{} \langle \mathfrak{A}^o_{p_{\theta\gamma}}, C_{p_{\theta\gamma}}\rangle \text{ cofinally.}$$

We can use the argument of Case 1 to get $r_\delta = p_{\theta\delta}$, i.e. setting $r_i = \pi^{-1}(p_{i\gamma})$ for $i \in I$, use §2.5 Fact 1 and the collapsing lemma (2.12) to get $r_i = p_{i\delta}$ and hence $r_\delta = \bigcup_{i\in I} r_i = \bigcup_{i\in I} p_{i\delta} = p_{\theta\delta}$.

$p_{\theta\delta} \in \mathbb{S}^*_\delta$ and $p_\delta \le p_{\theta\delta}$ so Ω_δ exists and $|p_\delta| \ge |p_{\theta\delta}| = \Omega_\delta$. All that remains is to show that $D \cap \delta \in \mathfrak{A}_{p_\delta}$. In fact we will have shown $D \cap \delta \in \mathfrak{A}_{p_{\theta\delta}}$.

$\langle p_i \upharpoonright \delta \mid i < \theta \rangle \in \mathfrak{A}_{p_{\theta\delta}}$ and by Lemma 2.4

$$D_{p_\gamma} \cap \delta = D_{r_\delta} , \ \langle \Pi_\nu^{r_\delta} \mid \nu \in D_r \rangle \text{ and } \langle r_\nu \mid \nu \in D_r \rangle \in \mathfrak{A}_{p_{\theta\delta}} .$$

In addition $\Pi_\delta^{p_\gamma} \circ \Pi_\nu^{r_\delta} = \Pi_\nu^{p_\gamma}$ for $\nu \in D_{r_\delta}$.

$\langle \Pi_\nu^{p_\gamma} " |r_\nu| \mid \nu \in D_{r_\delta} \rangle = \langle \Pi_\nu^{r_\delta} " |r_\nu| \mid \nu \in D_{r_\delta} \rangle \in \mathfrak{A}_{p_{\theta\delta}}$, since $\Pi_\delta^{p_\gamma \upharpoonright \delta} = \mathrm{id} \upharpoonright \delta$ and $|r_\nu| < \delta$ for $\nu \in D_{r_\delta}$. Hence $D \cap \delta \in \mathfrak{A}_{p_{\theta\delta}} \subseteq \mathfrak{A}_{p_\delta}$.

Q.E.D. (Case 2)

<u>Case 3</u> Case 1 and Case 2 fail.

Then $\forall \delta \in D_{p_\gamma} (\sup(I_\delta) < \lambda)$ and $\exists \delta \in D_{p_\gamma}$ s.t. $\sup(\Pi_\delta^{p_\gamma}) " |r_\delta| = |p_\gamma|$. If $ot(C_{p_\gamma}) = \gamma$ then for all $\delta \in D_{p_\gamma}$,

$$\Pi_\delta^{p_\gamma} : \langle \mathfrak{A}^o_{r_\delta}, C_{r_\delta} \rangle \xrightarrow[\Sigma_o]{} \langle \mathfrak{A}^o_{p_\gamma}, C_{p_\gamma} \rangle$$

and $(\Pi_\delta^{p_\gamma}) " C_{r_\delta}$ is an initial segment of C_{p_γ} (by the remarks after Lemma 2.4). But then $ot(C_{r\delta}) \le \delta < \gamma = ot(C_{p_\gamma})$ and it must be that $\sup(\Pi_\delta^{p_\gamma}) " |r_\delta| < |p_\gamma|$ and Case 2 would hold. So $ot(C_{p\gamma}) < \gamma$.

$\sup_{i<\lambda}(\mu^o_{p_{i\gamma}}) = \mu^o_{p_\gamma}$ so $cf(\lambda) = cf(\mu^o_{p_\gamma})$. If $cf(\lambda) = \omega$, then by the remarks after Lemma 2.4 (since $cf(\mu^o_{p_\gamma}) = \omega < cf(\gamma)$) imply that Case 1 holds. Hence

$\mathrm{cf}(\lambda) > \omega$. Set $\tilde{C} = \{\mu^{o}_{p_{i\gamma}} \mid i < \lambda\}$. $\tilde{C}$ and C_{p_γ} are both cub in $\mu^{o}_{p_\gamma}$ so $(\mathrm{cf}(\lambda) > \omega)$ $C = \tilde{C} \cap C_{p_\gamma}$ is cub in $\mu^{o}_{p_\gamma}$. $\mathrm{otp}(C) < \mathrm{otp}(C_{p_\gamma}) < \gamma$.

<u>Note</u> We deliberately refrain from using $\lambda < \gamma$ here in order for the proof to be applicable to other canonical constructions, in particular Lemma 3.25 below.

Set

$I = \{i < \lambda \mid \mu^{o}_{p_{i\gamma}} \in C\}$ and for $i \in I$ set

$B_i = \{\bar{\gamma} < \gamma \mid \bar{\gamma} = \gamma \cap X_{p_{i\gamma},\bar{\gamma}} \wedge p_i/\gamma, C_{p_\gamma} \cap |p_{i_\gamma}| \in X_{p_{i\gamma},\bar{\gamma}}\}$.

Then B_i is cub in γ . Set $B = \bigcap_{i \in I} B_i$.

<u>Claim 1</u> B is bounded in γ .

Proof Suppose by negation that B is unbounded in γ . Then both B and D_{p_γ} are cub in γ and we can choose a $\delta \in B \cap D_p$ s.t. (Case 2 fails) $\sup(\Pi^{p_\gamma}_{\delta})''|r_\delta| = |p_\delta|$. But then $(\Pi^{p_\gamma}_{\delta})''C_{r_\delta} = C_{p_\gamma}$ and hence $C \subseteq C_{p_\gamma} \subseteq \mathrm{rng}(\Pi^{p_\gamma}_{\delta})$. Since $\Pi^{p_\gamma}_{\delta}$ is cofinal it is Σ_1 and by Lemma 2.3 $\Pi^{p_\gamma}_{\delta}(r_\delta) = p_\delta$. Set $\pi = \Pi^{p_\gamma}_{\delta}$ and for $i \in I$ set $r_i = \pi^{-1}(p_{i\gamma})$. Apply §2.5 Fact 1 to get $\mathfrak{A}_{r_i} = \mathfrak{A}_{p_{i\gamma,\delta}}$ and $\pi/\mathfrak{A}_{r_i} = \Pi_{p_{i\gamma,\delta}}$. Since $\delta \in B$,

$$\forall i \in I(p_i/\gamma \in \mathrm{rng}(\Pi_{p_{i\gamma,\delta}}) \subseteq \mathrm{rng}(\pi)) .$$

But then Case 1 applies so we have a contradiction.

Q.E.D. (Claim 1)

<u>Claim 2</u> For $j \in I$, $\bigcap_{i<j} B_i$ is cub in γ .

Proof Suppose not, let θ be the least s.t. $\bigcap_{i<\theta} B_i$ is bounded in γ . θ must be a limit ordinal. Set

$\beta_o = \sup \bigcap_{i<\theta} B_i$ and

$$\alpha_i = \min(\bigcap_{j<i} (B_i \setminus \beta_o)) \quad \text{for} \quad i < \theta .$$

Then $\alpha_i \leq \alpha_{i+1}$ and $\sup_{i<\theta} \alpha_i = \gamma$. We have $\langle p_i \wedge \gamma \mid i < \theta \rangle \in \mathfrak{A}_{p_{\theta\gamma}}$, $|p_i \wedge \gamma| = |p_{i\gamma}|$ and $p_{i\gamma} = p_{\theta\gamma} \wedge |p_i \wedge \gamma|$. Hence $\langle |p_{i\gamma}| \mid i < \theta \rangle \in \mathfrak{A}_{p_{\theta\gamma}}$. It is also true that $\langle C_{p_\gamma} \cap |p_{i\gamma} \wedge \gamma| \mid i < \theta \rangle = \langle C_{p_{\theta\gamma}} \cap |p_{i\gamma}| \mid i < \theta \rangle \in \mathfrak{A}_{p_{\theta\gamma}}$ since $C_{p_\gamma} \cap \mu^o_{p_{\theta\gamma}} = C_{p_{\theta\gamma}} \in \mathfrak{A}_{p_{\theta\gamma}}$. This allows us to define $\langle B_i \mid i < \theta \rangle$ in $\mathfrak{A}_{p_{\theta\gamma}}$ and hence $\langle \alpha_i \mid i < \theta \rangle \in \mathfrak{A}_{p_{\theta\gamma}}$. This implies $\mathfrak{A}_{p_{\theta\gamma}} \models$ " γ is singular". Thus $\Omega_\gamma^{p_\gamma} \leq |p_{\theta\gamma}|$ which is a contradiction since we assumed $\Omega_\gamma^{p_\gamma} = |p_\gamma| > |p_{\theta\gamma}|$.

Q.E.D. (Claim 2)

Now let $\langle \eta_i \mid i < ot(C) \rangle$ enumerate I . Define a normal sequence $\langle \delta_i \mid i < ot(C) \rangle$ by

$$\delta_o = \text{the least } \delta \text{ s.t. } ot(C), \sup(B) < \delta < \gamma .$$

For $i > 0$,

$$\delta_i = \text{the least } \delta \in \bigcap_{j<i} B_{\eta_j} \text{ s.t. } \delta > \delta_j \text{ for } j < i .$$

Clearly $\sup_{i<ot(C)} (\delta_i) = \gamma$.

Set D = the limit points of $\langle \delta_i \mid i < ot(C) \rangle$. $D \in \mathfrak{A}_{p_\gamma}$ by the method we showed $\langle \alpha_i \mid i < \theta \rangle \in \mathfrak{A}_{p_\theta}$ in Claim 2) and D is cub in γ .

We shall show D has the desired properties. Let $\delta = \delta_\theta \in D$. Since $\delta \in \bigcap_{i<\theta} B_{\eta_i}$, we have $p_{\eta_i} \wedge \gamma \in X_{p_{\eta_i\gamma},\delta}$ for $i < \theta$. We may apply the collapsing lemma (2.12) to get $\pi_{p_{\eta_i\gamma,\delta}}(p_{\eta_i\delta}) = p_{\eta_i\gamma}$ for $i < \theta$. Set $\pi = \bigcup_{i<\theta} \pi_{p_{\eta_i\gamma,\delta}}$. π is well defined by §2.5 Fact 5

Then $(\pi^{-1})''(p_\gamma) = \bigcup_{i<\theta} \pi^{-1}(p_{\eta_i\gamma}) = \bigcup_{i<\theta} p_{\eta_i\delta} = p_{\theta\delta}$

$$\pi: \mathfrak{A}^o_{p_{\theta\delta}} \xrightarrow[\Sigma_1]{} \mathfrak{A}^o_{p_{\theta\gamma}} \text{ cofinally.}$$

Moreover, using the note of Lemma 2.14.5, Case 2.1, Claim 1, setting $\bar{C} = \bigcup_{i<\theta} \pi^{-1}(C_{p_\gamma} \cap |p_{\eta_i\delta}|)$ we have

$$\pi: \langle \mathfrak{A}^o_{p_{\theta\delta}}, \bar{C} \rangle \xrightarrow[\Sigma_1]{} \langle \mathfrak{A}^o_{p_{\theta\gamma}}, C_{p_{\theta\gamma}} \rangle \text{ cofinally.}$$

By Lemma 2.3 $C = C_{p_{\theta\delta}} = C_{\pi^{-1}(p_{\theta\gamma})}$.

As before $\langle p_{\eta_j} \restriction \delta \mid j < \theta \rangle$, $\langle p_{\eta_j\delta} \mid j < \theta \rangle \in \mathfrak{A}_{p_{\theta\delta}}$ and $\langle C_{p_{\theta\delta}} \cap p_{\eta_j\delta} \mid j < \theta \rangle \in \mathfrak{A}_{p_{\theta\delta}}$.

If we repeat the definition of $\langle \delta_i \mid i < \theta \rangle$ in $\mathfrak{A}_{p_{\theta\delta}}$ as we defined $\langle \delta_i \mid i < \mathrm{ot}(C) \rangle$ in $\mathfrak{A}_{p_\gamma}$ we get the same sequence using the properties of π . In particular,

$$\pi: \langle \mathfrak{A}^o_{p_{\theta\delta}}, C_{p_{\theta\delta}} \rangle \xrightarrow[\Sigma_1]{} \langle \mathfrak{A}^o_{p_{\theta\gamma}}, C_{p_{\theta\gamma}} \rangle \text{ cofinally,}$$

$\pi \restriction \delta = \mathrm{id} \restriction \delta$, $\pi(\delta) = \gamma$, $\pi(p_{\eta_i\delta}) = p_{\eta_i\gamma}$, $\pi(p_{\eta_i} \restriction \delta) = p_{\eta_i} \restriction \gamma$

and $\pi''C_{p_{\theta\delta}} = C_{p_{\theta\gamma}}$ for $i < \theta$.

Since $\sup_{i<\theta}(\delta_i) = \delta$ and $\theta < \mathrm{ot}(C) < \delta$ (by definition of the sequence δ_i) we have

$$\Omega_\delta = |p_{\theta\delta}| < |p_\delta| \text{ and}$$

$$D \cap \delta = \text{limit points of } \langle \delta_i \mid i < \theta \rangle \in \mathfrak{A}_{p_{\theta\delta}} .$$

Q.E.D. (Lemma 3.13)

Note The argument of Case 3 actually shows that it is impossible since we conclude that for some $\delta \in D_{p_\gamma}$ and some $\xi = |p_{\theta\delta}| < |p_\delta|$ that $p_{\theta\delta} = p_\delta{\restriction}\xi = r_\delta{\restriction}\xi \in \mathbf{S}^*_\delta$ and this is impossible since $r_\delta \in \mathbf{S}^*_\delta$.

3.5 THE INACCESSIBLE CASE

Lemma 3.25 *Let α be inaccessible. Then the conclusion of Theorem 3.2 holds at α .*

Proof The proof is similar to that of Lemma 3.13. We will only point out the differences.

Remember that $s \in \mathbf{S}_\alpha$, $\tau \leq \alpha$ and X , $\langle \Delta_\nu \mid \nu \in X \rangle \in \mathfrak{A}^1_s$. $X \subseteq \text{card} \cap [\tau, \alpha)$ s.t. X is thin in $\mathfrak{A}^1_s$ and Δ_ν is dense in $\mathbb{P}^s_\nu$. We may assume that X is unbounded in α , otherwise take $X' = X \cup \{\gamma{+} \mid \sup(X) < \gamma < \alpha\}$ and for $\gamma > \sup(X)$, $\Delta_{\gamma+} = \mathbb{P}^s_{\gamma+}$. Let $X = \langle \eta_i \mid i < \alpha \rangle$ be an enumeration of X . We must show that Δ is dense in $\mathbb{P}^s_\tau$ where

$$\Delta = \{p \in \mathbb{P}^s_\tau \mid \forall i < \alpha (p{\restriction}[\eta_i, \alpha] \in \Delta_{\eta_i}) \quad .$$

Let $p_o \in \mathbb{P}^s_\tau$. We must find a $p \leq p_o$ and $p \in \Delta$. We may assume that $|p_o{\restriction}\gamma| = |p_{o\gamma}|$ for $\gamma \in (\tau, \alpha]$ a limit cardinal.

The construction of p_i , Y^γ_i and f_i are slightly different. We have to define p_i for $i \leq \alpha$ and put $p_{i+1}{\restriction}[\eta_i, \alpha] \in \Delta_{\eta_i}$. This means that if $\gamma \in \text{card} \cap [\tau, \alpha)$, $p_{i\gamma}$ will be defined for $i > \gamma$. We must ensure that the cardinality of $p_{\lambda\gamma}$ does not grow larger than γ at some limit λ . This cannot happen if $\text{card}(\lambda) = \gamma$, so it suffices to require that $(p_{i+1})^{\eta_i} = (p_i)^{\eta_i}$ (hence $p_{i\gamma} = p_{\gamma+1,\gamma}$ for $i > \gamma$, since $\gamma < \eta_{\gamma+1}$).

By induction on $i < \rho \leq \alpha$ we define $p_i \in \mathbb{P}^s_\tau$, $Y^\gamma_i \prec \mathfrak{A}^1_s$ for $\gamma \in \text{card} \cap [\eta_i,\alpha]$ and $f_i \in \mathbb{F}(p_i)$ as follows

p_o is given.

For $\gamma \in \text{card} \cap [\eta_i,\alpha]$

Y^γ_i = the smallest $Y \prec \mathfrak{A}^1_s$ s.t. $\gamma \subseteq Y$ and

$p_j, Y^\gamma_j, X, \langle \Delta\eta_i \mid i < \alpha \rangle \in Y$ for $j < I$.

Hence $\lim(\lambda) \longrightarrow Y^\gamma_\lambda = \bigcup_{i<\lambda} Y^\gamma_i$.

for $\gamma \in \text{card} \cap [\eta_\lambda,\alpha]$

$$f_i(\gamma) = \begin{cases} Y^\gamma_i \cap \mathfrak{A}^1_{p_{i\gamma+}} & \text{if } \gamma+ \in Y_i \\ \text{undefined otherwise} \end{cases}$$

for $\gamma \in \text{card} \cap [\eta_i,\alpha)$.

$\text{dom}(f_i)$ is thin as in Lemma 3.13.

Define f^*_i by replacing each $\Delta \in f_i(\gamma)$ s.t. Δ is pre-dense in $\mathbb{P}^{p_{i\gamma+}}$ by

$$\Delta^* = \{q \in \mathbb{P}^{p_{i\gamma+}}_{\eta_i} \mid \Delta^q \text{ is dense in } \mathbb{P}^{q_{\eta_i}}_\tau\} .$$

q_{i+1} = the $\mathfrak{A}^1_s$-least $q \leq (p_i)_{\eta_i}$ s.t.

(a) $q \in \Delta\eta_i$

(b) $q \in \sum^{s,(p_i)_{\eta_i}}_{f^*_i}$

(c) $q(\alpha) \in \bigcap \{\Delta \in Y^\alpha_i \mid \Delta \text{ is dense in } \mathbb{R}^s\}$

(d) $|q/\gamma| = |q_\gamma|$ for limit cardinals $\gamma \in [\eta_i,\alpha]$.

$p_{i+1} = q_{i+1} \cup p_i$.

Note that q_{i+1} and p_i/η_i must be compatible since

$q_{i+1,\eta_i} \leq p_{i,\eta_i}$.

For $\lim(\lambda)$ set $p\colon \mathrm{card} \cap [\tau,\alpha] \longrightarrow V^2$ s.t.

$$p(\gamma) = \langle \bigcup_{i<\lambda} \dot{p}_{i\gamma}, \bigcup_{i<\lambda} p_{i\gamma} \rangle$$

$$p_\lambda = \begin{cases} p & \text{if } p \in \mathbb{P}^s_\tau \\ \text{undefined otherwise.} \end{cases}$$

We will show that p_i is defined for $i \leq \alpha$ and Lemma 3.25 will follow since $p_\alpha \leq p_o$ and $p_\alpha \restriction [\eta_i,\alpha] \in \Delta_{\eta_i}$.

Define $\eta^*_\lambda = \sup_{i<\lambda} \eta_i$ (η_i is not continuous).
For $\delta \in [\eta^*_\lambda, \eta_\lambda)$ we define

$$Y^\delta_\lambda = \bigcup_{i<\lambda} Y^\delta_i .$$

Lemma 3.26 *Let* $\eta^*_\lambda \leq \delta \leq \gamma \leq \beta < \alpha$. *Let* $\Delta \in Y^\delta_\lambda$ *be pre-dense in* $\mathbb{P}^{p_{i\beta+}}_\gamma$ *for some* $i < \lambda$. *Then there is a* $j < \lambda$ *s.t.* $(p_j)^{\beta+}_\gamma$ *meets* Δ .

Proof Same as Lemma 3.14.

Lemma 3.27 *Same as Lemma 3.15.*

Definition 1 We define $\sigma = \sigma^\gamma_\lambda$ and $b = b^\gamma_\lambda$ only for $\gamma \in \mathrm{card} \cap [\eta^*_\lambda, \alpha]$. We also define $X' = \sigma^{-1}(X)$. Otherwise the same as in the definition following Lemma 3.15.

Lemmas 3.28-3.34 *The same as Lemmas 3.16-3.22 except that all the lemmas refer only to* $\eta^*_\lambda \leq \gamma$.

Corollary 3.34.1 *If* $\mathfrak{N}_{p_\gamma} \models$ " γ is regular" *then* $X \cap \gamma \in \mathfrak{N}_{p_\gamma}$ *and* $\mathfrak{N}_{p_\gamma} \models$ " $X \cap \gamma$ is thin".

Proof $\mathfrak{N}_{p_\gamma} \models$ " γ is regular" $\longrightarrow b \models$ " γ is regular" .

Clearly X' is thin in b, $X' \in b$. Let $C \subseteq \text{card} \cap [\tau,\gamma)$ s.t. $b \models$ " C is cub in γ and $C \cap X' = 0$". But $\sigma^{-1}(C) = C$ and $X' \cap \gamma = X \cap \gamma \in \mathfrak{N}_{p_\gamma}$ so

$$\mathfrak{N}_{p_\gamma} \models \text{" } C \text{ is cub in } \gamma \text{ and } C \cap X \cap \gamma = 0 \text{ "} .$$

Q.E.D.

Lemma 3.35 $p(\alpha) \in \mathbb{R}^s$ *and* $p(\gamma) \in \mathbb{R}^{p_{\gamma+}}$ *for* $\gamma \in \text{card} \cap [\tau,\alpha]$

Proof For $\eta^*_\lambda \leq \gamma$ the same as Lemma 3.23. For $\gamma < \eta^*_\lambda$ $\exists i < \lambda$ s.t. $\gamma \leq \eta_i$ and then $p(\gamma) = p_i(\gamma) \in \mathbf{R}^{p_{i\gamma+}}$ and $\gamma^+ < \eta^*_\lambda$ so $p_{i\gamma+} = p_{\gamma+}$.

Lemma 3.36 $p{\restriction}\gamma \in \mathbb{P}_\tau^{p_\gamma}$ *for* $\gamma \in \text{card} \cap [\tau,\alpha]$.

Proof For $\gamma < \eta^*_\lambda$ $\exists i < \lambda$ s.t. $p{\restriction}\gamma = p_i{\restriction}\gamma \in \mathbb{P}_\tau^{p_{i\gamma}}$ and $p_{i\gamma} = p_\gamma$. For $\gamma \in [\eta^*_\lambda,\alpha]$ there are two differences.

(1) If $\lambda = \eta^*_\lambda = \gamma$ and $\mathfrak{N}_{p_\lambda} \models$ " λ is regular", then to show (ii) of the definition of $\mathbb{P}^s$ we define

D_i = the $\mathfrak{N}_{p_{i\lambda}}$ -least $D \subseteq \lambda$ s.t. D is cub in λ and $\delta \in D \longrightarrow \delta$ is a limit cardinal, $|p_i{\restriction}\delta| = |p_{i\delta}|$ and $\dot{p}_{i\delta} = 0$.

Let $D_1 = \{\delta < \lambda \mid \delta \in \bigcap_{\eta_i < \delta} D_i\}$,

$D_2 = \mathfrak{N}_{p_\gamma}$ -least $D \subseteq \gamma$ s.t. D is cub in γ and $D \cap X \cap \gamma = 0$,

and $D_3 = \{\delta < \lambda \mid \delta = \eta^*_\delta\}$.

Set $D = D_1 \cap D_2 \cap D_3$. Then $D \in \mathfrak{N}_{p_\gamma}$ and D is cub in γ .

For $\delta \in \text{card} \cap [\tau,\eta^*_\lambda)$, $p_\delta = \bigcup_{\eta_i \leq \delta} p_{i\delta}$, since for $\eta_i > \delta$, p_i does not contribute to p_δ .

$\delta \in D \longrightarrow \delta = \eta^*_\gamma$ and $\eta^*_\delta < \eta_\delta$ ($\eta^*_\delta \not\in X$) . So for $\delta \in D$ we have:

$|p \upharpoonright \delta| \geq \sup_{\eta_i < \delta}(|p_i \upharpoonright \delta|) = \sup_{\eta_i < \delta}(|p_{i\delta}| = |p_\delta|)$. This proves (ii).

(2) The verification of (iv) in Lemma 3.24 can be repeated because $\lambda < \gamma$ was not used.

3.6 THE SINGULAR CASE

Theorems 3.1 and 3.2 are now proven for regular α . There only remains the case when α is singular and then we only have to prove Theorem 3.1. The ω-cofinal case can be handled exactly as the inaccessible case using:

Lemma 3.37 *Let* $cf(\alpha) = \omega$. *Let* $f \in \mathbf{F}(p)$ s.t. $dom(f) \subseteq \alpha$. *Then* Σ_f *is dense in* $\mathbb{P}^s_\tau$.

Proof Let $q \leq p$. We must find a $q' \leq q$ s.t. $q \in \Sigma_f$. By the extension lemma (2.14) we can assume w.l.o.g. that $|q \upharpoonright \gamma| = |q_\gamma|$ for limit cardinals $\gamma \in (\tau, \alpha]$, Ω_α exists and $f \in \mathfrak{A}_{q_\alpha}$.

Let $\langle \alpha_i \mid i < \omega \rangle \in \mathfrak{A}_{q_\alpha}$ be a monotone sequence from card $\cap [\tau, \alpha)$ converging to α . For $i < \omega$ define f^*_i by setting $dom(f^*_i) = dom(f) \cap [\alpha^+_i, \alpha^+_{i+1})$ and $f^*_i(\gamma)$ is defined by replacing each $\Delta \in f(\nu)$ which is predense in $\mathbb{P}^{q_{\nu^+}}_\tau$ by

$$\{q' \in \mathbb{P}^{q_{\nu^+}}_{\alpha^+_i} \mid \Delta^{q'} \text{ is dense in } \mathbb{P}^{q'_{\alpha^+_i}}_\tau\} .$$

Define

$$q_o = \text{the } \mathfrak{A}^1_{q_{\alpha^+_o}}\text{-least } q' \text{ s.t. } q' \leq (q)^{\alpha^+_o} \text{ and}$$

$$q' \in \Sigma^{q_{\alpha^+_o}, q \upharpoonright \alpha^+_o}_{f \upharpoonright \alpha^+_o}$$

q_{i+1} = the $\mathfrak{A}^1_{q_{\alpha^+_{i+1}}}$ -least q' s.t. $q' \le (q)^{\alpha^+_{i+1}}_{\alpha^+_i}$ s.t.

$$q' \in \sum_{f^*_i}^{q_{\alpha^+_{i+1}}, (q)^{\alpha^+_{i+1}}_{\alpha^+_i}}$$

Set $q' = \bigcup_{i<\omega} q_i$. Then $q' \le q/\alpha$ and $q' \cup \{\langle \alpha, |q| \rangle\} \in \Sigma_f$ by methods similar to Lemma 3.12.

Q.E.D.

Lemma 3.38 *Let* $cf(\alpha) = \omega$. *Then* $\mathbb{P}^s_\tau$ *is* τ*-distributive in* $\mathfrak{A}^1_s$.

Proof Exactly as Lemma 3.13 except that we use Lemma 3.37 instead of Lemma 3.12.

Next we deal with the case that α is singular and $cf(\alpha) > \omega$. For this we need a somewhat altered version of Lemma 3.12. We first define a variation of Σ_f .

Definition Let $\omega < cf(\alpha) < \alpha$ and $\tau \in card \cap \alpha$. Let $\langle \alpha_i \mid i < \rho \rangle$ be the monotone enumeration of

$$\{\tau\} \cup \{\gamma_j \mid j < \lambda_\alpha \cap \gamma_j > \tau\}$$

(remember that $C_\alpha \backslash ot(C_\alpha) = \{\gamma^\alpha_i \mid i < \lambda_\alpha\}$) . Let $s \in \mathbf{S}_{\alpha+}$, $p \in \mathbf{P}^s_\tau$ and $f \in \mathbb{F}(p)$ s.t. $dom(f) \subseteq \alpha$.

$\Pi_f = \Pi^{s,p}_f$ = the set of $q \in \mathbf{P}^s_\tau$ s.t. q/p or $q \le p$ and for all $i < \lambda_\alpha$ and $\nu \in card \cap [\alpha_i, \alpha_{i+1})$,

$$(q)^{\nu^+}_{\alpha_i} \in \sum_{f/[\alpha_i,\nu]}^{q_{\nu^+}, (q)^{\nu^+}_{\alpha_i}} .$$

Lemma 3.39 *Let* $\omega < cf(\alpha) < \alpha$, $p \in \mathbb{P}^s_\tau$, $f \in \mathbb{F}(p)$, $dom(f) \subseteq \alpha$. *Then* $\Pi^{s,p}_f$ *is dense in* $\mathbb{P}^s_\tau$.

Proof Let $q \leq p$. We must find a $q' \leq q$, $q \in \Pi_f$.

Let $\langle \alpha_i \mid i < \rho \rangle$ be as in the preceding definition. By the extension lemma (2.14) we may assume w.l.o.g. that $|q \upharpoonright \gamma| = |q_\gamma|$ for limit cardinals $\gamma \in (\tau, \alpha)$, $f \in \mathfrak{A}_{q_\alpha}$, $f \upharpoonright \alpha_i \in \mathfrak{A}_{q_{\alpha_i}}$ for $i < \rho$ and $\Omega_{q_{\alpha_i}}$ exists for $i > 0$ (hence $\langle \alpha_j \mid j < i \rangle \in \mathfrak{A}_{q_{\alpha_i}}$). $cf(\alpha_{i+1}) = \omega$ by the note on p.42. Hence by the proof of Lemma 3.37 we may define q_i for $i < \rho$ s.t.

q_i = the $\mathfrak{A}_{q_{\alpha_{i+1}}}$-least q' s.t. $q' \leq (q)^{\alpha_{i+1}}_{\alpha_i}$ in $\mathbb{P}^{q_{\alpha_{i+1}}}_{\alpha_i}$ and

$$q' \cup \{\alpha_{i+1}, q(\alpha_{i+1})\} \in \sum\nolimits^{q_{\alpha^+_{i+1}}, (q)^{\alpha^+_{i+1}}_{\alpha_i}}_{f \upharpoonright [\alpha_i, \alpha_{i+1})} .$$

Set $q' = \bigcup_{i<\rho} q_i$. Then $q' \leq q \upharpoonright \alpha$ in $\mathbb{P}^{q_\alpha}_\tau$ and

$$q' \cup \{\langle \alpha, q(\alpha) \rangle\} \in \Pi_f .$$

Q.E.D.

By Lemma 3.11 in order to prove that $\mathbb{P}^s_\tau$ is τ-distributive over $\mathfrak{A}^1_s$, it is sufficient to prove that $\mathbb{P}^s_{\gamma^+}$ is γ^+-distributive for some γ with $\tau \leq \gamma \leq \alpha$.

Hence Theorem 3.1 will be proven if we prove:

Lemma 3.40 *Let* $\omega < cf(\alpha) < \alpha$. *Then* $\mathbb{P}^s_\tau$ *is* τ*-distributive in* $\mathfrak{A}^1_s$ *for* $\tau > ot(C_\alpha)$.

Proof We only point out the differences with Lemma 3.13. Let $\langle \alpha_i \mid i < \rho \rangle$ be as in the preceding definition.

Let $\langle \Delta_\nu \mid \nu < \tau \rangle$, p_o be as in Lemma 3.13.
We construct p_i, Y^γ_i, f_i for $i \leq \tau$ and $\gamma \in card \cap [\tau, \alpha]$ exactly as in Lemma 3.13 except that we replace

$$p_{i+1} \in \Sigma^{s,p_i}_{f_i} \quad \text{with} \quad p_{i+1} \in \Pi^{s,p_i}_{f_i}$$

As before we must show $p = \bigcup_{i<\lambda} p_i \in \mathbf{P}^s_\tau$ for $\lim(\lambda)$.

Lemma 3.40.1 *Let* $i < \rho$, $\alpha_i \le \delta \le \gamma \le \beta < \alpha_{i+1}$. *Let* $\Delta \in Y^\delta_\lambda$ *s.t.* Δ *is predense in* $\mathbf{P}^{p_{j\beta^+}}_\gamma$ *for some* $j < \lambda$. *Then* $\exists k < \lambda$ *s.t.* $(p_k)^{\beta^+}_\gamma$ *meets* Δ.

Proof $p_{k+1} \in \Pi^{s,p_k}_{f_k}$ implies that $(p)^{\beta^+}_{\alpha_i} \in \Sigma^{(p_k)_{\beta^+},(p_k)^{\beta^+}_{\alpha_i}}_{f_k \restriction [\alpha_i,\beta]}$.
With this fact the proof is the same as Lemma 3.14.

Q.E.D.

Lemma 3.40.2 *Same as Lemma 3.15.*

Definition 2 For $\gamma \in \text{card} \cap [\tau,\alpha]$ define $\sigma = \sigma^\gamma_\lambda$ and $b = b^\gamma_\lambda$ as in definition 1 in Lemma 3.13. Note that $\langle \alpha_i \mid i < \rho \rangle$ is $\mathfrak{N}^1_s$ definable from τ; hence $\forall i < \rho$ $(\alpha_i \in X^\gamma_\lambda)$. Set $\alpha'_i = \sigma^{-1}(\alpha_i)$ for $i < \rho$.

Lemma 3.40.3 *Let* $i < \rho$. *Let* $\alpha'_i \le \gamma \le \eta \le \delta < \alpha'_{i+1}$ *or* $\gamma \le \alpha'_i \le \eta \le \delta < \alpha'_{i+1}$ *s.t.* $\eta, \delta \in (\text{card})_b$. *Let* $\sigma((\delta^+)_b) = \beta^+$ *and* $r = \sigma^{-1}(p_{i\beta^+})$ *for some* $i < \lambda$. *Then* $D' \cap [\eta,(\delta^+)_b)$ *is* $\mathbf{P}^r_\eta$*-generic over* $(\mathfrak{N}^1_r)_b$.

Proof The same as Lemma 3.16.

Corollary $D' \cap [\alpha'_i,(\delta^+)_b)$ *is* $\mathbf{P}^{(\delta^+)_b}_{\alpha'_i}$*-generic over* $\mathfrak{N}^1_{(\delta^+)_b}$.

Lemma 3.40.4 *The same as Lemma 3.17.*

Lemma 3.40.5 *Let* $\delta \in (\text{card})_b \cap [\alpha'_i,\alpha'_{i+1})$. *Then* $L_{\alpha'_{i+1}}[A \cap \alpha'_{i+1}, D' \cap [\alpha'_i,\alpha'_{i+1})] \models$ "δ is a cardinal".

Proof The same as Lemma 3.18.

Lemma 3.40.6 *The same as Lemma 3.19.*

Before going on we need the following fact.

Fact $\langle \alpha_i' \mid i < \rho\rangle \in L_{\mu'}[A \cap \gamma]$.

Proof Set $X = \{\alpha_i \mid i < \rho\} \in \mathfrak{A}_s^1$ and $X' = \sigma^{-1}(X) = \{\alpha_i' \mid i < \rho\} \in b$.

By §2.2 definition 2 (p.28) $\mathfrak{A}_s^1 \models \exists Y \in L(Y \supseteq X \wedge \mathrm{card}(Y) = \mathrm{card}(X))$. Pick the ${}_s^1$-least such Y . $\mathrm{card}(Y) = \mathrm{card}(X) \leq \rho < \tau \leq \gamma$ $(\rho = \mathrm{ot}(C_\alpha \backslash \mu^o_{\mathrm{ot}(C_\alpha)}))$. Let $f \in L_{\mu_s'}$ enumerate Y , i.e. $f: \xi \longrightarrow Y$, where $\xi < \gamma$. Let $u = (f^{-1})''Y$. Then $u \in L_\gamma[A \cap \gamma]$ since $H_\gamma = L_\gamma[A \cap \gamma]$. Let $f' = \sigma^{-1}(f)$. $\sigma(u) = u$ so $(f')''u = X' \in L_{\mu'}[A \cap \gamma]$.

Q.E.D. (Fact)

Lemma 3.40.7 $A', s' \in L_{\mu'}[A \cap \gamma, p_\gamma]$.

Proof Instead of Claim 1 as in Lemma 3.20 we have for $i < \rho$:

<u>Claim (i)</u> $A' \cap \alpha_i' , D' \cap \alpha_i' \in L_{\mu'}[A \cap \gamma, p_\gamma]$.

Proof Let i_o be the least s.t. $\gamma < \alpha_{i_o}$. We define γ_j , $a_j = A' \cap \gamma_j$ and $d_j = D'_{\gamma_j}$, starting with $\gamma_o = \gamma$, $a_o = A' \cap \gamma$, and $d_o = \tilde{p}_\gamma$. Assume we already have Claim (i). $A' \cap \alpha_i' , D' \cap \alpha_i' \in L_{\mu'}[A \cap \gamma, p_\gamma]$. We define D'_{α_i} as in Lemma 3.20, Case (2). Then we proceed to define $A' \cap \alpha_{i+1}'$, $D' \cap \alpha_{i+1}'$ as in Lemma 3.20, Case (1). The only difference being that when we define $\mathbb{S}'_{\gamma_j}$ for $\gamma_j \in [\alpha'_i, \alpha'_{i+1})$ we use $L_{\alpha'_{i+1}}[a_j], L_{\alpha'_{i+1}}$ in place of $L_{\alpha'}[a_j], L_{\alpha'}$.

Now assume Claim (i) for $i < \theta$, $\lim(\theta)$. Since by the above fact $\langle \alpha'_i \mid i < \theta \rangle \in L_{\mu'}[A \cap \gamma]$, we get

$$A' \cap \alpha'_\theta = \bigcup_{i<\theta} A' \cap \alpha'_i \,,\quad D' \cap \alpha'_\theta = \bigcup_{i<\theta} D' \cap \alpha'_i \in L_{\mu'}[A \cap \gamma, p_\gamma] \,,$$

which is Claim (θ).

Following the process we get Claim (ρ),

$$A' \cap \alpha' \,,\; D' \cap \alpha' \in L_{\mu'}[A \cap \gamma, p_\gamma]$$

since $\sup_{i<\rho} \alpha'_i = \alpha'$. So Claim (ρ) gives us Claim (1) of Lemma 3.20. Claims $\rho + 1$ and $\rho + 2$ are the same as Claims 2 and 3 of Lemma 3.20.

Q.E.D.

Lemmas 3.40.8-3.40.11 *The same as 3.21-3.24.*

Q.E.D. (Lemma 3.40)

This completes Theorems 3.1 and 3.2.

3.7 DISTRIBUTIVITY OF $\mathbb{P}_\tau$

Theorem 3.3 $\mathbb{P}_\tau$ *is τ-distributive.*

Definition $\mathbb{P}^*_\tau = \{p \in \mathbb{P}_\tau \mid \exists\alpha(\mathrm{dom}(p) = \mathrm{card} \cap [\tau, \alpha]\}$.

We denote the above α by $\alpha = \alpha_p$ for $p \in \mathbb{P}^*_\tau$.

This is the same as the definition at the beginning of this chapter and it was shown that $\mathbb{P}^*_\tau$ is dense in $\mathbb{P}_\tau$ so it suffices to prove τ-distributivity for $\mathbb{P}^*_\tau$.

Clearly we have $\mathbb{P}_\tau^{\alpha^+} = \{p \in \mathbb{P}_\tau^* \mid \alpha = \alpha_p\}$.

Let Δ_i be a dense subclass of $\mathbb{P}_\tau^*$ for $i < \tau$ s.t.

$\{\langle i,p\rangle \mid p \in \Delta_i\}$ is $\Sigma_m(\langle M,A\rangle)$.

We claim that $\bigcap_{i<\tau} \Delta_i$ is dense in $\mathbb{P}_\tau^*$.

Let $p_o \in \mathbb{P}_\tau^*$. We may assume w.l.o.g. that $|p_o{\restriction}\gamma| = |p_{o\gamma}|$ for limit points of $\mathrm{card} \cap [\tau,\alpha_{p_o}]$. Let $\{\langle i,p\rangle \mid p \in \Delta_i\}$ be $\Sigma_m(\langle M,A\rangle)$ in the parameter e. Let $n > m$ be "sufficiently large" (in a sense described later).

We construct $p_i \in \mathbb{P}_\tau^*$ and $Y_i^\gamma \prec_{\Sigma_n} \langle M,A\rangle$ for $\gamma \in \mathrm{card} \cap [\tau,\infty)$ and $i \leq \tau$, by induction on i as follows:

p_o is given

Y_i^γ = the smallest $X \prec_{\Sigma_n} \langle M,A\rangle$ s.t. $\gamma \subseteq X$, $e \in X$,

$p_j, \langle Y_j^\gamma \mid \gamma \in \mathrm{card} \cap [\tau,\alpha_{p_j}]\rangle \in X$ for $j < i$.

We define $f_i \in \mathbb{F}(p_i)$ s.t. $\mathrm{dom}(f_i) \subseteq \alpha_{p_i}$ by

$$f_i(\nu) = \begin{cases} Y_i^\nu \cap \mathfrak{A}^1_{p_{i\nu^+}} & \text{if } \nu \in Y_i^\nu \\ \text{undefined otherwise} & \end{cases}$$

for $\nu \in \mathrm{card} \cap [\tau,\alpha_{p_i})$.

$\mathrm{dom}(f_i)$ is thin in $\langle M,A\rangle$ since for inaccessible δ $\{\gamma \mid \gamma = Y_i^\gamma \cap \delta\}$ is cub in δ.

p_{i+1} = the L[A]-least $p \in \mathbb{P}_\tau^*$ s.t. $p{\restriction}\alpha_{p_i}^+ \leq p_i$,

$p{\restriction}\alpha_{p_i}^+ \in \sum_{f_i}^{\alpha_{p_i}^+,p_i}$.

$p \in \Delta_i$ and $|p{\restriction}\gamma| = |p_\gamma|$ for limit cardinals $\gamma \in \mathrm{dom}(p)$

and $\alpha_p \geq \sup(Y_i^{\alpha_{p_i}} \cap \mathrm{On})$.

For $\lim(\lambda)$ set $\alpha_\lambda = \sup_{i<\lambda} \alpha_{p_i}$. Define a function p s.t. $\mathrm{dom}(p) = \mathrm{card} \cap [\tau, \alpha_\lambda]$ as follows:

$$p(\nu) = \langle \bigcup_{i<\lambda} \dot{p}_{i\nu}, \bigcup_{i<\lambda} p_{i\nu} \rangle \quad \text{for} \quad \nu < \alpha \quad \text{and} \quad p(\alpha) = \langle 0,0 \rangle .$$

If $p \in \mathbf{P}^*_\tau$ we set $p_\lambda = p$. Otherwise p_λ is undefined. We shall show p_i is defined for $i \leq \tau$, which will show that $p_\tau \leq p_0$ and $p_\tau \in \bigcap_{i<\tau} \Delta_i$.

<u>Claim</u> For $\gamma \leq \alpha_\lambda$, $\sup(Y^\gamma_\lambda \cap \mathrm{On}) = \alpha_\lambda$.

Proof For limit λ, $Y^\gamma_\lambda = \bigcup_{i<\lambda} Y^\gamma_i$ and

$$\alpha_{p_i} \leq \sup(Y^\gamma_{i+1} \cap \mathrm{On}) \leq \alpha_{p_{i+1}} .$$

Lemma 3.41 *Let* $\tau \leq \delta \leq \gamma \leq \beta < \alpha_\lambda$. *Let* $\Delta \in Y^\delta_\lambda$ *be predense in* $\mathbf{P}^{p_{i\beta+}}_\gamma$ *for some* $i < \lambda$. *Then* $\exists\, j < \lambda$ *s.t.* $(p_j)^{\beta+}_\gamma$ *meets* Δ.

Proof The same as Lemma 3.14.

There is no lemma corresponding to Lemma 3.15 since $p(\alpha_\lambda) = \langle 0,0 \rangle$ and has nothing to code.

Definition For $\gamma \in \mathrm{card} \cap [\tau, \alpha_\lambda)$ define $\sigma = \sigma^\gamma_\lambda$ and $b = b^\gamma_\lambda$ by $\sigma: b \cong Y^\gamma_\lambda$ and b is transitive.

We further define:

(a) $A' = (\sigma^{-1})''(A \cap \alpha_\lambda)$.

(b) $\mu' = \mathrm{On} \cap b$.

(c) $e' = \sigma^{-1}(e)$.

(d) $p'_i = \sigma^{-1}(p_i)$ for $i < \lambda$.

(e) For $\delta \in (\mathrm{card})_b \cap [\gamma, \mu')$ set

$$D'_\delta = \bigcup_{i<\lambda} \tilde{p}'_{i\delta}, \quad D' = \bigcup_{\delta \in (\mathrm{card})_b} D'_\delta .$$

The rest is the same as in the definition 1 following Lemma 3.15.

Lemma 3.42 *Let* $\gamma \leq \eta \leq \delta < \mu'$, $\eta, \delta \in (\mathrm{card})_b$. *Let* $\sigma((\delta^+)_b) = \beta^+$ *and* $r = \sigma^{-1}(p_{i\beta^+})$ *for some* $i < \lambda$. *Then* $D' \cap [\eta, (\delta^+)_b)$ *is* $\mathbb{P}^r_\eta$*-generic over* $(\mathfrak{A}^1_r)_b$.

Proof The same as Lemma 3.16.

The lemma corresponding to Lemma 3.17 is meaningless.

Lemma 3.43 *Let* $\delta \in (\mathrm{card})_b$. *Then* $L_{\mu'}[A', D'] \models$ " δ is a cardinal".

Proof The same as for Lemma 3.18 but note that it holds all the way up to μ'.

Lemma 3.44 *The same as Lemma 3.19.*

Lemma 3.45 A' *and* D' *are definable in* $L_{\mu'}[A \cap \gamma, p_\gamma]$.

Proof Note A' and D' are classes in b. Since we have a stronger version of 3.18 in 3.43, we only need the argument of the claim to show:

$\langle A' \cap \delta \mid \delta \in (\mathrm{card})_b \rangle$, $\langle D' \cap \delta \mid \delta \in (\mathrm{card})_b \rangle$ are definable in $L_{\mu'}[A \cap \gamma, p_\gamma]$ and $A' \cap \delta$, $D' \cap \delta \in L_{\mu'}[A \cap \gamma, p_\gamma]$ for $\delta \in (\mathrm{card})_b$.

Q.E.D.

Lemma 3.46 b, $\langle p'_i \mid i < \lambda \rangle \in L_{\mu'+\omega}[A \cap \gamma, p_\gamma]$.

Proof By Lemma 3.45, A' and D' are definable in $L_{\mu'}[A \cap \gamma, p_\gamma]$.

So

$$b = \langle L_{\mu'}[A',D'],A',D'\rangle \in L_{\mu'+\omega}[A \cap \gamma, p_\gamma] .$$

$\langle p'_i \mid i < \lambda\rangle$ can be defined from b, using e' as $\langle p_i \mid i < \lambda\rangle$ was defined from e in $\langle M,A\rangle$.

Note that we are assuming that all definability mentioned is at most Σ_n.

Q.E.D.

Lemma 3.47 (a) $p_\gamma \in \mathbf{S}_\gamma$. (b) $\gamma \geq \omega_2 \longrightarrow b \in \mathfrak{A}^1_{p_\gamma}$.

Proof (a) is the same as for Lemma 3.22. (b) is even easier since,

$$L_{\mu'}[A \cap \gamma, p_\gamma] \models |p_\gamma| \text{ is a cardinal.}$$

But if $L_{\mu'_{p_\gamma}}[A \cap \gamma, p_\gamma] \models \text{card}(|p_\gamma|) = \gamma$, then $\mu' < \mu^1_{p_\gamma}$.

Lemma 3.48 $p(\gamma) \in \mathbf{R}^{p_{\gamma+}}$ *for* $\gamma \in \text{card} \cap [\tau, \alpha_\lambda)$ *and* $p(\alpha_\lambda) = \langle 0,0\rangle \in \mathbf{R}^{\alpha^+_\lambda}$.

Lemma 3.49 $p/\gamma \in \mathbb{P}^{p_\gamma}_\tau$ *for* $\gamma \in \text{card} \cap [\tau, \alpha_\lambda]$.

Proof For $\gamma < \alpha_\lambda$ the same as Lemma 3.24. Let $\gamma = \alpha_\lambda$. We must show $p/\alpha_\lambda \in \mathbb{P}^{\alpha_\lambda}_\tau$.

$$X^{\alpha_\lambda}_\lambda = \langle L_{\alpha_\lambda}[A], A \cap \alpha_\lambda\rangle \prec_{\Sigma_n} \langle M,A\rangle .$$

We can define $\langle p_i \mid i < \lambda\rangle$ in $X^{\alpha_\lambda}_\lambda$ so $\langle p_i \mid i < \lambda\rangle$ is definable in $\langle L_{\alpha_\lambda}[A \cap \alpha_\lambda], A \cap \alpha_\lambda\rangle$.

But $\mathfrak{A}^1_{\alpha_\lambda} = \langle L_\mu[A \cap \alpha_\lambda], A \cap \alpha_\lambda \rangle$ for $\mu > \alpha_\lambda$ so

$$p = \bigcup_{i<\lambda} p_i \in \mathfrak{P}^1_{\alpha_\lambda} .$$

The rest of the proof is the same.

Q.E.D. (Theorem 3.3)

4 . The Dénouement

In this chapter we will complete Theorem 0.1 and prove Theorem 0.2. Let $D \subseteq [\tau,\infty)$ be $\mathbb{P}_\tau$-generic over $\langle M,A\rangle$. By the argument of Lemma 3.5, for $\gamma \in \text{card} \cap [\tau,\infty)$, $D \cap [\gamma^+,\infty)$ is $\mathbb{P}_{\gamma^+}$-generic over $\langle M,A\rangle$ and $D \cap \gamma^+$ is $\mathbf{P}_\tau^{D_{\gamma^+}}$-generic over $\mathfrak{A}^1_{D_{\gamma^+}}$. In particular $D \cap \gamma^+$ is $\mathbf{P}_\tau^{\gamma^+}$-generic over $\mathfrak{A}^1_{\gamma^+}$. This is the sense in which $\mathbb{P}$ is "Easton".

Lemma 4.1 *Let* $D \subseteq [\tau,\infty)$ *be* $\mathbb{P}_\tau$*-generic over* $\langle M,A\rangle$. *Let* $N = M[D]$. *Then:*

(i) $N \models ZF + GCH + \neg 0^{\#}$.

(ii) *Cardinals and cofinalities are preserved in* N.

(iii) $N \models$ "$V = L[A \cap \tau, D_\tau]$".

(iv) A *and* D *are* N*-definable in the parameters* $A \cap \tau$, D_τ.

Proof

<u>Claim 1</u> Let α be an $\langle M,A\rangle$ cardinal, then α is a cardinal in N, $cf_N(\alpha) = cf_M(\alpha)$ and $H_\alpha = L_\alpha[A,D]$.

Proof For M-cardinals $\beta \geq \alpha$, $D \cap \beta^+$ is $\mathbf{P}_\tau^{\beta^+}$-generic over $\mathfrak{A}^1_{\beta^+}$. By Lemma 3.6, α is a cardinal in $N_{\beta^+} = \mathfrak{A}_{\beta^+}[D \cap \beta^+]$ and $cf_{N_{\beta^+}}(\alpha) = cf_M(\alpha)$. If $\alpha \leq \tau$, then $H_\alpha = L_\alpha[A,D]$ in N_{β^+} since $\mathbf{P}_\tau^{\beta^+}$ is τ-distributive over $\mathfrak{A}^1_{\beta^+}$. By Lemma 3.7, $N_{\beta^+} = \mathfrak{A}^1_{\beta^+}[A \cap \tau, D_\tau]$ so for $\tau < \alpha \leq \beta^+$, $H_\alpha = L_\alpha[A,D]$ in N_{β^+}. But this holds for arbitrarily large β and $N = \bigcup_{\beta \in (\text{card})_M} N_{\beta^+}$.

Q.E.D. (Claim 1)

Claim 2 $N \models GCH + \neg 0^{\#}$ and all of ZF except for replacement.

Proof GCH, $\neg 0^{\#}$ and the power set axiom follow from cardinal preservation and $H_\alpha = L_\alpha[A,D]$ for all $\alpha \in$ card. The rest of ZF (except for replacement) are trivial.

Claim 3 $N \models$ "$V = L[A \cap \tau, D_\tau]$" and A and D are N-definable from $A \cap \tau$ and D_τ.

Proof Repeat of Lemma 3.7.

Let $\Vdash$ be the Schoenfeld forcing relation [SH] and let $\Vdash_{\Sigma_n}$ be $\Vdash$ restricted to Σ_n formula. We adopt the following notation. Let $M \models ZF^-$ and $\mathbf{P}$ a forcing relation over M. Then $M^{\mathbf{P}}$ denotes the terms of the forcing language. If D is $\mathbf{P}$-generic over M, we denote by $M^{\mathbf{P}}/D$ the model generated by the generic interpretation. In the case that $\mathbf{P}$ is a set we know that $M^{\mathbf{P}}/D = M[D] =$ the minimal model generated by M and D.

Claim 4 $\Vdash_{\Sigma_n}$ is $\langle M,A \rangle$-definable for $n < \omega$ and $M^{\mathbf{P}}/D = M[D]$.

Proof First we show that $\Vdash_{\Sigma_0}$ is definable. Let $p \in \mathbf{P}_\tau$, $\vec{t} \in M^{\mathbf{P}}$ and $\phi(\vec{x})$ a Σ_0 formula. Choose $\alpha \in$ card s.t. $p, \vec{t} \in L_\alpha[A]$. Then $\hat{p} \restriction \alpha^+ \in \mathbf{P}_\tau^{\alpha^+}$ (where $\hat{p}(\gamma) = p(\gamma)$ if $\gamma \in \text{dom}(p)$ and $\hat{p}(\gamma) = \langle 0,0 \rangle$ otherwise) and $\vec{t} \in (\mathfrak{N}^1_{\alpha^+})^{\mathbf{P}_\tau^{\alpha^+}}$. $D \cap \alpha^+$ is $\mathbf{P}_\tau^{\alpha^+}$-generic over $\mathfrak{N}^1_{\alpha^+}$ and $p \in G_D \longleftrightarrow \hat{p} \restriction \alpha^+ \in G_{D \cap \alpha^+}$.

Moreover $\vec{t}$ has the same interpretation in $M^{\mathbb{P}_\tau}/D$ as in $(\mathfrak{A}^1_{\alpha^+})^{\mathbb{P}^{\alpha^+}_\tau}/D \cap \alpha^+ = \mathfrak{A}^1_{\alpha^+}[D \cap \alpha^+]$. Since this holds for all $\mathbb{P}_\tau$-generic D we conclude:

$$p \Vdash_{\mathbb{P}_\tau} \phi(\vec{t}) \text{ iff } \exists\alpha(p, \vec{t} \in L_\alpha[A] \wedge \hat{p}/\alpha^+ \Vdash_{\mathbb{P}^{\alpha^+}_\tau} \phi(\vec{t})) .$$

Hence:

$$p \Vdash_{\Sigma_0} \phi(\vec{t}) \text{ iff } \exists\alpha(p, \vec{t} \in L_\alpha[A] \wedge \hat{p}/\alpha^+ \Vdash_{\mathbb{P}^{\alpha^+}_\tau} \phi(\vec{t}))$$

for Σ_0 formulas ϕ . Hence $\Vdash_{\Sigma_0}$ is definable. $\Vdash_{\Sigma_n}$ definability follows by standard methods.

Q.E.D. (Claim 4)

<u>Claim 5</u> $N \models ZF$.

Proof We need only verify the replacement axiom. It suffices to prove the following assertion.

Assertion If γ is a cardinal in N and $f: \gamma \longrightarrow On$ is N-definable, then f is bounded in $On \cap N$.

For $\gamma \leq \tau$ this is trivial since $\mathbb{P}_\tau$ is τ-distributive. Now let $\gamma > \tau$. The above assertion holds for $N' = M[D \cap [\gamma^+, \infty)]$ since $D \cap [\gamma^+, \infty)$ is $\mathbb{P}_{\gamma^+}$-generic over M and $\mathbb{P}_{\gamma^+}$ is γ^+-distributive. N' preserves γ^+ and γ^{++} so $N' \models$ "$\mathbb{P}^{D_\gamma}_\tau$ is γ^{++}-A.C.". But then the assertion holds for $N'[[D \cap [\tau, \gamma^+)] = M[[D \cap \gamma^+, \infty)], D \cap [\tau, \gamma^+)] = M[D] = N$ since $D \cap [\tau, \gamma^+)$ is $\mathbb{P}^{D_\gamma}_\tau$-generic over N' .

Q.E.D.

4.2 LARGE CARDINAL FACTS

We have now proven all of Theorem 0.1 except for the preservation of large cardinals. We recall the definitions of these cardinals and state some facts whose proofs can be found in Baumgartner [BA], Boos [BO] or Drake [DR].

Definition 1 κ is Mahlo iff $\{\tau < \kappa \mid \tau \text{ is regular}\}$ is stationary in κ.

Fact 4.2.1 If κ is Mahlo, then κ is inaccessible and $\{\tau < \kappa \mid \tau \text{ is inaccessible}\}$ is stationary in κ.

Definition 2 Let $m,n < \omega$. κ is π^n_m-indescribable iff whenever $A \subseteq V_\kappa$ and $\phi(x)$ is a π^n_m formula s.t. $V_\kappa \models \phi[A]$, then $\exists \tau < \kappa\ V_\tau \models \phi[A \cap V_\tau]$.

Fact 4.2.2 If κ is π^n_m-indescribable and A, ϕ are as above, then $\{\tau < \kappa \mid V_\tau \models \phi[A \cap V_\tau]\}$ is stationary (hence κ is Mahlo for $m,n \geq 1$).

Fact 4.2.3 κ is weakly compact iff κ is π^1_1-indescribable.

Definition 3 $X \subseteq \kappa$ is subtle in κ iff whenever $C \subseteq \kappa$ is cub in κ and $\langle A_\nu \mid \nu \in X\rangle$ is s.t. $A_\nu \subseteq \nu$, then

$$\exists \nu, \tau \in C \cap X (\nu < \tau \wedge A_\nu = \nu \cap A_\tau) .$$

Fact 4.2.4 Let $X \subseteq \kappa$ be subtle in κ. Then:

(a) X is stationary in κ.

(b) If $f: X \longrightarrow \kappa$ is regressive, then there is a subtle $Y \subseteq X$ s.t. $f \restriction Y$ is constant. Hence:

(c) If X is partitioned into $< \kappa$ parts, then one part is subtle in κ.

(d) For $n,m<\omega$, $\{\tau\in X \mid \tau$ is π^n_m-indescribable$\}$ is subtle in κ.

Fact 4.2.5 Let $X\subseteq\kappa$. The following are equivalent:

(1) X is subtle.

(2) For every $C\subseteq\kappa$, C is cub in κ and $\langle A_\nu \mid \nu\in X\rangle$ s.t. $A_\nu\subseteq H_\nu$,

$$\exists\nu,\tau\in C\cap X(\nu<\tau\wedge A_\nu = H_\nu\cap A_\tau).$$

We did not notice a proof of the following fact in the literature so we include one.

Fact 4.2.6 Let $X\subseteq\kappa$. The following are equivalent:

(a) X is subtle in κ.

(b) If $A_\alpha\subseteq\alpha$ for $\alpha\in X$, then there are arbitrarily large $\alpha\in X$ s.t. $A_\alpha = \alpha\cap A_\beta$ for arbitrarily large β.

Proof Assume X is subtle and (b) does not hold. Then there is a γ s.t. for $\alpha\in X\backslash\gamma$ there is a ξ_α s.t. $\beta\geq\xi_\alpha\longrightarrow A_\alpha\neq\alpha\cap A_\beta$. Define a normal sequence $\langle\gamma_i \mid i<\kappa\rangle$ as follows:

$\gamma_0=\gamma$ and for $\alpha\in X\cap\gamma_i$, $\gamma_{i+1}\geq\xi_\alpha$ and $\gamma_\lambda = \lim_{i<\lambda}(\gamma_i)$ for $\lim(\lambda)$. $C=\{\gamma_i \mid i<\kappa\}$ is cub in κ and for $\alpha,\beta\in C\cap X$, $\alpha<\beta\longrightarrow A_\alpha\neq\alpha\cap A_\beta$. This is a contradiction.

It is easy to see (b) $\longrightarrow X$ is stationary (if C is a counterexample take $A_\alpha=\sup(C\cap\alpha)$ for $\alpha\in X$ and reach a contradiction).

Now assume (b). Let C be cub in κ, $A_\alpha\subseteq\alpha$ for $\alpha\in X$. Define:

$$A'_\alpha = \begin{cases} A_\alpha & \text{if } \alpha \in C \\ \sup(\alpha \cap C) & \text{if } \alpha \notin C . \end{cases}$$

Let $\alpha \in X \cap C$ s.t. $A'_\alpha = \alpha \cap A'_\beta$ for arbitrarily large β. Clearly $A'_\alpha = \alpha \cap A'_\beta \longrightarrow \beta \in C$ for $\beta \geq \min(C \setminus \alpha)$.

Q.E.D.

Definition 4 κ is subtle iff κ is subtle in itself.

Fact 4.2.7 Every subtle cardinal is Mahlo.

Definition 5 κ is subtly π^n_m-indescribable iff whenever $A \subseteq \kappa$ and $\phi(x)$ is a π^n_m formula s.t. $V_\kappa \models \phi[A]$, then $\{\tau < \kappa \mid V_\tau \models \phi[A \cap V_\tau]\}$ is subtle in κ.

Definition 6 κ is (weakly) ineffable iff whenever $\langle A_\nu \mid \nu < \kappa \rangle$ is s.t. $A_\nu \subseteq \nu$. Then there is $A \subseteq \kappa$ s.t. $\{\nu \mid A_\nu = A \cap \nu\}$ is stationary (unbounded) in κ.

Fact 4.2.8 κ is weakly ineffable iff κ is subtly π^1_1-indescribable.

Fact 4.2.9 κ is ineffable iff κ is subtly π^1_2-indescribable.

Definition 7 $\kappa \longrightarrow (\alpha)^{<\omega}$ iff for all $f: [\kappa]^{<\omega} \longrightarrow 2$ there is an $H \subseteq \kappa$ of order type α s.t. $f \upharpoonright [H]^{<\omega}$ is constant. Such an H is called homogeneous for f.

Definition 8 Let $\alpha \leq \kappa$, $\lim(\alpha)$. κ is α-Erdos iff whenever C is cub in κ and $f: [C]^{<\omega} \longrightarrow \kappa$ is regressive $(f(\{a\}) < \min\{a\})$, then f has a homogeneous set of order type α.

Fact 4.2.10 The least κ s.t. $\kappa \longrightarrow (\alpha)^{<\omega}$ is α-Erdos.

Fact 4.2.11 Every α-Erdos cardinal is Mahlo.

Definition 9 Let $\mathfrak{A}$ be a transitive $\in$-model s.t. $\kappa \subseteq \mathfrak{A}$. $I \subseteq \kappa$ is a good set of indiscernibles for $\mathfrak{A}$ iff $\gamma \in I \longrightarrow I \setminus \gamma$ is a set of indiscernibles for $\langle \mathfrak{A}, \nu < \gamma \rangle$.

Fact 4.2.12 Let $\alpha \leq \kappa$, $\lim(\alpha)$. The following are equivalent:

(i) κ is α-Erdos.

(ii) Let $C \subseteq \kappa$ be cub in κ. Let $\mathfrak{A}$ be a transitive $\in$-model s.t. $\kappa \subseteq \mathfrak{A}$. Then $\exists I \subseteq C$, $\mathrm{ot}(I) = \alpha$ and I is a good set of indiscernibles for $\mathfrak{A}$.

Corollary *Let* κ *be* α*-Erdos and* $\mathfrak{A} = \langle L_\kappa[A], A \rangle$. *Then* $\mathfrak{A}$ *has a good set of indiscernibles s.t.*

$$\gamma \in I \longrightarrow \mathfrak{A} \restriction L_\gamma[A] \prec \mathfrak{A} \text{ (where } \mathfrak{A} \restriction L_\gamma[A] = \langle L_\gamma[A], A \cap \gamma \rangle\text{)}.$$

Proof Take $C = \{\gamma \mid \mathfrak{A} \restriction \gamma \prec \mathfrak{A}\}$.

Q.E.D.

We wish to show that the properties: Mahlo, subtle, (subtly) π^1_n indescribable for $n < \omega$ and α-Erdos for $\alpha < \omega_1$, are preserved in $N = M[G]$. In §2.1 we started with a ground model $\langle M, A \rangle$ and performed a preliminary extension to $\langle M'.A' \rangle$ by adding a Cohen generic subset $A^1_\alpha \subseteq [\alpha+\alpha, \alpha^+)$ for every cardinal $\alpha \geq \omega$.

$$A^o_\alpha = \{\alpha + \nu \mid \nu \in A \cap \alpha\} \text{ and}$$

$$A'_\alpha = A^o_\alpha \cup A^1_\alpha,\ A' = \bigcup_{\alpha \in \mathrm{card}} A'_\alpha \text{ and } M' = M[A].$$

Let **Q** be the class of conditions for executing such an

extension. We must show that **Q** preserves the above cardinal properties. In addition we need some facts about $\langle M', A'\rangle$ to show that **P** preserves some of the above cardinal properties. The following three facts are all that is needed for proving the preservation properties of **P** . These facts are proved in the appendix since the proofs are independent of the coding conditions.

Fact 4.2.13 $\langle M', A'\rangle \models ZF + GCH + \neg 0^{\#}$ in which cardinals and cofinalities are preserved. $H_\alpha = L_\alpha[A]$ for infinite cardinals α . The cardinal properties Mahlo, subtle, (subtly) π^1_m-indescribable $(m < \omega)$ and α-Erdos are preserved.

Fact 4.2.14 Let κ be (subtly) π^1_m-indescribable in M $(m \geq 1)$. Let $\xi < \kappa^+$ and let ϕ be a π_m formula s.t.

$$\langle H_{\kappa^+}, A \cap \kappa^+\rangle \models \phi(\xi) .$$

For $\alpha \leq \kappa$ set:

$$X_\alpha = \text{the least } X \prec \langle H_{\kappa^+}, A \cap \kappa^+\rangle \text{ s.t. } \alpha \cup \{\xi\} \subseteq X .$$

Set: $\sigma_\alpha : \langle L_{\delta_\alpha}[A'_\alpha], A'_\alpha\rangle \cong X_\alpha$ and $\xi_\alpha = \sigma_\alpha^{-1}(\xi)$.

Set: $D = \{\alpha < \kappa \mid \alpha = \kappa \cap X_\alpha \wedge A'_\alpha = A \cap \delta_\alpha \wedge \langle H_{\alpha^+}, A \cap \alpha^+\rangle \models \phi[\xi_\alpha]\}$.

Then D is stationary (subtle) in κ .

Fact 4.2.15 Let $\kappa, \xi, X_\alpha, \sigma_\alpha, \delta_\alpha$ be as in Fact 4.2.14. Set:

$$D = \{\alpha < \kappa \mid \alpha = \kappa \cap X_\alpha \wedge A'_\alpha = A \cap \delta_\alpha \wedge \langle L_{\delta_\alpha}[A], A \cap \delta_\alpha\rangle \prec_{\Sigma_m} \langle L_{\alpha^+}[A], A \cap \alpha^+\rangle\} .$$

Then D is stationary (subtle) in κ .

4.3 PRESERVATION OF LARGE CARDINALS

We are now ready to prove large cardinal property preservation. Assume w.l.o.g. that we are working over a model $\langle M,A\rangle$ s.t. $\langle M,A\rangle \models ZF + GCH + \neg 0^{\#}$ in which Facts 4.2.14 and 4.2.15 hold (14 and 15 for short). We also have (3) of §2.1 but it is not needed here. Facts 14 and 15 will only be used in proving preservation of (subtly) π^1_m-indescribable cardinals.

Lemma 4.3 *Let* κ *be inaccessible in* M *and let* $B \in N$, $B \subseteq \kappa$, B *cub in* κ. *Then there is a* $B' \subseteq B$, $B' \in M$ *s.t.* B' *is cub in* κ.

Proof We know that $B \in L_{\kappa^+}[D] \subseteq \mathfrak{A}^1_{\kappa^+}[D \cap \kappa^+]$ where $D \cap \kappa^+$ is $\mathbf{P}^{\kappa^+}$-generic over $\mathfrak{A}^1_{\kappa^+}$. Let $f \in \mathfrak{A}^1_{\kappa^+}[D \cap \kappa^+]$ be the monotone enumeration of B. Let $\mathring{B}, \mathring{f} \in (\mathfrak{A}^1_{\kappa^+})^{\mathbf{P}^{\kappa^+}}$ and let $p_0 \in \mathbf{P}^{\kappa^+} \cap G_D$ be s.t. B and f interpret $\mathring{B}$ and $\mathring{f}$ and $p_0 \Vdash$ "$\mathring{B}$ is cub in $\check{\kappa}$ and $\mathring{f}$ is the monotone enumeration of $\mathring{B}$".

Let $\Delta_\tau = \{p \in \mathbf{P}^{\kappa^+} \mid p/p_0 \vee \exists \nu(p \Vdash f(\check{\tau}) = \check{\nu})\}$.

By Lemma 3.4

$$\Delta^*_\tau = \{q \in \mathbf{P}^{\kappa^+}_{\tau^+} \mid \Delta^q_\tau \text{ is dense in } \mathbf{P}^{q_{\tau^+}}\} \text{ is dense in } \mathbf{P}^{\kappa^+}_{\tau^+}.$$

Then by Theorem 3.2,

$$\Delta = \{p \in \mathbf{P}^{\kappa^+} \mid \forall \tau (p)_{\tau^+} \in \Delta^*_\tau\} \text{ is dense in } \mathbf{P}^{\kappa^+}.$$

Hence we may take a $p \in \Delta \cap G$ s.t. $p \leq p_0$. For $r \in \Delta^{(p)_{\tau^+}}_\tau$ set:

$$\gamma_r = \text{that unique } \gamma \text{ s.t. } r \cup (p)_{\tau^+} \Vdash \mathring{f}(\check{\tau}) = \check{\gamma}.$$

Set $\gamma_\tau = \sup\limits_{r \in \Delta^{(p)_{\tau^+}}_\tau} (\gamma_r)$.

Set $B' = \{\alpha < \kappa \mid \forall\tau < \alpha(\gamma_\tau < \alpha)\}$. Clearly $B' \in M$ and B' is cub in κ . $p \Vdash \forall\tau < \check{\kappa}(\mathring{f}(\check{\tau}) \leq \check{\gamma}_\tau)$. Hence, for $\alpha \in B'$ $p \Vdash$ "$B' \cap \check{\alpha}$ is unbounded in $\check{\alpha}$" and thus, $p \Vdash \check{B}' \subseteq \mathring{B}$.

Q.E.D.

Corollary 4.3.1 Let κ be inaccessible. If $X \subseteq \kappa$, $X \in M$ is stationary in M , then X is stationary in N .

Corollary 4.3.2 If κ is Mahlo in M then κ is Mahlo in N .

Technical Fact 4.4 If $s \in \mathbf{S}_{\gamma^+}$, $\langle \dot{r}, |r| \rangle \in \mathbb{R}^s$ and $\xi \in [\gamma^+, |s|)$, then $\langle \dot{r}^\xi, |r| \rangle \in \mathbb{R}^{s \restriction \xi}$ where $\dot{r}^\xi = \dot{r}^o \cup (\dot{r}^1)^\xi$ and $(\dot{r}^1)^\xi = \{\langle b,\eta \rangle \in \dot{r}^1 \mid b = b_{s \restriction \delta}$ for $\delta < \xi\}$.

Proof Follows directly from the definition of $\mathbb{R}^s$.

Notation *For* $r = \langle \dot{r}, |r| \rangle \in \mathbb{R}^s$ *set* $r^\xi = \langle \dot{r}^\xi, |r| \rangle \in \mathbb{R}^{s \restriction \xi}$.

Lemma 4.5 *Let* $s \in \mathbf{S}_{\kappa^+}$, $\tau \leq \kappa$. *Let* ϕ *be a formula in the forcing language for* $\mathbb{P}^s_\tau$ *over* $\mathfrak{N}^1_s$. *Let* $\Delta = \{p \in \mathbb{P}^s_\tau \mid p \mid\mid \phi\}$. *Let* $\gamma < \kappa$ *and* $q \in \mathbb{P}^s_\tau$ *s.t.* $\Delta^{(q)_{\gamma^+}}$ *is dense in* $\mathbb{P}^{q_{\gamma^+}}_\tau$. *Let* $h \leq q$, $h \in \mathbb{P}^s_\tau$ *s.t.* $h \Vdash \phi$ $(h \Vdash {\sim}\phi)$. *Then* $h^* \cup (q)_{\gamma^+} \Vdash \phi$ $(\Vdash {\sim}\phi)$ *where* $h^* \restriction \gamma = h \restriction \gamma$ *and* $h^*(\gamma) = h^{|q_{\gamma^+}|}(\gamma) \in \mathbb{R}^{q_{\gamma^+}}$, $h^* \in \mathbb{P}^{q_{\gamma^+}}_\tau$.

Proof We show the case $h \Vdash \phi$ ($h \Vdash {\sim}\phi$ is analogous). By negation assume, ${\sim}(h^* \cup (q)_{\gamma^+} \Vdash \phi)$. Then there is a $p \leq h^* \cup (q)_{\gamma^+}$ s.t. $p \Vdash {\sim}\phi$. Define $p^* \in \mathbb{P}^{(q)_{\gamma^+}}_\tau$ as h^* was defined. Since $\Delta^{(q)_{\gamma^+}}$ is dense in $\mathbb{P}^{(q)_{\gamma^+}}_\tau$ there is a $r \leq p^*$, $r \in \Delta^{(q)_{\gamma^+}}$. Then $r \cup (q)_{\gamma^+} \in \Delta$. Hence

$r \cup (q)_{\gamma^+} \Vdash \phi$. But $r \cup (q)_{\gamma^+}$ is compatible both with h and p , $h \Vdash \phi$ and $p \Vdash {}^{\sim}\phi$, so $r \cup (q)_{\gamma^+}$ cannot decide ϕ . This is a contradiction.

Q.E.D.

<u>Note</u> As a consequence of the proof $h^* \in \Delta^{(q)_{\gamma^+}}$.

Lemma 4.6 *Let* κ *be inaccessible. If* $X \subseteq \kappa$, $X \in M$ *and* X *is subtle in* M *, then* X *is subtle in* N .

Proof Let $X \subseteq \kappa$ be non-subtle in N . It is sufficient to prove that there is a $X' \supseteq X$, $X' \in M$ s.t. X' is non-subtle in M . Since X is non-subtle in N there are C , $\langle A_\nu \mid \nu \in C \cap X \rangle = A \in N$ s.t. C is cub in κ , $A_\nu \subseteq \nu$ and $\forall \tau, \nu \in C \cap X (\nu < \tau \longrightarrow A_\nu \neq \nu \cap A_\tau)$. By Lemma 4.3 we may assume w.l.o.g. that $C \in M$. Then $C \in \mathfrak{N}^1_{\kappa^+}$ and X , $\langle A_\nu \mid \nu \in C \cap X \rangle \in \mathfrak{N}^1_{\kappa^+}[D \cap \kappa^+]$. Let $p_0 \in \mathbf{P}^{\kappa^+} \cap G_D$ be s.t. $p_0 \Vdash$ " $\mathring{A}$ is defined on $\check{C} \cap \mathring{X}$ and

$$\forall \nu \in \check{C} \cap \mathring{X} (\mathring{A}_\nu \subseteq \nu \wedge \forall \nu \forall \tau (\nu < \tau \longrightarrow \mathring{A}_\nu \neq \nu \cap \mathring{A}_\tau))" .$$

Since κ is inaccessible we may assume that $C \subseteq \mathrm{card} \cap \kappa$. For $\tau, \nu \in C$, $\nu < \tau$ set:

$$\Delta_{\tau\nu} = \{p \in \mathbb{P}^{\kappa^+} \mid p/p_0 \vee (p \,||\, \check{\tau} \in \mathring{X} \wedge (p \Vdash \check{\tau} \in \mathring{X} \longrightarrow p \,||\, \check{\nu} \in \mathring{A}_\tau))\}.$$

By Lemma 3.4,

$$\Delta^*_\tau = \{p \in \mathbf{P}^{\kappa^+}_{\tau^+} \mid \forall \nu < \tau \; \Delta^P_{\tau\nu} \text{ is dense in } \mathbf{P}^{P_{\tau^+}}\} \text{ is dense in } \mathbb{P}^{\kappa^+}_{\tau^+} .$$

By Lemma 3.2,

$$\Delta = \{q \in \mathbb{P}^{\kappa^+} \mid \forall \tau \in C ((q)_{\tau^+} \in \Delta^*_\tau)\} \text{ is dense in } \mathbb{P}^{\kappa^+} .$$

Let $q \in \Delta \cap G_D$. We may assume, by the extension lemma (2.14), that $|q/\gamma| = |q_\gamma|$ for limit cardinals $\gamma \leq \kappa$.

Set $X' = \{\nu < \kappa \mid {}^{\sim}q \Vdash \check{\nu} \notin \mathring{X}\}$. Then $X' \supseteq X$ and $X' \in M$.

<u>Claim</u> X' is not subtle in M.

Proof It suffices to show that

$X'' = \{\tau \in X' \mid$ is inaccessible$\}$ is not subtle in M.

For $\tau \in X'' \cap C$ choose a $p^\tau \in \mathbb{P}^{(q)_{\tau^+}} \cap G_{D \cap \tau^+}$ s.t. $p^\tau \leq q/\tau^+$ and

(a) $p^\tau \cup (q)_{\tau^+} \Vdash \check{\tau} \in \mathring{X}$.

(b) $\forall \nu < \tau \, \exists \gamma < \tau (\Delta_{\tau\nu}^{(q)_{\tau^+}})^{(p)_{\gamma^+}}$ is dense in $\mathbb{P}^{(p^\tau)_{\gamma^+}}$.

(c) $|p^\tau/\tau| = |(p^\tau)_\tau|$.

(b) is possible by Lemma 3.8. For $\tau \in X'' \cap C$ set:

$U_\tau^1 = \{\langle \nu,\gamma,r \rangle \mid \nu,\gamma<\tau \wedge r \in \mathbb{P}^{(p^\tau)_{\gamma^+}} \wedge r \cup (p^\tau)_{\gamma^+} \cup (q)_{\tau^+} \Vdash \check{\nu} \in \mathring{A}_{\check{\gamma}}\}$,

$U_\tau^0 = \{\langle \nu,\gamma,r \rangle \mid \nu,\gamma<\tau \wedge r \in \mathbb{P}^{(p^\tau)_{\gamma^+}} \wedge r \cup (p^\tau)_{\gamma^+} \cup (q)_{\tau^+} \Vdash \check{\nu} \notin \mathring{A}_{\check{\gamma}}\}$

and $V_\tau = \{\bar{\tau} < \tau \mid \bar{\tau} = \tau \cap X_{(p^\tau)_{\tau,\bar{\tau}}} \wedge p^\tau/\tau \in X_{(p^\tau)_{\tau,\bar{\tau}}}\}$.

Clearly V_τ is cub in τ. Now we assume, by negation, that X'' is subtle in M. Then there is a $\tau,\rho \in X'' \cap C$, $\tau < \rho$ s.t. $p^\tau/\tau = p^\rho/\tau$, $U_\tau^0 = H_\tau \cap U_\rho^0$, $U_\tau^1 = H_\tau \cap U_\rho^1$ and $V_\tau = \tau \cap V_\rho$. Since $V_\tau = \tau \cap V_\rho$ and V_τ is cub in τ we have:

$$\tau \in V_\rho, \text{ i.e. } \tau = \rho \cap X_{(p^\rho)_{\rho,\tau}} \text{ and } p^\rho/\rho \in X_{(p^\rho)_{\rho,\tau}}.$$

We also know by (c) that $|p^\rho/\rho| = |(p^\rho)_\rho|$, so we may use the collapsing lemma (2.12) to get:

$$\mathfrak{A}(p^\rho)_\tau = \mathfrak{A}(p^\rho)_{\rho,\tau}, \quad (p^\rho)_\tau = (\pi_{(p^\rho)_{\rho,\tau}})^{-1}[(p^\rho)_\rho],$$

$$|p^\rho/\tau| = |(p^\rho)_\tau| \text{ and } (\mathring{p}^\rho)_\tau = 0.$$

We have $p^\tau \upharpoonright \tau = p^\rho \upharpoonright \tau$ and $|p^\tau \upharpoonright \tau| = |(p^\tau)_\tau|$, hence $|(p^\rho)_\tau| = |(p^\tau)_\tau|$ and $(p^\rho)_\tau = (p^\tau)_\tau$ since $p^\tau \upharpoonright \tau$ and $p^\rho \upharpoonright \tau$ will code them the same way.

Set $q' = p^\rho \cup (q)_{\rho^+}$. Since $(\mathring{p}^\rho)_\tau = 0$ and $(p^\rho)_\tau = (p^\tau)_\tau$ we can define $q'' \leq q$ by $q''(\tau) = p^\tau(\tau)$ and $q''(\nu) = q'(\nu)$ otherwise. The only change made was in $(p^\rho)_\tau$, since $(\mathring{p}^\rho)_\tau = 0$ and $(\mathring{p}^\tau)_\tau \neq 0$. But $(\mathring{p}^\tau)_\tau$ correctly restricts $(p^\tau)_\tau = (p^\rho)_\tau$. Hence $q'' \leq q$, $q'' \upharpoonright \tau^+ = p^\tau$, $q'' \upharpoonright \rho^+ \leq p^\rho$ and $q'' \in G_D$.

<u>Subclaim</u> $q'' \Vdash \mathring{A}_{\check{\tau}} = \check{\tau} \cap \mathring{A}_{\check{\rho}}$.

Proof Let $h \leq q''$ and $\eta < \tau$ s.t. $h \Vdash \check{\eta} \in \mathring{A}_{\check{\tau}}$. Using the notation and results of Lemma 4.5,

$$h^* \cup (p^\tau)_{\gamma^+} \cup (q)_{\tau^+} \Vdash \check{\eta} \in \mathring{A}_{\check{\tau}} \text{ and } h^* \in (\Delta_{\tau\eta}^{(q)_{\tau^+}})^{(p^\tau)_{\gamma^+}}.$$

(This is not the exact statement of Lemma 4.5 but obviously can be deduced by the methods of Lemma 4.5 using a tedious argument. We, for instance, have to use the fact that $p^\tau \cup (q)_{\tau^+} \Vdash \check{\tau} \in \mathring{X}$. Hence $h^* \cup (p^\tau)_{\gamma^+} \cup (q)_{\tau^+} \Vdash \check{\eta} \in \mathring{A}_{\check{\tau}}$.)
But then $h^* \in U^1_\tau$ and $U^1_\tau = H_\tau \cap U^1_\rho$, hence

$$h^* \cup (p^\rho)_{\gamma^+} \cup (q)_{\rho^+} \Vdash \check{\eta} \in \mathring{A}_{\check{\rho}}.$$

$h \leq h^* \cup (p^\rho)_{\gamma^+} \cup (q)_{\rho^+}$ so $h \Vdash \check{\eta} \in \mathring{A}_\rho$. By the same method $h \Vdash \check{\eta} \notin \mathring{A}_{\check{\tau}}$ implies $h \Vdash \check{\eta} \notin \mathring{A}_{\check{\rho}}$ (using $U^0_\tau = H_\tau \cap U^0_\rho$).

Q.E.D. (Subclaim)

Now we also have $q'' \Vdash \check{\tau}, \check{\rho} \in \mathring{X}$ and $q'' \in G_D$. Hence $N \models A_\tau = \tau \cap A_\nu \wedge \tau, \nu \in X \cap C$. This contradicts the assumption that X is non-subtle in N.

Q.E.D. (Lemma 4.6)

Corollary 4.6.1 *If* κ *is subtle in* M, *it is subtle in* N.

Before proceeding further there is a need for three more facts.

Fact 4.7 $\mathbb{P}^{\kappa^+} \subseteq L_{\kappa^+}[A]$ and is $\Delta_1(\langle L_{\kappa^+}[A], A \cap \kappa^+\rangle)$ uniformly in κ.

Proof Clear.

Fact 4.8 Let

$L_\kappa = \{\phi(\mathring{X}_1,\ldots,\mathring{X}_n,\mathring{D}) \mid \phi$ is Σ_o, $\mathring{X}_1,\ldots,\mathring{X}_n \in L_{\kappa^+}[A]$ terms of the forcing language for $\mathbf{P}^{\kappa^+}$ and $\mathring{D}$ the canonical predicate for D $\}$.

Then there is a relation $\rhd \subseteq \mathbf{P}^{\kappa^+} \times L_\kappa$ s.t.

(a) $p \rhd \phi \longrightarrow p \Vdash_{\mathbf{P}^{\kappa^+}} \phi$.

(b) $\{p \mid p \rhd \phi \vee p \rhd {}^{\sim}\phi\}$ is dense in $\mathbb{P}^{\kappa^+}$.

(c) $\rhd$ is $\Delta_1(\langle L_{\kappa^+}[A], A \cap \kappa^+\rangle)$ uniformly in κ.

Proof The proof is similar to that of Lemma A.11 in the appendix. Note that $\mathbf{P}^{\kappa^+} = \mathbf{P}^{\emptyset_{\kappa^+}}$ where $\emptyset_{\kappa^+} \in \mathbf{S}_{\kappa^+}$ is the empty function. Hence for $p \in \mathbf{P}^{\kappa^+}$, $p(\kappa) = \langle \dot{p}_\kappa, p_\kappa \rangle \in \mathbf{R}^{\emptyset_{\kappa^+}}$. Thus $p(\kappa) \in L_{\kappa^+}[A]$. Following the proof of Lemma A.11 in the appendix set:

$$C = \{p \in \mathbb{P}^{\kappa^+} \mid \dot{p}^{\phi}_\kappa \subseteq (A \cap |p_\kappa|) \times [\alpha, |p_\kappa|)$$
$$\wedge\ \dot{p}^1_\kappa \subseteq \mathbb{P}_{<\alpha^+}(\alpha^+ \cap |p_\kappa|) \times [\alpha, |p_\kappa|) \wedge \langle L_{|p_\kappa|}[\tilde{p}_\kappa], \tilde{p}_\kappa\rangle \models ZF^-\}.$$

The proof now proceeds as in Lemma **A**.11 of the appendix.

Fact 4.9 Let ϕ be π_n $(1 \leq n < \omega)$. There is a π_n

formula ϕ^* s.t. for all $\mathbb{P}^{\kappa^+}$ terms $\mathring{X} \in L_{\kappa^+} A$:

$$[p \Vdash_{\mathbb{P}^{\kappa^+}} (\langle L_{\kappa^+}[\mathring{B}], \mathring{B}\rangle \models \phi(\mathring{X}))] \longleftrightarrow [\langle L_{\kappa^+}[A], A\rangle \models \phi^*(p, \mathring{X})] .$$

Proof The same as Corollary A.11.1 in the appendix.

Lemma 4.10 *Let* κ *be (subtly)* π^1_n*-indescribable in* M . *Then it remains so in* N .

Proof Let $B \subseteq \kappa$ and let ϕ be a π^1_n formula s.t. $N \models$ "$V_\kappa \models \phi(B)$" . Set:

$$E = \{\tau < \kappa \mid V_\tau \models \phi(B \cap \tau)\} \text{ (defined in N).}$$

<u>Claim</u> E is stationary (subtle).

Proof Clearly $B \in L_{\kappa^+}[D] = H_{\kappa^+}$ in N . Let $f = f_D : \kappa^+ \xrightarrow{\text{Onto}} L_{\kappa^+}[D]$ be uniformly primitive recursive in D . Let $\xi < \kappa^+$ be s.t. $B = f(\xi)$. The statement: "$f_D(\xi) \subseteq \kappa \wedge V_{\kappa^+} \models \phi[f_D(\xi)]$" is $\pi_n(\langle L_{\kappa^+}[D], D \cap \kappa^+\rangle)$ uniformly in D , κ and ξ for inaccessible κ . By Fact 4.9 there is a $p \in \mathbb{P}^{\kappa^+} \cap G_D$ s.t.

$$p \Vdash_{\kappa} \text{"}(V_\kappa \models \phi(f_{\mathring{D}}(\check{\xi}))) \wedge (f_{\mathring{D}}(\check{\xi}) \subseteq \check{\kappa})\text{"}$$

and the statement is $\pi_n(\langle L_{\kappa^+}[A], A \cap \kappa^+\rangle)$ uniformly in κ .

We may assume that:

(a) $|p/\kappa| = |p_\kappa|$.

(b) For $\nu < \kappa$, there is a $\gamma < \kappa$ s.t. $\Delta_\nu^{(p)_{\gamma^+}}$ is dense in $\mathbb{P}^{P_{\gamma^+}}$, where $\Delta_\nu = \{q \in \mathbb{P}^\kappa \mid \check{q} \Vdash \nu \in f_{\mathring{D}}(\check{\xi})\}$.

(a) can be assumed by the extension lemma (2.14) and (b) by Lemma 3.4. By Facts 4.8 and 4.9 it is clear that conditions

(a) and (b) on p are $\Delta_1(\langle L_{\kappa^+}[A], A\cap\kappa^+\rangle)$.

For $\alpha<\kappa$ let X_α, σ_α, δ_α, A'_α and ξ_α be as in Fact 4.2.14. Set:

$$J=\{\alpha<\kappa \mid \alpha=\kappa\cap X_\alpha \wedge p\in X_\alpha \wedge A'_\alpha = A\cap\delta_\alpha \wedge \langle L_{\delta_\alpha}[A'_\alpha],A'_\alpha\rangle \prec_{\Sigma_m} \langle L_{\alpha^+}[A],A\cap\alpha^+\rangle\} .$$

Then J is stationary (subtle) by Fact 4.2.15. Let $\bar\kappa\in J$, $\bar\kappa$ inaccessible. Let $\bar p = \sigma_{\bar\kappa}^{-1}(p)$ and $\bar\xi = \xi_{\bar\kappa} = \sigma_{\bar\kappa}^{-1}(\xi)$. Then $\bar p\in \mathbb{P}^{\bar\kappa^+}$ by Fact 4.7 and

$$\langle L_{\delta_{\bar\kappa}}[A'_{\bar\kappa}],A'_{\bar\kappa}\rangle \prec \langle L_{\bar\kappa^+}[A],A\cap\bar\kappa^+\rangle$$

by Fact 4.9,

$$p \Vdash_{\mathbb{P}^{\bar\kappa^+}} [(V_{\bar\kappa} \models \phi[f^\circ_D(\check{\bar\xi})]) \wedge (f^\circ_D(\check{\bar\xi}) \subseteq \check{\bar\kappa})]$$

and for $\nu<\bar\kappa$, there is a $\gamma<\bar\kappa$ s.t. $\bar\Delta_\nu^{(\bar p)_{\gamma^+}}$ is dense in $\mathbb{P}^{\bar p_{\gamma^+}}$, where $\bar\Delta_\nu = \{q\in\mathbb{P}^{\bar\kappa^+} \mid q \mid\mid \check\nu\in f^\circ_D(\check{\bar\xi})\}$. But $\bar\Delta^{(\bar p)_{\gamma^+}} \in L_{\bar\kappa}[A'_{\bar\kappa}]$ $(\gamma<\bar\kappa)$ and $\sigma_{\bar\kappa}\restriction L_{\bar\kappa} = id\restriction L_{\bar\kappa}$. Thus

$$\Delta_\nu^{(p)_{\gamma^+}} = \sigma(\bar\Delta_\nu^{(\bar p)_{\gamma^+}}) = \bar\Delta_\nu^{(\bar p)_{\gamma^+}} .$$

We have that if $r\in\bar\Delta_\nu^{(\bar p)_{\gamma^+}} = \Delta_\nu^{(p)_{\gamma^+}}$ then,

(1) $[r\cup(\bar p)_{\gamma^+} \Vdash_{\mathbb{P}^{\bar\kappa^+}} \check\nu\in f^\circ_D(\check{\bar\xi})] \longrightarrow [r\cup(p)_{\gamma^+} \Vdash_{\mathbb{P}^{\kappa^+}} \check\nu\in f^\circ_D(\check\xi)]$

or

(2) $[r\cup(\bar p)_{\gamma^+} \Vdash_{\mathbb{P}^{\bar\kappa^+}} \check\nu\notin f^\circ_D(\check{\bar\xi})] \longrightarrow [r\cup(p)_{\gamma^+} \Vdash_{\mathbb{P}^{\kappa^+}} \check\nu\notin f^\circ_D(\check\xi)]$.

$\sigma_{\bar\kappa}(\bar p_{\bar\kappa}) = p_\kappa$, hence $\sigma_{\bar\kappa}(\mathfrak{A}_{p_{\bar\kappa}}) = \mathfrak{A}_{p_\kappa}$. As in §2.5 Fact 1 it can be shown that $\sigma_{\bar\kappa}{}''\mathfrak{A}_{\bar p_{\bar\kappa}} = X_{p_{\kappa,\bar\kappa}}$. Hence $\sigma_{\bar\kappa}\restriction\mathfrak{A}_{\bar p_{\bar\kappa}} = \pi_{p_{\kappa,\bar\kappa}}$. The collapsing lemma (2.12) gives us:

$$\dot p_{\bar\kappa} = 0 , \quad p_{\bar\kappa} = \Pi^{-1}_{p_{\kappa,\bar\kappa}}(p_\kappa) = p_{\bar\kappa} .$$

Note that $\dot{p}_{\bar\kappa} = 0$ and $\dot{\bar{p}}_{\kappa} \neq 0$, so for any $q \le p$ s.t. $\dot{q}_{\bar\kappa} \supseteq \dot{\bar{p}}_{\bar\kappa}$ it is true that $q/\bar\kappa^{+} \le \bar{p}$ in $\mathbb{P}^{\bar\kappa^{+}}$. If $q/\bar\kappa^{+} \Vdash_{\mathbb{P}^{\bar\kappa^{+}}} \check\nu \in f^{\circ}_{D}(\check{\bar\xi})$ then by Lemma 4.5, there is a $\gamma < \bar\kappa$ s.t. $q^{*} \cup (\bar{p})_{\gamma^{+}} \Vdash_{\mathbb{P}^{\bar\kappa^{+}}} \check\nu \in f^{\circ}_{D}(\check{\bar\xi})$, $q^{*} \in \bar\Delta_{\nu}^{(\bar{p})_{\gamma^{+}}}$.
By (1) above $q^{*} \cup (p)_{\gamma^{+}} \Vdash_{\mathbb{P}^{\kappa^{+}}} \check\nu \in f^{\circ}_{D}(\check\xi)$. $q \le q^{*} \cup (p)_{\gamma^{+}}$, so $q \Vdash_{\mathbb{P}^{\kappa^{+}}} \check\nu \in f^{\circ}_{D}(\check\xi)$. The same holds for $q/\kappa^{+} \Vdash_{\mathbb{P}^{\bar\kappa^{+}}} \check\nu \notin f^{\circ}_{D}(\check{\bar\xi})$.
We have shown for such a q, that:

$$q/\bar\kappa^{+} \Vdash_{\mathbb{P}^{\bar\kappa^{+}}} \check\nu \in f^{\circ}_{D}(\check{\bar\xi}) \longleftrightarrow q \Vdash_{\mathbb{P}^{\kappa^{+}}} \check\nu \in f^{\circ}_{D}(\check\xi) .$$

<u>Subclaim 1</u> For any $q \le p$ s.t. $\dot{q}_{\bar\kappa} \supseteq \dot{\bar{p}}_{\bar\kappa}$,

$$q \Vdash_{\mathbb{P}^{\kappa^{+}}} f^{\circ}_{D}(\check{\bar\xi}) = \check{\bar\kappa} \cap f^{\circ}_{D}(\check\xi) .$$

Proof Let G be $\mathbb{P}^{\kappa^{+}}$-generic over $\mathfrak{N}^{1}_{\kappa^{+}}$ s.t. $q \in G$.
We show that

$$\mathfrak{N}^{1}_{\kappa^{+}}[D_G] \models f_{D_G}(\bar\xi) = \bar\kappa \cap f_{D_G}(\xi) .$$

Let $\bar{B} = f_{D_G}(\bar\xi)$, $\bar{B} \in \mathfrak{N}^{1}_{\bar\kappa^{+}}[D_G \cap \bar\kappa^{+}]$ $(q \Vdash f^{\circ}_{D}(\check\xi) \subseteq \check\kappa)$.
$D_G \cap \bar\kappa^{+}$ is $\mathbb{P}^{\bar\kappa^{+}}$-generic so there is a $q' \le q/\bar\kappa^{+}$ in $\mathbb{P}^{\bar\kappa} \cap G$ s.t. $q' \Vdash_{\mathbb{P}^{\bar\kappa^{+}}} \check{\bar{B}} = f^{\circ}_{D}(\check{\bar\xi})$.
We have $q' \cup q \in \mathbb{P}^{\kappa} \cap G$ s.t. $(q' \cup q)_{\bar\kappa} \supseteq \dot{p}_{\bar\kappa}$ so for every $\nu \in \bar{B}$, $q' \cup q/\bar\kappa^{+} \Vdash_{\mathbb{P}^{\bar\kappa^{+}}} \check\nu \in f^{\circ}_{D}(\check{\bar\xi}) \longleftrightarrow q' \cup q \Vdash_{\mathbb{P}^{\kappa^{+}}} \check\nu \in f^{\circ}_{D}(\check\xi)$.
Hence $\mathfrak{N}^{1}_{\kappa^{+}}[D_G] = f_{D_G}(\bar\xi) = \bar\kappa \cap f_{D_G}(\xi)$.

Q.E.D. (Subclaim 1)

Remember that we are in the middle of a claim that E is stationary (subtle) in κ. Suppose, by negation, that this is not true. By the Claim of Lemma 4.6 there is an $E' \supseteq E$, $E' \in M$ and E' is not stationary (subtle) in κ.

For $\bar{\kappa} \in J$ set $\Delta_{\bar{\kappa}} = \{q \leq p \mid \dot{q}_{\bar{\kappa}} \supseteq \dot{\bar{p}}_{\bar{\kappa}}\}$.

<u>Subclaim 2</u> $\Delta = \{q \in \mathbb{P}^{\kappa^+} \mid q/p \vee q \in \bigcup_{\bar{\kappa} \in J \setminus E'} \Delta_{\bar{\kappa}}\}$ is dense in $\mathbb{P}^{\kappa^+}$.

Proof Let $q \leq p$. Set: $K_q = \{\nu \in \text{card} \cap \kappa \mid \dot{q}_\nu \neq 0\}$. By the definition of $\mathbb{P}^s_\tau$, K_q is the compliment of a cub set, so K_q is not stationary in κ. Hence $Q = (J \setminus E') \setminus K_q$ is stationary (subtle) in κ. So we may choose an inaccessible $\bar{\kappa} \in Q$ and as before set $\bar{p} = \sigma^{-1}_{\bar{\kappa}}(p)$. Since $\dot{q}_{\bar{\kappa}} = 0$ surely there is a $q' \leq q$ s.t. $\dot{q}'_{\bar{\kappa}} \supseteq \dot{\bar{p}}_{\bar{\kappa}}$ and $q' \in \Delta_{\bar{\kappa}}$.

Q.E.D. (Subclaim 2)

Now choose a $q \leq p$ s.t. $q \in \Delta \cap G_D$. There is a $\bar{\kappa} \in J \setminus E'$ s.t. $q \in \Delta_{\bar{\kappa}}$. But then it is known:

$$\bar{p} \geq q/\bar{\kappa}^+ \Vdash (V_{\bar{\kappa}} \models \phi[f^\circ_D(\check{\bar{\xi}})]) \text{ and } q \Vdash f^\circ_D(\check{\bar{\xi}}) = \check{\bar{\kappa}} \cap f^\circ_D(\xi) .$$

Hence $V_{\bar{\kappa}} \models \phi[B \cap \bar{\kappa}]$ in N and $\bar{\kappa} \in E \subseteq E'$ contradicting $\bar{\kappa} \in J \setminus E'$.

Q.E.D. (Lemma 4.10)

The following lemma will complete the proof of Theorem 0.1.

Lemma 4.11 *Let* κ *be* α*-Erdos for* $\alpha < \omega_1$ *in* $\langle M, A \rangle$. *Then* κ *is* α*-Erdos in* N.

We start by showing that $\mathbb{P}_\omega$ preserves α-Erdos. Assume $D \subseteq [\omega, \infty)$ is $\mathbb{P}_\omega$-generic over $\langle M, A \rangle$ and $N = M[D]$. Let $C, f \in N$ s.t. C is cub in κ and $f: [C]^{<\omega} \longrightarrow \kappa$ is regressive. By Lemma 4.3 we may assume $C \in \langle M, A \rangle$, since the lemma gives a $C' \in \langle M, A \rangle$, $C' \subseteq C$, C' cub in κ and we could show that $f/[C']^{<\omega}$ has a homogeneous set of type α. $f \in \mathfrak{A}^1_{\kappa^+}[D \cap \kappa^+]$ and there is a term

$\mathring{f} \in (\mathfrak{A}^1_{\kappa^+})^{\mathbb{P}^{\kappa^+}_\omega}$ s.t. f interprets $\mathring{f}$. Let $q_o \in \mathbb{P}^{\kappa^+}_\omega$ be s.t. $q_o \Vdash$ "$\mathring{f}[\check{C}]^{<\check{\omega}} \longrightarrow \check{\kappa}$ is regressive".

Define a well ordering on $[C]^{<\omega}$ by maximum difference. For $a,b \in [C]^{<\omega}$ define:

$$d(a,b) = \max(\nu \mid \chi_a(\nu) \neq \chi_b(\nu)\} .$$

Then $a \prec b$ iff $\chi_a(d(a,b)) < \chi_b(d(a,b))$.

$\prec$ is a well ordering of type κ. Note that we define $\max(0) = 0$. Set $e(a,b) = \mathrm{card}(\max(b \cap d(a,b))$. We define $p_a \leq q_o$ in $\mathbb{P}^{\kappa^+}_\omega$ by induction on $\prec$ so that

$p_a \leq \bigcup_{b<a} (p_b)_{e(a,b)}$.

$p_o = q_o$.

Assume p_b is defined for $b \prec a$ and $\bigcup_{b<a} (p_b)_{e(a,b)} \in \mathbb{P}^{\kappa^+}_\omega$.

Set $q_a = \bigcup_{b<a} (p_b)_{e(a,b)}$. For $\gamma \in \mathrm{card} \cap [\omega,\kappa]$ define:

X^γ_a = the least $X \prec \mathfrak{A}^1_{\kappa^+}$ s.t. $\gamma \subseteq X$ and

$a, c, \mathring{f}, b, (p_b)_{e(a,b)}, X^\gamma_b \in X$ for $b \prec a$ and $e(a,b) \leq \gamma$

and

$$f_a(\gamma) \begin{cases} X^\gamma_a \cap \mathfrak{A}^1_{q_{a\gamma^+}} & \text{if } \gamma^+ \in X^\gamma_a \\ \text{undefined otherwise.} \end{cases}$$

p_a = the $\mathfrak{A}^1_{\kappa^+}$-least $p \in \mathbb{P}^{\kappa^+}_\omega$ s.t. $p \leq q_a$ and

(1) $p \in \sum^{\kappa^+, q_a}_{f_a}$

(2) $p(\kappa) \in \bigcap \{\Delta \in X^\kappa_a \mid \Delta \text{ is dense in } \mathbb{R}^{\kappa^+}\}$

(3) $|p \restriction \gamma| = |p_\gamma|$ for limit cardinals $\gamma \in (\omega,\kappa]$

(4) $\exists \xi (p \Vdash \mathring{f}(\check{a}) = \check{\xi})$.

We will show that $q_a \in \mathbb{P}^{\kappa^+}_\omega$ by induction on $\prec$.

Case 1 a is a successor of a^-. It is easy to see that for $b \prec a^- \prec a$, $e(a^-,b) = e(a,b)$. Hence

$$\bigcup_{b\prec a} (p_b)_{e(a,b)} = \bigcup_{b\prec a^-} (p_b)_{e(a^-,b)} \cup (p_{a^-})_{e(a,a^-)} .$$

Thus q_a is a condition by the induction hypothesis.

Case 2 a is a limit point of $\prec$.

The proof of case 2 basically follows that of Lemma 3.13 and is similar to Lemma 3.25. Note that we are dealing with $(p_b)_{e(a,b)}$ instead of all p_b. $\mathrm{card}(\{b \mid b \prec a, e(a,b)=\gamma\}) \le \gamma$ for $\omega \le \gamma \le \kappa$, thus there is no cardinality problem at γ since $\mathrm{card}(\mathrm{dom}(q_{a\gamma})) \le \gamma$. We only mention the difference with Lemma 3.13.

Lemma (analog. to 3.14): Let $\omega \le \delta \le \gamma \le \beta < \kappa$. Let $\Delta \in X_a^\delta$ be predense in $\mathbb{P}_\gamma^{(p_b)\beta^+}$ for some $b \prec a$, $e(a,b) \le \gamma$. Then there is a c, $b \prec c \prec a$ s.t. $(p_c)_\gamma^{\beta^+}$ meets Δ and $e(a,c) \le \gamma$.

Proof Suppose that for every $c \prec a$, $(p_c)_\gamma^{\beta^+}$ does *not* meet Δ. Let $b \prec a_o \prec a$ s.t. $\Delta \in X_{a_o}^\delta$ and $e(a,a_o) \le \gamma$ (note that by definition $X_a^\delta = \bigcup_{\substack{b\prec a\\ e(a,b)\le\delta}} X_b^\delta$ and a is a limit point of $\prec$ implies that a is a limit point of $\prec \restriction \{b \mid e(a,b) \le \delta\}$ for all $\delta \le \gamma$). $\gamma, \beta \in X_{a_o}^\delta$ as in Lemma 3.14.

Let a_o^+ be the successor of a_o. It is easy to see that $e(a,a_o^+) = e(a,a_o) \le \gamma$. $e(a_o^+,b) \le \gamma$ since for $x \prec y \prec z$, $e(z,x)$, $e(z,t) \le \gamma$ implies $e(y,x) \le \gamma$. $p_{a_o^+} \in \sum_{f_{a_o^+}}^{\kappa^+, q_{a_o^+}}$ and $q_{a_o^+} \le (p_b)_\gamma$. Δ is predense in $\mathbb{P}_\gamma^{(p_b)\beta^+}$ implies Δ is predense in $\mathbb{P}_\gamma^{(q_{a_o^+})\beta^+}$ by Lemma 2.11. Hence $\sum_{f_{a_o^+}}^{\kappa^+, q_{a_o^+}}$ takes care of Δ and the rest of the proof follows similarly.

We collapse X_a^γ to B, $\sigma: B \cong X_a^\gamma$ as in Definition 1 inside Lemma 3.13 with the following changes:

$p'_b = \sigma^{-1}((p_b)_{e(a,b)})$ for $b \prec a$, $b' = \sigma^{-1}(b)$ for $b \prec a$,

$D'_\delta = \bigcup_{\substack{b<a \\ e(a,b)\le\delta}} \tilde{p}'_{b\delta}$ etc.

The rest of the proof is the same just using $(p_b)_{e(a,b)}$ instead of p_i and noting that $\{b \mid b<a, e(a,b) \le \gamma\}$ is definable in X_a^γ.

Hence p_a and q_a are well defined for $a \in [C]^{<\omega}$.

Set $E = \{\langle \xi, p_a, a\rangle \mid a \in [C]^{<\omega}, p_a \Vdash \mathring{f}(\check{a}) = \check{\xi}\}$ and $L = \langle L_{\kappa^+}[A], A \cap \kappa^+, E\rangle$.

By Fact 4.2.12 there is a good set of indiscernibles, I, for L s.t. $I \subseteq C$ and $\mathrm{ot}(I) = \alpha$. Let I be the $\mathfrak{A}^1_{\kappa^+}$ least such good set. Clearly for $a,b \in [I]^n$ we have: $p_a \Vdash \mathring{f}(\check{a}) = \check{\xi}$ iff $p_b \Vdash \mathring{f}(\check{b}) = \check{\xi}$.

Lemma 4.11 will be proven if it is shown that $\bigcup_{a\in[I]^{<\omega}} p_a \in \mathbf{P}_\omega^{\kappa^+}$, since setting $p^* = \bigcup_{a\in[I]^{<\omega}} p_a$, we will have $p_o \ge p^* \Vdash$ "I is homogeneous for $\mathring{f}$".

Jensen in [J2] showed, in similar circumstances, that all the p_a are compatible. In our case we must show that $\bigcup_{a\in[I]^{<\omega}} p_a$ can be embedded in a canonical construction (of Lemma 3.13), besides showing their compatibility. We will in fact show that $\bigcup_{a\in[I]^{<\omega}} p_a = \bigcup_{i<\omega} p_{b_i}$ for a sequence $\langle b_i \mid i<\omega\rangle$ cofinal in $[I]^{<\omega}$ and $\bigcup_{i<\omega} p_{b_i}$ will be canonical. First we need some lemmas from [J2].

Lemma 4.11.1 *If* $a,b \in [I]^n$, $r \cap a = r \cap b$ *then* $(p_a)^r = (p_b)^r$.

Proof Let $a = \gamma_o,\ldots,\gamma_i,\beta_o,\ldots,\beta_{n-i-2} = \vec{\gamma}\vec{\beta}$

$b = \gamma_o,\ldots,\gamma_i,\beta'_o,\ldots,\beta'_{n.i.2} = \vec{\gamma}\vec{\beta}'$

and let γ' be the next indiscernible above γ_i. Then $p_{\vec{\beta}\vec{\gamma}}$ is L-definable from $\vec{\gamma}\vec{\beta}$. It is therefore $L' = \langle L,0,1,\ldots,\xi_{\xi<\gamma'}\rangle$ -definable from $\vec{\beta}$. But $I\backslash\gamma'$ is good for L'. So $(p_{\vec{\gamma}\vec{\beta}})^{\gamma'} = (p_{\vec{\gamma}\vec{\beta}'})^{\gamma'}$, since (the ordinal code of) $(p_{\vec{\gamma}\vec{\beta}})^{\nu}$ is one of the named constants $\forall\nu < \gamma'$.

Lemma 4.11.2 *If* $a,b \in [I]^n$, $\max(a) \le \min(b)$ *then* $p_b \le p_a$.

Proof Let $a = \{\nu_1,\ldots,\nu_n\}$, $b = \{\tau_1,\ldots,\tau_n\}$.

$(p_a)^{\nu_1} = (p_b)^{\nu_2}$ by Lemma 4.11.1.

$(p_a)^{\nu_{i+1}} = p^{\nu_{i+1}}_{\{\nu_1,\ldots,\nu_i,\tau_{i+1},\ldots,\tau_n\}}$ by Lemma 4.11.1.

$p_b \le (p\{\nu_1,\ldots,\nu_i,\tau_{i+1},\ldots,\tau_n\})_{\nu_1}$ by definition.

Hence $p_b \le ((p_a)^{\nu_{i+1}})_{\nu_i}$ for $1 \le i \le n$.

$p \le (p_a)_{\nu_n}$ by definition.

Putting the above together the lemma is proved.

Q.E.D.

Let $\langle\beta_i \mid i<\omega\rangle$ be the $\mathfrak{A}^1_{\kappa^+}$-least ω cofinal sequence in I. Set $I^* = \{\beta_i \mid i<\omega\}$ and $\langle a_i \mid i<\omega\rangle = \langle I^*, \prec\rangle$.

<u>Claim</u> $\bigcup_{a\in[I]^{<\omega}} p_a = \bigcup_{i<\omega} p_{a_i}$.

Proof For $a \in [I]^n$, $1 \le n < \omega$ chose an $i<\omega$ s.t. $\max(a) < \min(a_i)$ and $a_i \in [I^*]^{<\omega}$. By Lemma 4.11.2 $p_{a_i} \le p_a$.

Q.E.D. (Claim)

Actually $\langle p_{a_i} \mid i<\omega\rangle$ is not our final sequence. We will eliminate all subsequences for which there is a drop in the length of a_i. First we make two observations.

Observation 1 If $a_i \in (I^*)^n$ and j is the least $i < k$ s.t. $a_k \in (I^*)^n$ for some n, then $p_{a_j} \le p_{a_i}$.

Proof Let $a_i = \{\gamma_o, \ldots, \gamma_{n-1}\}$. It is easy to see that there must be a k, $0 \le k \le n-1$ s.t. if $\gamma_k = \beta_\ell$ then $a_j = \{\gamma_o, \ldots, \gamma_{k-1}, \beta_{\ell+1}, \gamma_{k+1}, \ldots, \gamma_{n-1}\}$. By Lemma 4.11.2 $(p_{a_j})^{\beta_\ell} = (p_{a_i})^{\beta_\ell}$. $e(a_j, a_i) = \beta_\ell$ by definition; hence $(p_{a_j}) \le (p_{a_i})_{\beta_\ell}$.

Q.E.D.

A consequence of the observation is that if $a_i, a_{i+1} \in (I^*)^n$ then $p_{a_{i+1}} \le p_{a_i}$.

Observation 2 If $a_i \in (I^*)^n$ and $a_{i+1} \in (I^*)^{n+1}$ then $p_{a_{i+1}} \le p_{a_i}$.

Proof Let $a_i = \{\gamma_o, \ldots, \gamma_{n-1}\}$. Then it must be that $a_{i+1} = \{\beta_o, \gamma_o, \ldots, \gamma_{n-1}\}$. Clearly $e(a_{i+1}, a_i) = \emptyset$ and $p_{a_{i+1}} \le p_{a_i}$.

Q.E.D.

Define a sequence $\langle b_i \mid i < \omega \rangle$, $b_i \in [I^*]^{<\omega}$ as follows:

$b_o = a_o$

Assume b_i is defined and $b_i = a_k$ for $i \le k$. Let $a_k \in (I^*)^n$. If $a_{k+1} \in I^n$ or $a_{k+1} \in I^{n+1}$ then define $b_{i+1} = a_{k+1}$. Otherwise let j_o be the least j s.t. $j > k$ and $a_j \in I^n$. Then define $b_{i+1} = a_{j_o}$.

Clearly $\langle b_i \mid i < \omega \rangle$ is cofinal in $[I^*]^{<\omega}$ (and hence $I^{<\omega}$).

From the observations it is clear that $b_{i+1} \leq b_i$, $i < \omega$.

Define $p_i = p_{b_i}$. We must show that $\bigcup_{i<\omega} p_i = \bigcup_{i<\omega} p_{a_i}$, i.e. that nothing is lost by omitting the subsequences of a_i's.

<u>Claim</u> If $b_k = a_i$, $b_{k+1} = a_j$ and $i < j$ then

$$p_{k+1} = p_{a_j} \leq \bigcup_{i\leq\ell<j} p_{a_\ell} .$$

Proof It is sufficient to show that for any $i \leq \ell < j$ there is an m, $\ell < m \leq j$ s.t. $p_{a_m} \leq p_{a_\ell}$.
If $a_\ell \in (I^*)^n$ and $a_{\ell+1} \in (I^*)^n$ or $a_{\ell+1} \in (I^*)^{n+1}$ then $p_{a_{\ell+1}} \leq p_{a_\ell}$ by Observations 1 and 2 respectively. Note that this includes the case $p_{a_j} \leq p_{a_{j-1}}$ since if $a_j \in (I^*)^{n_o}$ then $a_{j-1} \in (I^*)^{n_o-1}$.

The only case left is $a_\ell \in (I^*)^n$ and $a_{\ell+1} \in (I^*)^h$ for $h < n$ (a drop).

Note that for $i < \ell < j$, if $a_\ell \in (I^*)^n$ and $a_i, a_j \in (I^*)^{n_o}$ then $n < n_o$, since a_j is the next set of cardinality n_o by definition. Thus there must be an m, $\ell < m < j$ s.t. $a_m \in (I^*)^n$. If we take a least such m, by Observation 1 we have $p_{a_m} \leq p_{a_\ell}$.

Q.E.D. (Claim)

All that is left to prove is that

$$p^* = \bigcup_{i<\omega} p_i = \bigcup_{i<\omega} p_{b_i} = \bigcup_{i<\omega} p_{a_i} = \bigcup_{a\in[I]^{<\omega}} p_a \in \mathbb{P}_\omega^{\kappa^+} .$$

Set $X_I^\gamma = \bigcup_{a\in[I]^{<\omega}} X_a^\gamma$ for $\gamma \in \text{card} \cap [\omega,\kappa]$. $X_I^\gamma \prec \mathfrak{A}_{\kappa^+}^1$ and $\forall a \in [I]^{<\omega}$, $a \in X_I^\gamma$. If we define I in X_I^γ as we defined I in $\mathfrak{A}_{\kappa^+}^1$ we get the same I since $\langle \xi, p_b, b\rangle \in E \cap X_I^\gamma$ for all $b \in [I]^{<\omega}$ s.t. $p_b \Vdash \mathring{f}(\check{b}) = \check{\xi}$.

Hence $I \in X_I^\gamma$. We can also define $\langle \beta_i \mid i < \omega \rangle$, I^*, $\langle a_i \mid i < \omega \rangle$ and $\langle b_i \mid i < \omega \rangle$ all in X_I^γ, getting the same sets. The important fact to note here is that ultimately we can redefine the $\langle p_i \mid i < \omega \rangle$ sequence in X_I^γ.

Again we copy the proof of Lemma 3.13 though in the present case the proof is simple since there is only one limit point. We only point out differences.

Lemma (analog to 3.14) *Let* $\omega \le \delta \le \gamma \le \beta < \kappa$. *Let* $\Delta \in X_I^\delta$ *be predense in* $\mathbf{P}_\gamma^{(p_i)_{\beta^+}}$ *for some* $i < \omega$. *Then there is a* $j < \omega$ *s.t.* $(p_j)_\gamma^{\beta^+}$ *meets* Δ.

Proof Suppose for every $j < \omega$, $(p_j)_\gamma^{\beta^+}$ does not meet Δ. Let $i \le i_o < \omega$ be s.t. $\Delta \in X_{b_{i_o}}^\delta$. Let $b_{i_o} \in [I^*]^n$. Let j_o be the least $j > i_o$ s.t. $b_j \in [I^*]^{n+1}$. Then $e(b_{j_o}, b_{j_o - 1}) = \emptyset$ and $q_{b_{j_o}} \le p_{j_o - 1} \le p_{i_o}$. $p_{j_o} \in \sum_{f_{b_{j_o}}}^{\kappa^+, q_{b_{j_o}}}$ and by Lemma 2.11, Δ predense in $\mathbf{P}_\gamma^{(p_{i_o})_{\beta^+}}$ implies Δ predense in $\mathbf{P}_\gamma^{(q_{b_{j_p}})_{\beta^+}}$. Hence $\sum_{f_{b_{j_o}}}^{\kappa^+, q_{b_{j_o}}}$ takes care of Δ and the rest of the proof follows.

When collapsing X_I^γ to B, $\sigma_i B \cong X_I^\gamma$, we need images for I and I^* to show we can imitate the construction of $\langle p_{b_i} \mid i < \omega \rangle$ in X_I^γ to get $\langle p_i' \mid i < \omega \rangle \in B$, where $p_i' = \sigma^{-1}(p_{b_i})$.

The rest of the proof is the same.

Q.E.D.

We have shown that $\mathbf{P}_\omega$ preserves the α-Erdos property. Let $D \subseteq [\omega, \infty)$ be $\mathbf{P}_\omega$-generic over $\langle M, A \rangle$. We have to show that $\mathbf{P}^{D_\omega}$ preserves the α-Erdos property. It is known that small conditions $(\mathrm{card}(\mathbf{P}^{D_\omega}) < \kappa)$ preserve α-Erdos.

Q.E.D.

Lemma 4.11 implies that the property "κ is the least s.t. $\kappa \longrightarrow (\alpha)^{<\omega}$, $\alpha < \omega_1$" is preserved. Since if κ is α-Erdos then $\kappa \longrightarrow (\alpha)^{<\omega}$ holds and since $\alpha < \omega_1$ the statement "I is a set of indiscernibles for $f: [\kappa]^{<\omega} \longrightarrow 2$" is equivalent to the well-foundedness of a set of finite maps which is absolute. Thus we have:

If $\lambda < \kappa$ and $N \models \lambda \longrightarrow (\alpha)^{<\omega}$ then $M \models \lambda \longrightarrow (\alpha)^{<\omega}$.

4.4 $0^{\#}$ AND THEOREM 0.2

In this section we assume $M = L$, $A = 0$ and $0^{\#}$ exists. L satisfies all the requirements on $\langle M,A\rangle$ listed in §2.1. Let I be the canonical class of indiscernibles for L. I is uniquely characterized by:

(a) $I \subseteq On$ is closed in On.

(b) I is a class of indiscernibles for L.

(c) Every $x \in L$ is $\Sigma_1(L)$ in parameters from I.

Properties about I can be found in Silver [SI] and Kunen [K]. The following facts are used:

Fact 4.12.1 $I \setminus \gamma$ are indiscernibles for $\langle L, \nu < \gamma\rangle$ for $\gamma \in I$.

Notation $\langle \gamma_0, \ldots, \gamma_n\rangle = \vec{\gamma} \in I^{i+1}$ *is always assumed to be in ascending order.*

Fact 4.12.2 $\gamma \in I \longrightarrow L_\gamma \prec L$.

Fact 4.12.3 Let $X_{\gamma,\vec{\delta}}$ = the least $X \prec L$ s.t. $\gamma \cup \{\gamma\} \cup \{\delta_o,\ldots,\delta_n\} \subseteq X$, where $\vec{\delta} = \langle\delta_o,\ldots,\delta_n\rangle, \langle\gamma,\vec{\delta}\rangle \in I^{n+1}$. Let $\gamma^* = \min(I \setminus (\gamma+1))$. Then $X_{\gamma,\vec{\delta}} \cap L_{\gamma^*} = X_{\gamma,\vec{\eta}} \cap L_{\gamma^*}$ for $\langle\gamma,\vec{\delta}\rangle, \langle\gamma,\vec{\eta}\rangle \in I^{n+1}$.

Fact 4.12.4 Let $\langle\delta_o,\delta_1,\ldots,\delta_n,\ldots \mid n<\omega\rangle \in (I\setminus(\gamma+1))^\omega$ and let $\gamma^* = \min(I\setminus\gamma+1)$ for $\gamma \in I$ then

$$L_{\gamma^*} \subseteq X = \text{the least } Y \prec L \text{ s.t. } \gamma\cup\{\gamma\}\cup\{\delta_o,\ldots\} \subseteq Y .$$

We shall use I to construct a $\mathbb{P}_\omega$-"pseudo generic" $D \subseteq [\omega,\infty)$ over L. D is "pseudo generic" if I is the canonical set of indiscernibles for $\langle L[D],D\rangle$ and $\forall\alpha \in$ card $D \cap \alpha^+$ is $\mathbb{P}_\omega^{\alpha^+}$-generic over $\mathfrak{A}^1_{\alpha^+}$. D will be definable in $L[0^\#]$.

We shall define $\langle s^{i\nu} \in \mathbb{S}_{\nu^+} \mid \nu\in I\rangle$ and $\langle p^{i\nu} \in \mathbb{P}_\omega^{s^{i\nu}} \mid \nu\in I\rangle$ by induction on $i<\omega$, s.t.:

(a) $s^{i\nu} = p^{i\tau}_{\nu^+}$ for $\langle\nu,\tau\rangle \in I^2$ $(\nu<\tau)$.

(b) $p^{i\nu} \le p^{j\nu}$ in $\mathbb{P}_\omega^{s^{i\nu}}$ for $j \le i$.

(c) $|p^{i\nu}\restriction\gamma| = |p^{i\nu}_\gamma|$ for limit cardinals $\gamma \le \nu$.

(d) $p^{i\nu}$ is uniformly $\Sigma_1(L)$ in $\langle\nu,\tau_1,\ldots,\tau_i\rangle \in I^{i+1}$.

<u>Claim 4.13</u> (a), (b), (c) and (d) imply:

(e) $s^{i\nu} \le s^{j\nu}$ in $\mathbb{S}_{\nu^+}$ for $j \le i$.

(f) $p^{i\nu}\restriction\nu = p^{i\tau}\restriction\nu$ for $\nu<\tau$.

(g) $\dot{p}^{i\tau}_\nu = 0 \wedge p^{i\tau}_\nu = p^{i\nu}_\nu$ for $\nu<\tau$.

Proof (e) trivially follows from (a) and (b). We will show that (f) and (g) follow from (c) and (d). Let $\pi': I \longrightarrow I$ be order preserving s.t. $\pi'\restriction I\cap\nu = \mathrm{id}\restriction I\cap\nu$ and $\pi'(\nu) = \tau$. π' extends to a unique $\pi: L \xrightarrow[\Sigma_1]{} L$ s.t.

$\pi \upharpoonright L_\nu = \mathrm{id} \upharpoonright L_\nu$ and $\pi(\nu) = \tau$. By (d) $\pi(p^{i\nu}) = p^{i\tau}$ so, $p^{i\nu} \upharpoonright \nu = \pi(p^{i\nu}) \upharpoonright \nu = p^{i\tau} \upharpoonright \nu$. Thus (f) holds. Clearly, $\pi(p^{i\nu}_\nu) = p^{i\tau}_\tau$. Hence §2.5 Fact 1 implies,

$$\pi \upharpoonright \mathfrak{A}_{p^{i\nu}_\nu} = \Pi_{p^{i\tau}_\tau, \nu}, \quad \mathfrak{A}_{p^{i\nu}_\nu} = \mathfrak{A}_{p^{i\tau}_\tau, \nu} \quad \text{and}$$

$$p^{i\tau} \upharpoonright \tau = \pi(p^{i\nu} \upharpoonright \nu) \in X_{p^{i\tau}_\tau, \nu} .$$

By (c) $|p^{i\tau} \upharpoonright \tau| = |p^{i\tau}_\tau|$ so we may use the collapsing lemma (2.12) to get: $\dot{p}^{i\tau}_\nu = 0$ and $p^{i\tau}_\nu = p^{i\nu}_\nu$.

Q.E.D. (Claim)

Definition 1 We define $p^{i\nu}, s^{i\nu}$ by induction as follows:

$p^{o\nu}$ = the constant function $\langle 0,0 \rangle$ on $\mathrm{card} \cap [\omega, \nu)$.

$s^{o\nu} = 0$.

Now let $p^{i\nu}$ and $s^{i\nu}$ be defined s.t. (a), (b), (c) and (d) (hence (e), (f) and (g)) hold. For $\gamma \in \mathrm{card} \cap [\omega, \nu)$ and $\langle \nu, \vec{\tau} \rangle \in I^{i+1}$ where $\vec{\tau} = \tau_1 < \tau_2 \ldots < \tau_i$ set:

$Y^\gamma_{i\nu\vec{\tau}}$ = the least $Y \prec L$ s.t. $\gamma \cup \{\nu, \vec{\tau}\} \subseteq Y$.

For $\gamma < \nu$ set:

$$f_{i\nu}(\gamma) = \begin{cases} Y^\gamma_{i\nu\vec{\tau}} \cap \mathfrak{A}^1_{p^{i\nu}_\gamma} & \text{if } \gamma \in Y^\gamma_{i\nu\vec{\tau}} \\ \text{undefined otherwise.} \end{cases}$$

and set $f_{i\nu}(\nu) = Y^\nu_{i\nu\vec{\tau}} \cap \mathfrak{A}^1_{s^{i\nu}}$.

It is clear by Fact 4.12.3 that $f_{i\nu} \in \mathbf{F}(p^{i\nu})$ is independent of the choice of $\vec{\tau} \in I^i$.

<u>Claim</u> $f_{i\nu}$ is uniformly $\Sigma_1(L)$ in $\langle \nu, \tau_1, \ldots, \tau_{i+1} \rangle \in I^{i+2}$.

Proof

(A) $p^{i\nu}$ is uniformly $\Sigma_1(L)$ in $\nu,\tau_1,\dots,\tau_i$ (note this implies $p^{i\nu}$ is independent of $\tau_1,\dots,\tau_i$).

(B) $s^{i\nu}$ is uniformly $\Sigma_1(L)$ in $\nu,\tau_1,\dots,\tau_{i+1}$ since $s^{i\nu} = p^{i\tau_2}_{\nu+}$ and $p^{i\tau_1}$ is uniformly $\Sigma_1(L)$ in $\tau_1,\tau_2,\dots,\tau_{i+1}$.

(C) $\langle Y^{\gamma}_{i\nu\tau_1,\dots,\tau_i} \mid \gamma \le \nu\rangle$ is uniformly $\Sigma_1(L)$ in $\nu,\tau_1,\tau_2,\dots,\tau_{i+1}$ since $Y \prec L$ is the same as $Y \prec L_{\tau_{i+1}} \prec L$.

Q.E.D. (Claim)

By (A) $p^{i\nu} \in Y^{\gamma}_{i\nu}$.

Define:

$p^{i+1,\nu}$ = the $\mathfrak{A}^1_{s^{i\nu}}$-least $p \le p^{i\nu}$ s.t.

(1) $p \in \Sigma^{s^{i\nu},p^{i\nu}}_{f_{i\nu}}$.

(2) $p(\nu) \in \cap\{\Delta \mid \Delta \in f_{i\nu}(\nu),\ \Delta \text{ is dense in } \mathbb{R}^{s^{i\nu}}\}$.

(3) $|p{\restriction}\gamma| = |p_\gamma|$ for limit cardinals $\gamma \le \nu$.

Define $s^{i\nu}$ by (a). Then $s^{i\nu},p^{i\nu}$ clearly satisfy (a), (b), (c) and (d).

Set $p^i = \bigcup_{i \in I} p^{i\nu}{\restriction}\nu$.

For $\nu \in I$, $p^i{\restriction}\nu = p^{i\nu}{\restriction}\nu$ and $p^i_\nu = p^{i\nu}_\nu$ but $\dot{p}^i_\nu = 0$ and $\dot{p}^{i\nu}_\nu \ne 0$ (by (f) and (g)) for $i > 0$. Clearly, $s^{i\nu} = p^i_{\nu+}$ for $i > 0$. p^i is a proper class; intuitively $p^i \in \mathbb{P}^\infty_\omega$ since $p^i{\restriction}\nu \in \mathbb{P}^\nu_\omega$ and $L_\nu < L_\infty = L$ for $\nu \in I$.

Definition 2 Let $\alpha \in \text{card} \setminus \omega$.

G^i_α = the set of $q \in \mathbb{P}^{p^i_{\alpha+}}_\omega$ s.t. $q{\restriction}(\alpha^+ \setminus u) = p^i{\restriction}(\alpha^+ \setminus u)$, where $u \subseteq I \cap \alpha^+$ is finite and $q(\eta) = p^{i\eta}(\eta)$ for $\eta \in u$. The only difference between $q \in G^i_\alpha$ and p^i is that it may be that $\dot{q}_\eta \neq 0$ but always $\dot{p}^i_\eta = 0$ for $\eta \in u$. In any case $q_\eta = p^i_\eta$ for $\eta \in u$ so G^i_α is a set of mutually compatible conditions in $\mathbf{P}_\omega^{p^i_{\alpha^+}}$.

Lemma 4.14 *Let* $\Delta \in \mathfrak{A}^1_{p^i_{\alpha^+}}$ *be predense in* $\mathbf{P}_\omega^{p^i_{\alpha^+}}$. *Then there is a* $j \geq i$, $p \in G^j_\alpha$ s.t. p *meets* Δ.

Proof Suppose the conclusion to be false. Let α be the least counter example. For $i < \omega$ set $q^i = p^{i\nu}\restriction\alpha^+$ and $f_i = f_{i\nu}\restriction\alpha^+$ where $\nu = \min(I\backslash\alpha)$. Then $q^i \in G^i_\alpha$ since $q^i\restriction\alpha^+$ differs from $p^i\restriction\alpha^+$ at most at $\dot{q}^i_\nu$ (if $\alpha = \nu$). It is clear that $f_i \in F(q^i)$.
By Fact 4.12.4 $\mathfrak{A}^1_{p^i_{\alpha^+}} \subseteq \bigcup_{i\leq j<\omega, \vec{\tau}\in I^j} Y^\alpha_{j\nu,\vec{\tau}}$.

Hence for sufficiently large i, $\alpha \in \mathrm{dom}(f_i)$ and $\mathfrak{A}^1_{p^i_{\alpha^+}} \subseteq \bigcup_{j\geq i} f_j(\alpha)$.

Let $n \geq i$ s.t. $\Delta \in f_{n-1}(\alpha)$. Then by definition,

$$q^n \in \sum\nolimits_{f_{n-1}}^{p^{n-1}_{\alpha^+}, q^{n-1}}$$

Since q^n does not meet Δ, there is a $\gamma < \alpha$ s.t. $\Delta^{(q^n)_{\gamma^+}}$ is dense in $\mathbf{P}_\omega^{q^n_{\gamma^+}}$. But $q^n_{\gamma^+} = p^n_{\gamma^+}$. By the minimal choice of α, there is a $m \geq n$, $r \in G^m_\gamma$ s.t. r meets $\Delta^{(q^n)_{\gamma^+}}$. Set $p = r \cup (q^m)_{\gamma^+}$. Then $p \in G^m_\alpha$ and p meets Δ. We have reached a contradiction.

Q.E.D.

Definition 3 Set $D_\gamma = \bigcup_{i<\omega} \tilde{p}'_i$ for $\gamma \in \text{card}\setminus\omega$.
$D = \bigcup_{\gamma\in\text{card}} D_\gamma$.

Corollary 4.14.1 $D \cap \alpha^+$ *is* $\mathbb{P}_\omega^{p^i_{\alpha^+}}$*-generic over* $\mathfrak{A}^1_{p^{i\nu}}$.

Lemma 4.15 *Let* $\nu \in I$. *Then* $I \cap \nu$ *is a set of indiscernibles for* $\mathfrak{A}^1_{s^{i\nu}}[D \cap \nu^+]$.

Proof Let

$$G = \{q \in \mathbb{P}_\omega^{s^{i\nu}} \mid \exists p \in \bigcup_{i<\omega} G^i_\nu (p{\restriction}\nu = q{\restriction}\nu \wedge q_\nu = p_\nu \wedge \dot{q}_\nu = \dot{p}_\nu \cap \mathfrak{A}^1_{s^{i\nu}})\} .$$

Then G is a mutually compatible set in $\mathbb{P}^s_\omega$ s.t.

$$[\mathfrak{A}^1_{s^{i\nu}}[D \cap \nu^+] \models \phi(\vec{\eta})] \longleftrightarrow [\exists p \in G\ (p \Vdash \phi(\check{\vec{\eta}})] ,$$

where $\Vdash$ is the $\mathbb{P}^s_\omega$ forcing relation. It suffices to show:

<u>Claim</u> Let $\vec{\eta} \in I \cap \nu$ s.t. $\mathfrak{A}^1_{s^{i\nu}}[D \cap \nu^+] \models \phi(\vec{\eta})$.

Let $\pi' \colon I \cap \nu \longrightarrow I \cap \nu$ be order preserving. Then

$$\mathfrak{A}^1_{s^{i\nu}}[D \cap \nu^+] \models \phi[\pi'(\vec{\eta})] .$$

Proof Let $p \in G$ s.t. $p \Vdash \phi(\check{\vec{\eta}})$. By the definition of G and the properties of $p^{i\delta}$ (for $\delta \in I$), $p = f(\vec{\rho}, \nu, \vec{\tau})$ where $\vec{\rho} \in I \cap \nu$, $\vec{\tau} \in I \cap (\nu, \infty)$ are ascending m-tuples, f is $\Sigma_1(L)$ and $f(\vec{\rho}', \nu, \vec{\tau}) \in G$ for all ascending m-tuples $\vec{\rho}' \in I \cap \nu$. We may assume that s (hence $\Vdash$) is $\Sigma_1(L)$ in $\nu, \vec{\tau}$ (by making $\vec{\tau}$ longer if necessary. $\pi' \cup \text{id}{\restriction}(I\setminus\nu)$ extends to a unique $\pi \colon L \xrightarrow[\Sigma_1]{} L$. But then $\pi(s) = s$, $\pi(\Vdash) = \Vdash$ and $\pi(p) = f(\pi(\vec{\rho}), \nu, \vec{\tau}) \in G$. So $\pi(p) \Vdash \phi(\pi(\vec{\eta}))$ and thus $\mathfrak{A}^1_{s^{i\nu}}[D \cap \nu^+] \models \phi(\pi(\vec{\eta}))$.

Q.E.D.

Corollary 4.15.1 *For* $\eta, \eta' \in I$, $\eta < \eta'$

$\langle L_\eta[D], D \cap \eta\rangle \prec \langle L_{\eta'}[D], D \cap \eta'\rangle$ *and*

$\langle L_\eta[D], D \cap \eta\rangle \prec \langle L[D], D\rangle$.

Proof Take $\eta, \eta' < \nu$. Then $I \cap \nu$ are indiscernibles for $\mathfrak{A}^1_{s^{i\nu}}[D \cap \nu^+]$, hence

$\langle L_\eta[D], D\rangle \prec \langle L_{\eta'}[D], D \cap \eta'\rangle$.

Thus $\langle L[D], D\rangle$ satisfaction is definable from I and

$\langle L_\eta[D], D\rangle \prec \langle L[D], D\rangle$.

Corollary 4.15.1 I *is a class of indiscernibles for* $\langle L[D], D\rangle$.

Proof $I \cap \nu$ is a set of indiscernibles for $\langle L_\nu[D], D \cap \nu\rangle$ and $\langle L_\nu[D], D \cap \nu\rangle \prec \langle L[D], D\rangle$ for $\nu \in I$.

Lemma 4.16 $L[D] = L[D_\omega]$, D *is* $L[D_\omega]$ *definable in the parameter* D_ω *and* $L[D]$ *has all the preservation properties of Theorem 0.1.*

Proof It is sufficient to prove the lemma for $\langle L_\nu[D], D \cap \nu\rangle$ for $\nu \in I$ since $\langle L_\nu[D], D \cap \nu\rangle \prec \langle L[D], D\rangle$. By Corollary 4.14.1 $D \cap \nu^+$ is $\mathbf{P}_\omega^{p^i_{\nu^+}}$-generic over $\mathfrak{A}^1_{s^{i\nu}}$ ($s^{i\nu} = p^i_{\nu^+}$). Thus all the preservation theorems hold in $L_\nu[D]$ since they hold in $\mathfrak{A}^1_{s^{i\nu}} \supseteq L_\nu[D]$. By Lemma 3.7 $D \cap \nu \in L_{\mu 1_{s^{i\nu}}}[D_\omega]$ (even $D \cap \nu^+ \in L_{\mu 1_{s^{i\nu}}}[D_\omega]$). An examination of the proof of Lemma 3.7 shows that all the coding of $D \cap \nu$ occurs in $L_\nu[D_\omega]$ since ν is inaccessible.

Q.E.D.

Lemma 4.17 *We can construct a* $\mathbb{P}^{D_\omega}$*-generic* $a = D_o \subseteq \omega$ *(using* I*) over* $L[D_\omega]$ *s.t.:*

(a) I *is a class of indiscernibles for* $\langle L[a \cup D], D, a\rangle$.

(b) $L[a \cup D] = L[a]$ *and all assertions of Lemma 4.17 hold with* $a \cup D$ *and* a *in place of* D *and* D_ω .

Proof Let $\langle \nu_i \mid i < \omega\rangle$ enumerate the first ω elements of I .

Set:

X_I = the least $X \prec L$ s.t. $\nu_o, \ldots, \nu_{i-1} \in X$ and

$Y_i = X_I \cap L_{\omega_1}[D]$ (ω_1 of L) .

Each Y_i is countable and $L_{\omega_1}[D] = \bigcup_{i<\omega} Y_i$.

Set g_i = the L-least $g: \omega \longrightarrow Y_i$ which is surjective.

Set $g: \omega \xrightarrow{On} L_{\omega_1}[D]$ by $g(\{i,j\}) = g_i(j)$ and 0 otherwise.

$\Delta \subseteq \mathbb{P}^{D_\omega} \longrightarrow \Delta \in L_{\omega_1}[D]$.

Define $p_i \in \mathbb{P}^{D_\omega}$ by induction on i as follows:

p_o = the L[D]-least $p \in g(0)$ if $g(0)$ is dense in $\mathbb{P}^{D_\omega}$, otherwise $p_o = 0$.

p_{i+1} = the L[D]-least $p \leq p_i$ s.t. $p \in g(i+1)$ if $g(i+1)$ is dense in $\mathbb{P}^{D_\omega}$. Otherwise $p_{i+1} = p_i$.

Clearly $\bigcup_{i<\omega} p_i$ is $\mathbb{P}^{D_\omega}$-generic over $L[D_\omega]$. It is also clear that (a) holds since it makes no difference which indiscernibles we use to define g . (b) follows easily.

Q.E.D.

Lemma 4.18 $0^{\#} = a^{\#}$ *as Turing degrees.*

Proof $0^{\#}$ and $a^{\#}$ are the complete theories of

$\langle L, \nu_o, \nu_1, \ldots, \nu_i, \ldots \mid i < \omega \rangle$ and

$\langle L[a], a, \nu_o, \ldots, \nu_i, \ldots \mid i < \omega \rangle$ respectively.

Let T be the complete theory of

$\langle L[D], D, \nu_o, \ldots, \nu_i, \ldots \mid i < \omega \rangle$.

It is apparent from the construction that T is recursive in $0^{\#}$ and $a^{\#}$ is recursive in T . $0^{\#}$ is trivially recursive in $a^{\#}$.

Q.E.D.

Lemma 4.19 a *is not set generic over* L .

Proof a adds a Cohen generic to each successor cardinal.

Q.E.D.

Theorem 0.2 is now completed.

5 . Applications

5.1 A NEW VERSION OF SOLOVAY'S CONJECTURE

When Solovay's conjecture was proven false (in Chapter 0) a new version was introduced.

Conjecture *If* $0^{\#} \notin L[a]$, $a \subseteq \omega$, *then* a *is class generic over* L .

The notion "class generic over L " has to be clarified. Let $M \models ZF^-$. We require a class of conditions, $\mathbb{P}$, to be definable in M and that the class $\{\langle p,\vec{t}\rangle \mid p \Vdash \phi(\vec{t})\}$ be definable in M . If $\mathbb{P}$ is a class of conditions satisfing the above, $G \subseteq \mathbf{P}$ is $\mathbb{P}$-generic over M if G intersects all the M-definable dense classes in $\mathbf{P}$ (and the other requirements on genericity). When working over L , it may be interesting to use a class $\mathbb{P}$ which is not L-definable. In such a case we would use the model $\langle L[\mathbb{P}],\mathbb{P}\rangle$ where $\mathbf{P}$ is definable. Three notions of "class generic over L " are defined.

Definition 1

(1) G is "strong class generic over L" iff G is $\mathbb{P}$-generic over L for an L-definable class of conditions $\mathbb{P}$.

(2) G is "medium class generic over L " iff G is $\mathbb{P}$-generic over $\langle L,\mathbf{P}\rangle$ for a class of conditions $\mathbb{P}$, s.t. $\langle L,\mathbb{P}\rangle$ is amenable.

(3) G is "weak class generic over L " iff G is $\mathbb{P}$-generic over $\langle L[\mathbb{P}],\mathbf{P}\rangle$, where $\mathbf{P}$ is a class of conditions.

It is, of course, required that $\{\langle p,\vec{t}\rangle \mid p \Vdash \phi(\vec{t})\}$ is definable in the appropriate model. With respect to the above conjecture we are interested in a $\mathbb{P}$ s.t. $\langle L[\mathbb{P}],\mathbb{P}\rangle \models \neg 0^{\#}$. It is known that if G is strong or medium generic over L then $\langle L[G],\mathbb{P}\rangle \models \neg 0^{\#}$. The case of weak class generic is open.

Definition 2 Let I be the canonical class of indiscernibles for L. Let $\alpha,\beta \in I$, $\alpha < \beta$. Set $\pi_{\alpha\beta}: L \xrightarrow[\Sigma_1]{} L$ to the unique isomorphism determined by $\pi_{\alpha\beta}\restriction\alpha = \mathrm{id}\restriction\alpha$ and $\pi(\alpha) = \beta$.

Lemma 5.1 *Let* $\langle L,A\rangle$ *be amenable. Assume* $0^{\#}$ *exists and set* I *to be the canonical class of indiscernibles for* L. *Then*

$$\exists \nu \in I \ \forall \delta \in I \ (\nu \le \delta \longrightarrow \langle L_\delta, A\cap\delta\rangle \prec \langle L,A\rangle).$$

Proof For $\gamma \in I$, let $f_\gamma: \gamma \xrightarrow{On} \gamma^{+L}$ be canonically defined from γ and ω-many larger indiscernibles so that f is onto γ^{+L}. Since f_γ is uniformly defined in γ and ω-many larger indiscernibles, $\pi_{\gamma\gamma'}(f_\gamma) = f_{\gamma'}$ for $\gamma < \gamma'$, $\gamma,\gamma' \in I$. Define a regressive function $f: I \longrightarrow On$ by: $f(\gamma) = f_\gamma^{-1}(\xi)$, where ξ is the least L-code of $A\cap\gamma$ ($\langle L,A\rangle$ amenable means $A\cap\gamma \in L$ for every $\gamma \in On$ and ξ clearly lies between γ and γ^{+L}).

<u>Claim 1</u> There exists an unbounded $X \subseteq I$ s.t. $f\restriction X$ is constant.

Proof Assume $\forall \alpha \in On$ ($\{\gamma \in I \mid f(\gamma) < \alpha\}$ is bounded). Define $\gamma_i \in I$ for $i < \omega$ by:

γ_o is arbitrary.

γ_{i+1} = the least $\gamma \in I$ s.t. $\gamma > \sup\{\delta \in I \mid f(\gamma) < \gamma_i\}$.

$\gamma_\omega = \sup_{i<\omega} \gamma_i$.

f is regressive, hence there is an i s.t. $f(\gamma_\omega) < \gamma_i$ which contradicts the definition of the sequence γ_i .

Q.E.D. (Claim 1)

Let $X \subseteq I$ be unbounded and for $\gamma \in X$, $f(\gamma) = \delta_o$. For $\gamma \in I$ set $f_\gamma(\gamma_o) = \xi_\gamma$. For $\gamma \in I$, ξ_γ is the L-code of $A \cap \gamma$. For $\gamma, \gamma' \in X$, $\gamma < \gamma'$, $\pi_{\gamma\gamma'}(A \cap \gamma) = A \cap \gamma'$. Let $\gamma_o = \min(X)$.

<u>Claim 2</u> For $\delta \in I \setminus \gamma_o$, ξ_δ codes $A \cap \delta$.

Proof $\pi_{\gamma\delta}(A \cap \gamma) \subseteq A \cap \delta$ and $\pi^{-1}_{\delta\gamma'}(A \cap \gamma') \supseteq A \cap \delta$ by the definition of the $\pi_{\alpha\beta}$. Then

$$A \cap \delta \supseteq \pi_{\gamma\delta}(A \cap \gamma) = \pi^{-1}_{\delta\gamma'}(A \cap \gamma') \supseteq A \cap \delta$$

since $\pi_{\delta\gamma'} \circ \pi_{\gamma\delta} = \pi_{\gamma\gamma'}$. Hence $\pi_{\gamma\delta}(A \cap \gamma) = A \cap \delta$ and ξ_δ codes $A \cap \delta$.

Q.E.D. (Claim 2)

For $\gamma, \gamma' \in I \setminus \delta_o$, $\pi_{\gamma\gamma'}(A \cap \gamma) = A \cap \gamma'$ and $\pi_{\gamma\gamma'}(\langle L_\gamma, A \cap \gamma \rangle) = \langle L_{\gamma'}, A \cap \gamma' \rangle$, thus

$$\langle L_\gamma, A \cap \gamma \rangle \prec \langle L_{\gamma'}, A \cap \gamma' \rangle \quad \text{for } \gamma, \gamma' \in I \setminus \delta_o .$$

Q.E.D.

Corollary 5.1.1 *If* G *is* $\mathbf{P}$*-generic over* $\langle L, \mathbf{P} \rangle$ *and* $\langle L, \mathbf{P} \rangle$ *is amenable, then* $\langle L[G], \mathbf{P} \rangle \models \neg 0^{\#}$ *exists.*

Proof Assume $\langle L[G], \mathbf{P} \rangle \models 0^{\#}$ exists. Then there is a $p_o \in G$ s.t. $p_o \Vdash 0^{\#}$ exists. But then the satisfaction relation of $\langle L, \mathbf{P} \rangle$ can be defined in $\langle L, \mathbf{P} \rangle$ by (using Lemma 5.1):

$\langle L,\mathbf{P}\rangle \models \phi(x)$ iff $\exists \gamma \in I \ \exists p \in \mathbf{P} \ (x \in L_\gamma \wedge p \le p_0 \wedge$

$\wedge p \Vdash \forall \delta \in I(\delta \ge \gamma \quad \langle L_\gamma, \mathbf{P}\cap\gamma\rangle \prec \langle L_\delta, \mathbf{P}\cap\delta\rangle) \wedge \langle L_\gamma, \mathbf{P}\cap\gamma\rangle \models \phi(x)$.

This is impossible.

Q.E.D.

The $\mathbf{P}$-pseudo generic, a , of Theorem 0.2 (which is constructed in §4.4) was a candidate for refuting the above conjecture. We shall show that any $\mathbf{P}$-pseudo generic over L is weak class generic over some class of conditions. We recall the definition of pseudo generic.

Definition 3 Let $\mathbf{P}$ be the class of coding conditions. D is $\mathbf{P}$-pseudo over L if I remains the class of indiscernibles for L[D] and $\forall \alpha \in \text{card}(D \cap \alpha^+$ is $\mathbf{P}^{\alpha^+}$-generic over $\mathfrak{A}^1_{\alpha^+})$.

In §4.4 the pseudo generic, $a \subseteq \omega$, is coded in two steps. First by $\mathbf{P}_\omega$ to $D_\omega \subseteq [\omega,\aleph_1)$ and then by almost disjoint forcing to $a \subseteq \omega$. The last step is generic so it is enough to deal with $\mathbf{P}_\omega$ and show that the class D is generic for an appropriate set of conditions. Set $\mathbf{P} = \mathbf{P}_\omega$.

Definition 4 For κ inaccessible set ${}^\kappa\mathbf{P}_\tau = \mathbf{P}_\tau \cap L_\kappa$. All notation that applied to $\mathbf{P}_\tau$ will apply to ${}^\kappa\mathbf{P}_\tau$. Δ is dense in ${}^\kappa\mathbf{P}_\tau$ over L_κ iff $\Delta \subseteq {}^\kappa\mathbf{P}_\tau$ is dense and Δ is definable in L_κ . Note that such a Δ is a class in L_κ . G is ${}^\kappa\mathbf{P}_\tau$-generic over L_κ iff $G \cap \Delta \ne 0$ for all Δ dense in ${}^\kappa\mathbf{P}_\tau$ over L_κ . Note that there are only κ such Δ . ${}^\kappa\mathbf{P}_\tau$ is τ-distributive over L_κ iff for every sequence $\langle \Delta_\nu \mid \nu < \tau\rangle$ definable in L_κ s.t. each Δ_ν is dense in ${}^\kappa\mathbf{P}_\tau$ over L_κ , $\bigcap_{\nu\in\tau} \Delta_\nu$ is dense in ${}^\kappa\mathbf{P}_\tau$.

Lemma 5.2 *Let* κ *be inaccessible. Let* $\langle \Delta_\nu \mid \nu < \gamma^+ < \kappa\rangle$

be a L_κ-definable sequence of ${}^\kappa\mathbb{P}_\tau$ dense classes over L_κ $(\tau \le \gamma)$. Then $\Delta = \{p \in {}^\kappa\mathbb{P}_{\gamma+} \mid \forall \nu < \gamma^+ \; \Delta_\nu^p$ is dense in $\mathbb{P}_\tau^{p_{\gamma+}{}^\kappa}\}$ is dense in ${}^\kappa\mathbb{P}_{\gamma+}$.

Proof Similar to Lemma 3.4 except that it must be noted that everything is definable in L_κ. Let $p_o \in {}^\kappa\mathbb{P}_{\gamma+}$ and $D \subseteq [\gamma^+, \kappa)$ be ${}^\kappa\mathbb{P}_{\gamma+}$-generic over L_κ s.t. $p_o \in G_D$. We show $\Delta_\nu^D = \{q \in \mathbb{P}_\tau^{D_{\gamma+}{}^\gamma} \mid \exists p \in G_D (q \cup p \in \Delta_\nu)\}$ is dense in $\mathbb{P}_\tau^{D_{\gamma+}{}^\gamma}$ as in Lemma 3.4 noting that

$H = \{p \in {}^\kappa\mathbb{P}_{\gamma+} \mid p/q_o \vee \exists q \le q_o (q \cup p \in \Delta_\nu)\}$ is definable in L_κ for $q_o \in \mathbb{P}_\tau^{p_{\gamma+}{}^\gamma}$. Continue the proof as in 3.4 and note that $\langle p_{r\nu} \mid r \in X_\nu, \nu < \gamma^+\rangle \in L_\kappa$ and hence $\langle \Delta_{r\nu} \mid r \in X_\nu, \nu < \gamma^+\rangle$ is definable in L_κ. The proof is now completed as in 3.4.

Q.E.D.

Lemma 5.3 *There is a $D \subseteq [\omega, \infty)$ s.t. D satisfies all the properties of Lemmas 4.15 and 4.16 and D is $\mathbb{P}$-generic over L.*

Proof Let $\langle \Delta_\nu \mid \nu \in On\rangle$ be the L-definable sequence of dense classes in $\mathbb{P}$ s.t. Δ_ν is definable in L, defined by $\nu < \tau$ iff the Gödel number of the defining formula for Δ_ν is less than that of Δ_τ. For $\gamma \in I$ if we define a sequence in L_γ by the same definition as the above sequence we get $\langle \Delta_\nu \cap {}^\gamma\mathbb{P} \mid \nu < \gamma\rangle$, $\langle \Delta_\nu \cap {}^\gamma\mathbb{P} \mid \nu < \gamma\rangle$ is uniformly $\Sigma_1(L)$ in γ and $\langle \Delta_\nu \cap {}^\gamma\mathbb{P} \mid \nu < \gamma\rangle \in \mathfrak{A}_\gamma^1$.

We define $p_i^\nu = p_i \in {}^\nu\mathbb{P}^*$ by induction on $i < \nu$ as follows:

$p_o = 0$.

Assume $p_i \in {}^\nu\mathbb{P}^*$ is already defined. By definition $p_i \in {}^\nu\mathbb{P}^*$ implies that $p_i \in {}^\nu\mathbb{P}$ and $\mathrm{dom}(p_i) = [\omega, \alpha_{p_i}]$ for $\alpha_{p_i} < \nu$. Set $\alpha_i = \alpha_{p_i}$. Assume that $i \le \alpha_i < \nu$.

For $\gamma \in [\mathrm{card}(i)^+, \alpha_i]$ define,

$X^\gamma_{i\nu} = X^\nu_i$ = the least $X \prec \mathfrak{A}^1_\nu$ s.t. $\gamma \subseteq X$ and $\langle \Delta_\alpha \mid \alpha < \nu \rangle$, $p_j, X_j \in X$ for $j < i$.

Note that $X^\gamma_\lambda = \bigcup_{i<\lambda} X_i$ for $\lim(\lambda)$.

For $\gamma \in [\mathrm{card}(i)^+, \alpha_i)$ define,

$$f_{i\nu}(\gamma) = f_i(\gamma) = \begin{cases} X^\gamma_i \cap \mathfrak{A}^1_{(p_i)_{\gamma^+}} & \text{if } \gamma^+ \in X^\gamma_i \\ \text{undefined otherwise.} \end{cases}$$

Let $\beta = \mathrm{card}(i)^+$. Define f^*_i by $\mathrm{dom}(f^*_i) = \mathrm{dom}(f_i)$ and for $\gamma \in \mathrm{dom}(f_i)$, replace each $\Delta \in f_i(\gamma)$ s.t. Δ is dense in $\mathbf{P}^{p_{i\gamma^+}}$ by $\Delta^* = \{q \in \mathbf{P}_\beta^{p_{i\gamma+}} \mid \Delta^q \text{ is dense in } \mathbf{P}^{q_\beta}\}$.

q_{i+1} = the $\mathfrak{A}^1_\nu$-least $q \in {}^\nu\mathbf{P}^*_\beta$ s.t.

(1) $q \restriction \alpha^+_i \leq (p_i)_\beta$.

(2) $q \restriction \alpha^+_i \in \sum^{\alpha^+_i, p_i}_{f^*_i}$.

(3) $|q \restriction \gamma| = |q_\gamma|$ for limit cardinals $\gamma \in \mathrm{dom}(q)$.

(4) $\alpha_q \geq \sup(X^{\alpha_i}_i \cap \mathrm{On})$.

(5) $\Delta^{q_\beta}_j$ is dense in $\mathbf{P}^{(q)_\beta}$ for $j \leq i$.

The above clause (5) can be satisfied by Lemma 5.2.

Finally set:

$p_{i+1} = p_i \cup q_{i+1}$.

For $\lim(\lambda)$ set $\alpha_\lambda = \sup_{i<\lambda}(\alpha_i)$ and define p_λ on $\mathrm{card} \cap [\omega, \alpha_\lambda]$ by:

$p_\lambda(\gamma) = \langle \bigcup_{i<\lambda} \dot{p}_{i\gamma}, \bigcup_{i<\lambda} p_{i\gamma} \rangle$ for $\gamma < \alpha_\lambda$ and $p_\lambda(\alpha_\lambda) = \langle 0,0 \rangle$.

<u>Claim 1</u> $p_i \in {}^\nu\mathbf{P}^*$ for $i < \nu$ and $\bigcup_{i<\nu} p_i \in \mathbf{P}^\nu$.

The proof of the claim is a combination of techniques used in Lemma 3.25 and Theorem 3.3 (proved in §3.7). The proof will be sketched referring to the corresponding lemmas of Lemma 3.13. As in Theorem 3.3 $\sup(X_\lambda^\gamma \cap On) = \alpha_\lambda$ for $\lim(\lambda)$ and $\gamma \in card \cap \alpha_\lambda^+$. For $\lim(\lambda)$ set:

$$\lambda^* = \sup_{i<\lambda}(card(i)^+) .$$

We must show $p_\lambda \in {}^\nu P^*$ for $\lim(\lambda)$.

Lemma (corresponding to 3.14) *The same for* $\lambda^* \le \delta \le \gamma \le \beta < \alpha_\lambda$. *Note that* $\lambda^* = \alpha_\lambda$ *on a cub set in* ν.

Lemmas 3.15 and 3.17 *are meaningless.*

Definition For $\gamma \in [\lambda^*, \alpha_\lambda)$ define $\sigma = \sigma_\lambda^\gamma$, $b = b_\lambda^\gamma$ by $\sigma\colon b \cong X_\lambda^\gamma$. Set $\mu' = On \cap b$, $p_i' = \sigma^{-1}(p_i)$ etc. (there is no A to code).

Lemma 3.16 *Same for* $\lambda^* \le \gamma \le \eta \le \delta < \mu'$.

Lemmas 3.18-3.23 *As in Theorem 3.3 for* $\lambda^* \le \gamma < \alpha_\lambda$.

Lemma 3.24 *For* $\gamma < \lambda^*$ *and* $\gamma = \lambda = \lambda^* < \alpha_\lambda$ *as in Lemma 3.5. For* $\gamma = \alpha_\lambda$ *as in Theorem 3.3.*

To show $\bigcup_{i<\nu} p_i \in \mathbf{P}^\nu$ is trivial since for $\gamma < \nu$ $\exists i(p^\nu/\gamma = p_i/\gamma \in \mathbf{P}^{p_{i\nu}})$ and $\langle p_i \mid i < \nu\rangle$ was defined in $\mathfrak{A}_\nu^1$.

Q.E.D. (Claim)

Define $p^\nu = \bigcup_{i<\nu} p_i \in \mathbf{P}^\nu$. $\mathfrak{A}_\nu^1 = L_{\mu_\nu'}$ where μ_ν' = the least $\mu > \nu$ s.t. $L_\mu \models ZF^-$ and $\mathfrak{A}_\nu^1$ is uniformly $\Sigma_1(L)$ in ν. Hence it is clear that $p^\nu \in \mathbf{P}^\nu$ uniformly $\Sigma_1(L)$

in ν . $\pi_{\nu\tau}(p^\nu) = p^\tau$ for $\nu, \tau \in I$ $\nu < \tau$ hence $p^\nu = \pi_{\nu\tau}(p^\nu)/\nu = p^\tau/\nu$.

Set $p^{o\nu} = p^\nu \cup \{\nu, \langle 0,0\rangle\} \in \mathbb{P}^{\nu^+}$ and $s^{o\nu} = p^{o\tau}_{\nu^+}$ for $\nu < \tau \in I$. Then $|p^{o\nu}/\nu| = |p^{o\nu}_\nu| = |\emptyset_\nu| = \nu$ and $p^{o\nu}$ and $s^{o\nu}$ satisfy (a)-(d) of §4.4. We can repeat the definition on p. 148 starting from $p^{o\nu}$ and $s^{o\nu}$. All the theorems of §4.4 go through. If D is defined as in §4.4 we can prove:

<u>Claim 2</u> D is $\mathbb{P}$-generic over L .

Proof Let $\Delta \subseteq \mathbb{P}$ be dense in $\mathbb{P}$ and L-definable. Then there is an i s.t. $\Delta = \Delta_i$. By the construction $\Delta_i^{(p_{i+1})_\beta}$ is dense in $\mathbb{P}^{p_{i+1},\beta}$ where $\beta = \text{card}(i)^+$. By Lemma 4.14 $\exists p \in G^j_\beta$, $j < \omega$ s.t. $p \in \Delta_o^{(p_{i+1})_\beta}$. $p_{i+1} \in G_D$ hence $p \cup p_{i+1} \in \Delta_i \cap G_D$.

Q.E.D.

Definition 5 $p \in \mathbb{P}^\infty$ iff p is a class and p: card $\longrightarrow V \times V$ s.t. $\forall \alpha \in \text{card}(p/\alpha \in \mathbb{P}^{p_\alpha})$.

Lemma 5.4 *We can construct a* $\mathbb{P}$*-pseudo generic* D *with all the properties of Lemmas 4.15 and 4.16 s.t.* D *is <u>not</u>* $\mathbb{P}$*-generic over* L .

Proof Define $p^* \in \mathbb{P}^\infty$ as follows:

$p^*(\alpha) = \langle 0,0\rangle$ for limit cardinals.

$p^*(\beta^+) = \langle 0, \{<\beta^+, 0>\}\rangle$ for $\beta \in$ card .

Clearly $p^*/\beta \in \mathbb{P}^{p^*_\beta}$ for $\beta \in$ card by the argument of Lemma 2.9. p^* is an L-definable class. Define $p^{o\nu} = p^*/\nu^+$ and $s^{o\nu} = p^*_{\nu^+}$. $p^{o\nu}$ and $s^{o\nu}$ satisfy (a)-(d) of §4.4 and it is possible to continue the construction of §4.4 getting a $\mathbb{P}$-pseudo generic D satisfying the Theorems of §4.4.

Set $\Delta = \{q \in \mathbb{P} \mid q/p^* \wedge \sup(\mathrm{dom}(q))^+\}$. Δ is definable in L. Every $p \in G_D$ is compatible with p^*/α for every $\alpha \in$ card. Hence $G_D \cap \Delta = 0$ and G_D is not $\mathbb{P}$-generic over L.

Q.E.D.

Definition 6 Let $p \in \mathbb{P}^\infty$. Define:

$$(\mathbb{P}/p) = \{q \in \mathbb{P} \mid q \leq p \wedge \sup(\mathrm{dom}(q))^+\} .$$

$(\mathbb{P}/p)$ is definable in $\langle L,p\rangle$ and $\langle L,p\rangle$ is amenable. All previous notation is carried over to $(\mathbb{P}/p)$.

Lemma 5.5 *Let* $p \in \mathbf{P}^\infty$. *Then* $(\mathbf{P}/p)_\tau$ *is* τ*-distributive over* $\langle L,p\rangle$.

Proof The proof closely follows that of Theorem 3.3 (in §3.7). Set $\mathbf{Q} = (\mathbf{P}/p)$. We will prove that the dense subclass $\mathbf{Q}^*_\tau$ is τ-distributive. Let $\langle\Delta_i \mid i<\tau\rangle$ be a sequence of dense classes in $\mathbb{Q}^*_\tau$ s.t. $\{\langle i,q\rangle \mid q \in \Delta_i\}$ is $\Sigma_m(\langle L,p\rangle)$ in the parameter e. We claim $\bigcap_{i<\tau} \Delta_i$ is dense in $\mathbb{Q}^*_\tau$.

Let $p^o \in \mathbb{Q}^*_\tau$. Then $p^o \in \mathbf{P}_\tau^{\alpha^+_{p_o}}$ and $p^o \leq p/\alpha^+_{p_o}$. By the extension lemma (2.14) we may assume w.l.o.g. that $|p_o/\gamma| = |p_{o\gamma}|$ for all limit cardinals $\gamma \in \mathrm{dom}(p_o)$, since any $p'_o \leq p_o$, $p'_o \in \mathbb{P}_\tau^{\alpha^+_{p_o}}$ also satisfies $p'_o \leq p/\alpha^+_{p'_o}$ and $p'_o \in \mathbf{Q}^*_\tau$ $(\alpha_{p_o} = \alpha_{p'_o})$. Note that lemma 2.14 about $\mathbb{P}^{\alpha'_{p_o}}$ is used to get an extention in $(\mathbb{P}/p)^{\alpha^+_{p_o}} = \mathbb{Q}^{\alpha^+_{p_o}}$. Let $n > m$ be "sufficiently large". We construct $p_i \in \mathbf{Q}^*_\tau$ for $i \leq \tau$ as follows:

p_o is given.

Y_i^γ = the least $Y \prec_{\Sigma_n} \langle L,p\rangle$ s.t. $\gamma \subseteq Y$ and $e, p_j, Y_j^\gamma \in Y$ for $j < i$.

Define f_i as in Theorem 3.3. Set $\alpha_{p_i} = \alpha_i$.

p_{i+1} = the $\langle L,p\rangle$-least $q \in \mathbb{Q}_\tau^*$ s.t.

(1) $q \restriction \alpha_i^+ \le p_i$.

(2) $q \restriction \alpha_i^+ \in \sum_{f_i}^{\alpha_i^+, p_i}$.

(3) Let κ be inaccessible s.t. $\sup(Y_i^{\alpha_i} \cap \mathrm{On}) < \kappa$. Since $p \restriction \kappa \in \mathbf{P}^{p_\kappa}$, there is a $C \in \mathfrak{U}_{p_\kappa}$ s.t. C is cub in κ and $\gamma \in C \longrightarrow \gamma$ is a limit cardinal, $\dot{p}_\gamma = 0$ and $|\mathbf{P} \restriction \gamma| = |\mathbf{P}_\gamma|$. Let $\gamma_i = \min(C \setminus \sup(Y_i^{\alpha_i} \cap \mathrm{On}))$. Then we require $q \restriction \gamma_i^+ \le p_i \cup p \restriction \gamma_i^+$ and $|q \restriction \gamma| = |q_\gamma|$ for limit cardinals $\gamma \in [\tau, \gamma_i]$.

(4) $q \in \Delta_i$.

We must show that clauses (1)-(4) are dense in $\mathbb{Q}_\tau^*$. First take a $q' \in \mathbf{P}_\tau^{\alpha_i^+}$ s.t. $q' \le p_i$ and $q' \in \sum_{f_i}^{\alpha_i^+, p_i}$. Since $q' \le p_i \le p \restriction \alpha_i^+$ we get $q' \in \mathbb{Q}_\tau^*$. Note that we are using the fact that $\sum_{f_i}^{\alpha_i^+, p_i}$ is dense in $\mathbf{P}_\tau^{\alpha_i^+}$.

Next take $q'' \le q' \cup p \restriction \gamma_i$, $q'' \in \mathbb{P}_\tau^{p_{\gamma_i}}$ s.t. $|q''| = |p_{\gamma_i}|$ and $|q'' \restriction \gamma| = |q''_\gamma|$ for limit cardinals $\gamma \in [\tau, \gamma_i]$. Here the collapsing lemma (2.12) is used in $\mathbb{P}_\tau^{p_{\gamma_i}}$. Next set $q''' = q'' \cup \{\langle \gamma_i, \langle 0, p_{\gamma_i}\rangle\rangle\}$. Clearly $q''' \in \mathbf{Q}_\tau^*$.

Finally take, $q \le q'''$ and $q \in \Delta_i$ ($\Delta_i \subseteq \mathbb{Q}_\tau^*$). For $\lim(\lambda)$ define p_λ as in Theorem 3.3.

<u>Claim</u> $p_i \in \mathbf{Q}_\tau^*$ for all $i \le \tau$.

Proof Almost exactly as in Theorem 3.3. The difference being that the p_{i+1} defined in the present lemma are such that $|p_{i+1}\wedge\gamma| = |p_{i+1\gamma}|$ for all limit $\gamma \in [\tau,\gamma_i]$ but it may be that $\gamma_i < \sup(\mathrm{dom}(p_{i+1}))$. For $\lim(\lambda)$ it must be shown that for γ a limit cardinal, $\gamma \in [\tau,\alpha_\lambda]$, $|p_\lambda\wedge\gamma| = |p_{\lambda\gamma}|$. This is accomplished by:

$$|p_\lambda\wedge\gamma| = \sup_{i_o\le i<\lambda}(|p_i\wedge\gamma|) = \sup_{i_o\le i<\lambda}(|p_{i\gamma}|) = |p_{\lambda\gamma}|, \text{ where}$$

i_o = the least i s.t. $\alpha_{p_i} \ge \gamma$.

Q.E.D. (Lemma 5.5)

Corollary 5.5.1 *For inaccessible* κ, ${}^\kappa(\mathbb{P}/p)_\tau$ *is* τ-*distributive over* L.

Proof The same as above.

Lemma 5.6 *Let* κ *be inaccessible. Let* $\langle \Delta_\nu \mid \nu < \gamma^+ < \kappa\rangle$ *(for* $\tau \le \gamma$*) be an* L_κ*-definable sequence of* ${}^\kappa(\mathbb{P}/p)_\tau$ *dense classes over* L_κ. *Then*

$$\Delta = \{q \in {}^\kappa(\mathbb{P}/p)_{\gamma^+} \mid \forall\nu<\gamma^+\ \Delta^q \text{ is dense in } \mathbf{P}_\tau^{q_{\gamma^+}}\}$$

is dense in ${}^\kappa(\mathbf{P}/p)_{\gamma^+}$.

Proof The same as Lemma 5.2 (which in turn is similar to Lemma 3.4). Corollary 5.5.1 is used to get that ${}^\kappa(\mathbf{P}/p)_{\gamma^+}$ is γ^+-distributive.

Q.E.D.

Theorem 5.7 *For any* $\mathbb{P}$*-pseudo generic* D *over* L *there is a* $p \in \mathbf{P}^\infty$ *s.t.* D *is* $(\mathbf{P}/p)$*-generic over* $\langle L,p\rangle$.

Proof D is pseudo generic implies that I is the class of indiscernibles for $L[D]$ and $D \cap \alpha^+$ is $\mathbb{P}^{\alpha^+}$-generic over $\mathfrak{A}^1_{\alpha^+}$ for $\alpha \in$ card. Let $\kappa \in I$. Let $\dot{G}$ be the canonical predicate for a generic over $\mathbb{P}^{\kappa^+}$. It is clear that:

$$\Vdash_{\mathbb{P}^{\kappa^+}} \alpha \in \check{\kappa} \cap \widecheck{\mathrm{card}}(\mathring{G} \cap \mathbb{P}^{\check{\alpha}^+} \text{ is } \check{\mathbb{P}}^{\alpha^+}\text{-generic over } \check{\mathfrak{A}}^1_{\alpha^+}) .$$

Set:

$\Delta_\kappa = \{p \in \mathbb{P}^{\kappa^+} \mid \exists q \in \mathbb{P}^{p_\kappa} (p \Vdash_{\mathbb{P}^{\kappa^+}} \mathring{G} \cap \check{L}_\kappa$ is ${}^\kappa(\check{\mathbb{P}}/p)$-generic over $\langle L_\kappa, q\rangle \wedge p/\kappa \le q$ in $\mathbb{P}^{p_\kappa}\}$.

<u>Claim</u> Δ_κ is dense in $\mathbb{P}^{\kappa^+}$.

Proof Let $q \in \mathbb{P}^{\kappa^+}$. We may assume w.l.o.g. that $|q/\kappa| = |q_\kappa|$. Let $n < \omega$ be the least n s.t. $q/\kappa \in \mathfrak{A}^n_{|q_\kappa|}$. We wish to construct a $p \le q/\kappa$, $p \in \mathbb{P}^{q_\kappa}$ s.t. $\mathfrak{A}^n_{|q_\kappa|} \models$ "$p \cup q \Vdash \mathring{G} \cap \check{L}_\kappa$ is ${}^\kappa(\mathbb{P}/\check{q}/\kappa)$-generic over $\langle \check{L}, \check{q}/\check{\kappa}\rangle$".

Set $\mathbf{Q} = {}^\kappa(\mathbb{P}/q/\kappa)$ and $\mathfrak{A} = \mathfrak{A}^n_{|q_\kappa|}$.

Let $\langle \Delta_i \mid i < \kappa\rangle \in \mathfrak{A}$ be the canonical enumeration of the $\langle L_\kappa, q/\kappa\rangle$-definable dense classes in $\mathbf{Q}$ (defined as in Lemma 5.3). We follow the proof of Lemma 5.3 closely. We define $p_i \in \mathbf{Q}^*$ by induction on $i < \kappa$.

$p_o = 0$.

Assume p_i is defined, $\alpha_{p_i} = \alpha_i$, $\mathrm{card}(i) \le \alpha_i < \kappa$. Set $\beta = \mathrm{card}(i)^+$. For $\gamma \in [\beta, \alpha_i]$ define

X_i^γ = the least $X \prec \mathfrak{A}$ s.t. $\gamma \subseteq X$ and $\langle \Delta_\alpha \mid \alpha < \kappa\rangle, X_j^\gamma, p_j \in X$ for $j < i$. Define f_i and f_i^* as in Lemma 5.3.

q_{i+1} = the -least $h \in \mathbf{Q}^*_\beta$ s.t.

(1) $h/\alpha_i^+ \le (p_i)_\beta$.

(2) $h/\alpha_i^+ \in \sum_{f_i^*}^{\alpha_i^+, p_i}$.

(3) Let $C = \mathfrak{A}$-least $C \in \mathfrak{A}$ s.t. $C \subseteq \kappa$ is cub in κ, $\gamma \in C \longrightarrow \gamma$ is a limit cardinal, $\dot{q}_\gamma = 0$ and $|q/\gamma| = |q_\gamma|$. Set $\gamma_i = \min(C \setminus \sup(X_i^{\alpha_i} \cap On))$. Then $h/\gamma_i^+ \leq p_i \cup q/\gamma_i^+$ and $|h/\gamma| = |h_\gamma|$ for limit cardinals $\gamma \in (\beta, \gamma_i]$.

(4) Δ_j^h is dense in $\mathbf{Q}^h$ for $j \leq i$.

We must show that clauses (1)-(4) are dense in $\mathbf{Q}^*$. (1)-(3) are the same as in Lemma 5.5 and (4) follows from Lemma 5.6.

$p_{i+1} = p_i \cup q_{i+1}$.

Define p_λ for $\lim(\lambda)$ as in Lemma 5.3.

The proof that $p_i \in \mathbf{Q}^*$ for $i < \kappa$ is exactly the same as Lemma 5.3. $\bigcup_{i<\kappa} p_i \in \mathfrak{A}^n_{|q_\kappa|}$ since the whole definition of $\langle p_i \mid i < \kappa \rangle$ was carried out in $\mathfrak{A}^n_{|q_\kappa|}$.

Clearly $p = \bigcup_{i<\kappa} p_i \leq q/\kappa$ and $p \in \mathbf{P}^{q_\kappa}$.

$p \cup q \Vdash \mathring{G} \cap \check{L}$ is $\check{\mathbf{Q}}$-generic over $\langle \widetilde{L, q/\kappa} \rangle$ by the construction of p. Hence Δ_κ is dense in $\mathbf{P}^{\kappa^+}$ for each $\kappa \in I$. Δ_κ is uniformly $\Sigma_1(L)$ definable in κ.

Choose p_κ = the L[D]-least $p \in \Delta_\kappa \cap G$ and q_κ = L[D]-least $q \in \mathbf{P}^{p_\kappa}$ s.t. $p/\kappa \leq q$ and $p_\kappa \Vdash \mathring{G} \cap \check{L}_\kappa$ is $\check{\mathbf{Q}}$-generic over $\langle \widetilde{L_\kappa, q} \rangle$.

Then p_κ and q_κ are uniformly $\Sigma_1(L)$ in κ. $\pi_{\bar{\kappa}\kappa}(p_{\bar{\kappa}}) = p_\kappa$, $\pi_{\bar{\kappa}\kappa}(q_{\bar{\kappa}}) = q_\kappa$ for $\bar{\kappa} < \kappa$, $\bar{\kappa}, \bar{\kappa} \in I$. Hence $q_{\bar{\kappa}} = q_\kappa/\bar{\kappa}$ and $p_{\bar{\kappa}}/\bar{\kappa} = p_\kappa/\bar{\kappa}$ for $\bar{\kappa} < \kappa$, $\bar{\kappa}, \kappa \in I$.

Set $q = \bigcup_{\kappa \in I} q_\kappa \in \mathbf{P}^\infty$.

Since $p_\kappa \in G$ for all $\kappa \in I$ it is clear that G is

$\mathbb{P}$/q-generic over $\langle L,q\rangle$.

Q.E.D.

We complete this section with some discussion about the conjecture mentioned at the start. A more elementary question, but one that hints to a solution of the conjecture, is:

<u>Open Question</u>: Let $a \subseteq \omega$, $L[a] \models 0^{\#}$ does not exist. Is there a $b \in L[a]$, $a \notin L[b]$ and a is set generic over $L[b]$?

Presumably $b \subseteq \omega_1$ ($\subseteq \omega_2$?) and if we iterate the process could we get a backwards coding that finally produces an L-definable class for which a is generic? Earlier we mentioned that if G was weak class generic over L it was open whether

$$\langle L[G,\mathbb{P}],\mathbb{P},G\rangle \models \neg 0^{\#} .$$

We can connect this with the open question by starting with $L[0^{\#}]$ and defining a class $\mathbb{P}$ s.t. $\langle L[\mathbb{P}],\mathbb{P}\rangle \models \neg 0^{\#}$ but $0^{\#}$ is $\mathbb{P}$-generic over $\langle L[\mathbb{P}],\mathbb{P}\rangle$.

5.2 DESTROYING COUNTABLE MODELS OF ZF

In this section we shall prove:

Theorem 5.8 *Let* $\beta < \omega_1$ *be s.t.* $L_\beta \models ZF$. *Then there is an* $a \subseteq \omega$ *s.t.* $L_\beta[a] \models ZF$ *and* $\forall \alpha < \beta\ (L_\alpha[a] \not\models ZF)$.

The problem had previously been solved for ZF^- and admissible models [SA] and is problem 51 of Friedman [F]. Our strategy will be to start with $V = L_\beta$; perform a few class generic extensions to destroy ZF models for cardinals in L_β and then using the coding method, with a small change, to code the world down to a real and destroy the rest of the ZF models on the way. The proof will stretch out over several subsections.

5.2.1 Avoiding Inaccessibles

Let $\beta < \omega_1$ and $L_\beta \models ZF$. Working in L_β, we set:

$$C = \{\alpha < \beta \mid L_\beta \models \text{" } \alpha \text{ is inaccessible"}\} .$$

C is an L_β-definable class. We define a class of conditions $\mathbb{P}_1$ as follows: $p \in \mathbb{P}_1$ iff $p: \text{dom}(p) \longrightarrow \beta$ s.t.

(1) $\text{dom}(p) = \alpha + 1$ for some $\alpha < \beta$.

(2) p is a normal function.

(3) $\text{rng}(p) \subseteq (\text{card})^{L_\beta}$ and $\text{rng}(p) \cap C = 0$.

$\mathbb{P}_1$ is an L_β-definable class.

Claim 1 Let $\text{dom}(p) < \delta < \beta$, $p \in \mathbb{P}_1$. Then there is a $q \leq p$, $q \in \mathbb{P}_1$ s.t. $\text{dom}(q) = \delta + 1$.

Proof Let $\text{dom}(p) = \alpha + 1$. Define q by:

$$q(\alpha+1) = \text{card}(\delta)^{+L_\beta} .$$

$$q(\eta+1) = q(\eta)^{+L_\beta} \quad \text{for} \quad \eta \in [\alpha+2, \delta) .$$

$$q(\lambda) = \sup_{i<\lambda}(q(i)) \quad \text{for} \quad \lim(\lambda) , \lambda \in (\alpha, \delta] .$$

For $\lim(\lambda)$, $q(\lambda) \notin C$ since $\lambda < \gamma < \gamma^+ < q(\lambda)$. Clearly $q \in L_\beta$.

Q.E.D. (Claim 1)

Claim 2 $\mathbb{P}_1$ is α-distributive over L_β for every $\alpha < \beta$. Let $\langle \Delta_i \mid i < \alpha \rangle$ be an L_β-definable sequence of dense classes in $\mathbb{P}_1$. We must show that $\bigcap_{i<\alpha} \Delta_i$ is dense in $\mathbb{P}_1$. Let

$p_o \in \mathbb{P}_1$, $\mathrm{dom}(p_o) = \delta + 1$. Define p_i , $i \le \alpha$ as follows:

p_o - given.

p_1 = the L_β-least $p \le p_o$ s.t. $p(\delta) \ge (\mathrm{card}(\alpha))^+$ and $p \in \Delta_i$.

p_{i+1} = the L_β-least $p \le p_i$ s.t. $p \in \Delta_i$ for $i > 0$.

For $\lim(\lambda)$ set $\delta = \sup_{i<\lambda}(\mathrm{dom}(p_i))$ and $\gamma = \sup_{i<\lambda}(\max(\mathrm{rng}(p_i)))$.

Set: $p_\lambda = \bigcup_{i<\lambda} p_i \cup \{\langle \delta, \gamma \rangle\}$.

$\gamma > \mathrm{card}(\alpha)^+ > \lambda$ and $\langle \max(\mathrm{rng}(p_i)) \mid i < \lambda \rangle \in L_\beta$. Hence

$\gamma \notin C$ and $p_\lambda \in \mathbb{P}_1$. Clearly $p_\alpha \in \bigcap_{i<\alpha} \Delta_i$.

Q.E.D.

Claim 2 implies that any generic extension of L_β by $\mathbb{P}_1$ does not add any sets and satisfies the replacement axiom. Let G be $\mathbb{P}_1$-generic over L_β . Set

$$A_G = A_1 = \{\alpha \mid \exists p \in G(\alpha \in \mathrm{rng}(p))\} .$$

Then G is definable in $\langle L_\beta, A_1 \rangle$ and A_1 is a closed and unbounded class of cardinals s.t. $A_1 \cap C = 0$.

5.2.2 Destroying Inaccessibles

Set $N_1 = \langle L_\beta, A_1 \rangle \models ZF$. Now we work in N_1 . Let $\langle \alpha_i \mid i < \beta \rangle$ be the N_1 definable enumeration of A_1 . Define a class of conditions $\mathbb{P}_2$ by:

$p \in \mathbb{P}_2$ iff $p\colon \mathrm{On} \times \mathrm{On} \longrightarrow \mathrm{On}$ s.t.

(1) $p(0,\delta) = \gamma$ iff $\delta < \omega$ and $\gamma < \alpha_o$.

(2) $p(i+1,\delta) = \gamma$ iff $\delta < \alpha_i^+$ and $\gamma < \alpha_{i+1}$ for $i < \beta$.

(3) $\mathrm{card}(\mathrm{dom}(p) \cap (i+2) \times \alpha_i^+) < \alpha_i^+$.

(4) $\mathrm{dom}(p) \cap \{1\} \times \omega$ is finite.

$\mathbb{P}_2$ are the standard class of Easton conditions which collapse α_{i+1} to α_i^+ for $i < \beta$ and α_o to ω. Let G be $\mathbb{P}_2$-generic over N_1. Then $\langle L_\beta[G], A_1, G\rangle \models$ ZF+GCH+ " α is a cardinal iff $\alpha = \omega$, $\exists i(\alpha = \alpha_i^+)$ or α is a limit point of A_1" $N_1 \models$ "α is singular" $\longrightarrow \langle L_\beta[G], A_1, G\rangle \models$ " α is singular". Hence $\langle L_\beta[G], A_1, G\rangle \models$ "there are no inaccessibles". Set:

$$A^* = \{ \{\{i,\delta\},\gamma\} \mid \exists p \in G(p(i,\delta) = \gamma)\} ,$$

$$A_\alpha^o = \{\alpha + \nu \mid \nu \in A_1 \cap \alpha\} , \quad A_\alpha^1 = \{\alpha + \nu \mid \nu \in A^* \cap [\alpha, \alpha^+)\} ,$$

$$A_\alpha^* = A_\alpha^o \cup A_\alpha^1 \quad \text{and} \quad A_2 = \bigcup_{\alpha \in \text{card}} A_\alpha^* .$$

$L_\beta[A_2] = L_\beta[A_1, G]$ and A_1 is recursive in A_2. Set $N_2 = \langle L_\beta[A_2], A_2 \rangle$. $N_2 \models$ ZF+GCH+ "there are no inaccessibles"

5.2.3 Eliminating Singular Cardinal Models of ZF

We now work in N_2. Define $\mathbb{P}_3$ by:

$$p \in \mathbb{P}_3 \quad \text{iff} \quad p = \langle \alpha, x \rangle$$

where α is a singular cardinal, $x \in N_2$, $x \subseteq \alpha$ and $\forall \delta \leq \alpha$ (δ is a singular cardinal $\longrightarrow \langle L_\delta[A_2 \cap \delta, x \cap \delta] A_2 \cap \delta, x \cap \delta\rangle \not\models$ ZF). $\langle \alpha, x\rangle \leq \langle \gamma, y\rangle$ iff $\alpha \geq \gamma$ and $x \cap \gamma \supseteq y$.

<u>Claim 1</u> Let $\langle \alpha, x\rangle \in \mathbb{P}_3$ and $\delta > \alpha$ s.t. δ is a singular cardinal. Then there is a $\langle \delta, y\rangle \in \mathbb{P}_3$, $\langle \delta, y\rangle \leq \langle \alpha, x\rangle$.

Proof Let $\langle \delta_i \mid i < \beta\rangle \in N_2$ be an enumeration of the singular cardinals greater than α. We prove the claim by induction on i.

<u>Case 1</u> $\exists i(\delta = \delta_{i+1})$: By the induction hypothesis we may assume $\alpha = \delta_i$. Let $\text{cf}^{N_2}(\delta_{i+1}) = \gamma < \delta_{i+1}$. Let $D \subseteq [\delta_i, \delta_i + \gamma)$ code a cofinal $f: \gamma \longrightarrow \delta_{i+1}$. $\langle \delta_{i+1}, x \cup \chi_D\rangle \in \mathbb{P}_3$ and $\langle \delta_{i+1}, x \cup \chi_D\rangle \leq \langle \alpha, x\rangle$.

Case 2 $\delta = \delta_\lambda$ for $\lim(\lambda)$: Let $cf^{N_2}(\delta_\lambda) = \eta < \delta_\lambda$. Let $\langle \gamma_i \mid i < \eta \rangle \in N_2$ be s.t. $\sup_{i<\eta}(\gamma_i) = \delta_\lambda$ and $\gamma_o = \alpha$ and each γ_i is a singular cardinal. Choose $\langle \gamma_i, x_i \rangle \leq \langle \alpha, x \rangle$, $\langle \gamma_i, x_i \rangle \in \mathbb{P}_3$ by the induction hypothesis. Define $y \subseteq \delta_\lambda$ by:
$y \cap \alpha = x$, $y \cap Z_o \cap [\alpha, \gamma_i) = x_i$ for $i < \eta$ and
$y \cap Z_1 \cap [\alpha, \gamma_i) = \{\{\gamma_j, 1\} \mid j < i\}$ (where $\{\ ,\ \}$ is the Gödel pairing function defined in Chapter 1). ($Z_i = \{\{\alpha, i\} \mid \alpha \in On\}$.)

$\forall i < \lambda$ $(\langle L_{\delta_i}[A_2 \cap \delta_i , y \cap \delta_i], A_2 \cap \delta_i , y \cap \delta_i \rangle \models ZF$

since $y \cap \delta_i \cap Z_o = x_j \cap \delta_i$ for a j s.t. $\gamma_j \geq \delta_i$.

$\langle L_{\delta_\lambda}[A_2 \cap \delta_\lambda, y] A_2 \cap \delta_\lambda, y \rangle \models ZF$

since $y \cap Z_1 \cap [\alpha, \delta_\lambda) = \{\{\gamma_i, 1\} \mid i < \lambda\}$ and thus the sequence $\langle \gamma_i \mid i < \lambda \rangle$ is $\langle L_{\delta_\lambda}[A_2 \cap \delta_\lambda, y], A_2 \cap \delta_\lambda, y \rangle$ definable.

Q.E.D. (Claim 1)

Claim 2 $\forall \alpha < \beta$ $\mathbb{P}_3$ is α-distributive over N_2 .

Proof Let $\{\langle p, i \rangle \mid p \in \Delta_i , i < \alpha\}$ be $\Sigma_n(N_2)$ definable in a parameter $e \in L_\eta[A_2 \cap \eta]$ where $\eta > \alpha$ and each Δ_i is dense in $\mathbb{P}_3$. Let $p_o = \langle \theta, x_o \rangle \in \mathbb{P}_3$ and set $\delta = \max\{\theta, \eta\}$. We must find a $p \leq p_o$ s.t. $p \in \bigcap_{i<\alpha} \Delta_i$. Let $m > n$ be "large enough". Define γ_i by induction on $i \leq \alpha$ as follows:

γ_o = the N_2-least $\gamma > \delta$ s.t. γ is a limit cardinal and

$$\langle L_\gamma[A_2 \cap \gamma], A_2 \cap \gamma \rangle \prec_{\Sigma_m} N_2 ,$$

γ_{i+1} = the N_2-least $\gamma > \gamma_i$ s.t. γ is a limit cardinal and

$$\langle L_\gamma[A_2 \cap \gamma], A_2 \cap \gamma \rangle \prec_{\Sigma_m} N_2 \quad \text{and}$$

$\gamma_\lambda = \sup_{i<\lambda}(\gamma_i)$ for $\lim(\lambda)$ (all limit cardinals are singular in N_2).

Define p_i by induction on $i \le \alpha$ as follows:

p_o is given.

p_{i+1} = the N_2-least $p \le p_i$ s.t. $p \le \langle \gamma_i, x_i \rangle$ and $p \in \Delta_i$.

$p_\lambda = \bigcup_{i<\lambda} p_i$ for $\lim(\lambda)$.

We claim that $p_\lambda \in \mathbf{P}_3$ for $\lim(\lambda)$. Set $x_\lambda = x = \bigcup_{i<\lambda} x_i$ then $p_\lambda = \langle \gamma_\lambda, x \rangle$. For δ a singular cardinal, $\delta < \gamma_\lambda$ $\langle L_\delta[A_2 \cap \delta, x \cap \delta], A_2 \cap \delta, x \cap \delta \rangle = \langle L_\delta[A_2 \cap \delta, x_i \cap \delta], A_2 \cap \delta, x_i \cap \delta \rangle \not\models ZF$ for an i s.t. $\delta < \gamma_i$. The sequence $\langle \gamma_i \mid i < \lambda \rangle$ is definable in $\langle L_{\gamma_\lambda}[A_2 \cap \gamma], A_2 \cap \gamma \rangle$ by the same definition that $\langle \gamma_i \mid i < \alpha \rangle$ is N_2 definable. Note that

$$p_{i+1} \in \langle L_{\gamma_{i+1}}[A_2 \cap \gamma_{i+1}], A_2 \cap \gamma_{i+1} \rangle .$$

Hence $\langle L_{\gamma_\lambda}[A_2 \cap \gamma_\lambda], A_2 \cap \gamma_\lambda \rangle \not\models ZF$.

Q.E.D. (Claim 2)

Now let G be $\mathbf{P}_3$-generic over N_2 . By Claim 2, G does not add any sets.
Set $A_3^* = \{\alpha \mid \exists \gamma(\langle \gamma, x \rangle \in G \wedge \alpha \in x)\}$. Let A code A_3^* and A_2 . Set $N_3 = \langle L_\beta[A], A \rangle$. Then
$N_3 \models$ ZF+GCH+ $\neg 0^\#$+ "There are no inaccessible cardinals" + "for every singular cardinal α , $\langle L_\alpha[A], A \cap \alpha \rangle \not\models ZF$" .

5.2.4 Purging the Rest of the ZF Models

Now we will work in N_3 and use the coding conditions to code the class A by a subset $a \subseteq \omega$. $\mathbf{S}_\alpha$ is changed so that it kills ZF models in $[\alpha, \alpha^+)$.

5.2.4.1 New Definition of $\mathbf{S}_\alpha$:

Define $\mathbf{S}_\alpha$, $\alpha \in$ card by: $p \in \mathbf{S}_\alpha$ iff $p\colon [\alpha, |p|) \longrightarrow 2$ and

(1) $\xi \in [\alpha, |p|] \longrightarrow L[A \cap \alpha, p \restriction \xi] \models \mathrm{card}(\xi) \le \alpha$.

Let $\mu_\xi^* =$ the least μ s.t. $L_\mu[A\cap\alpha, p\restriction\xi] \models \mathrm{card}(\xi) \leq \alpha$.
Set $\mu^* = \mu^*_{|p|}$.

(2) $\forall\alpha \geq \omega_2 \quad L \models \mathrm{card}(\mu_\xi^*) \leq \xi$.

(3) $\forall\delta \in [\alpha, \mu_p^*] \ (\langle L_\delta[A\cap\alpha, p\restriction\delta^o], A\cap\alpha, p\restriction\delta^o\rangle \not\models \mathrm{ZF})$
where $\delta^o = \{\xi \mid \mu_\xi^* < \delta\}$.

Note From (3) one can infer that if $L_\gamma[A\cap\alpha, p\restriction\xi] \models \mathrm{ZF}$ then $L_\gamma[A\cap\alpha, p\restriction\xi] \models \mathrm{card}(\xi) \leq \alpha$.

We will show that the entire proof of Theorem 0.1 goes through with this new definition of $\mathbf{S}_\alpha$. We accomplish this by noting where the change of definitions affects the proof and changing the proof accordingly.

It must be shown that the conditions $\mathbf{S}_\alpha$ can be arbitrarily extended. This was originally shown in the proof of Theorem 1.4.

Claim 1 Let $\alpha \in \mathrm{card}$, $p \in \mathbf{S}_\alpha$. Let $\delta \in [|p|, \alpha^+)$. Then there is an $r \in \mathbf{S}_\alpha$, $r \leq p$ and $|r| = \delta$.

Proof The r chosen in Theorem 1.4 already satisfies the new definition of $\mathbf{S}_\alpha$ (assuming that p does). Let $|p| = \eta$. Then $r\restriction[\eta, \eta+\alpha)$ codes a well ordering of type δ. The μ_r chosen is $\delta+\omega$ (i.e. $L_{\delta+\omega}[A\cap\delta, r] \models$ "$\mathrm{card}(\delta) = \alpha$"). For $\beta \in (\eta, \delta+\omega]$, $L_\beta[A\cap\beta, r\restriction\beta] \not\models \mathrm{ZF}$ since for $\beta \in (\eta, \eta+\alpha] \cup (\delta, \delta+\omega]$, $L_\beta \not\models \mathrm{ZF}$ and for $\beta \in (\eta+\alpha, \delta]$, $r\restriction[\eta, \eta+\alpha)$ codes a well ordering of type δ.

Q.E.D. (Claim 1)

The auxiliary Lemmas (2.3, 2.4, 2.5) go through just as before. The α-distributivity of $\mathbf{S}_\alpha$ must be shown. In Theorem 1.4 the α-distributivity is proven for the $\mathbf{S}_\alpha$-like conditions over $V = L[B]$, where $B \subseteq \alpha^+$. In Lemma 2.6(3) the α-distributivity of $\mathbf{S}_\alpha$ is over $\mathfrak{N}_s^1[D]$, where $s \in \mathbf{S}_\alpha$ and $D \subseteq \alpha^+$ is a generic over $\mathfrak{N}_s^1$ that codes both $A\cap\alpha^+$ and s. Hence $\mathfrak{N}_s^1[D] \models$ "$V = L[D]$". There a slight variation of $\mathbf{S}_\alpha$ is used and it is noted that the proof in Theorem 1.4 goes through. The corresponding conditions for the new $\mathbf{S}_\alpha$ are defined as follows:

Set $D' = A \cap \alpha \cup D$. Forcing over $\mathfrak{A}^1_s[D]$ is done by the set of pairs $\langle r, |r| \rangle$ s.t.

(a) $r: [\alpha, |r|) \cap \bigcup_{i<1} Z_i \longrightarrow 2$ (where $|r| \in [\alpha, \alpha^+)$) .

(b) $\forall \delta \le |r| \ (L[D' \cap \delta, r/\delta] \models \mathrm{card}(\delta) \le \alpha)$.

define μ^*_ξ and μ^*_r as above.

(c) $\alpha \ge \omega_2 \quad L \models \mathrm{card}(\mu^*_\xi) \le \xi$.

(d) $\delta \in [\alpha, \mu^*_r] \ (\langle L_\delta[D' \cap \delta, r/\delta^o], D' \cap \delta, r/\delta^o \rangle \not\models ZF)$.

<u>Claim 2</u> In the above context $\mathbf{S}_\alpha$ is α-distributive over $\mathfrak{A}^1_s[D]$.

Proof We shall show that the original proof in Theorem 1.4 goes through with the new definition of $\mathbf{S}_\alpha$. We recall part of the proof. The problem is to show that $\bigcup_{i<\lambda} p_i = p \in \mathbf{S}_\alpha$ for $\lim(\lambda)$.

$\mu_p = \delta_\lambda + \delta_\lambda$ where,

$\sigma: \langle L_{\delta_\lambda}[D' \cap \alpha_\lambda], D' \cap \alpha_\lambda \rangle \cong X \prec \mathfrak{A}^1_s[D]$, $\sigma(\alpha_\lambda) = \alpha^+$ and $X \cap \alpha^+ = \alpha_\lambda$.

We have to show that:

$$\forall \eta \le \delta_\lambda + \delta_\lambda \ (\langle L_\eta[D' \cap \eta, p/\eta^o], D' \cap \eta, p/\eta^o \rangle \not\models ZF) .$$

Set $\mathfrak{A}_\eta = \langle L_\eta[D' \cap \eta, p/\eta^o], D' \cap \eta, p/\eta^o \rangle$.

For $\eta \in (\delta_\lambda, \delta_\lambda + \delta_\lambda]$, $\mathfrak{A}_\eta \not\models ZF$ because there is an L_{δ_λ}-definable function from δ_λ on η .

For $\eta < \alpha_\lambda$, $\mathfrak{A}_\eta \not\models ZF$ since there is an $i < \lambda$ s.t. $p_i/\eta = p/\eta$ and by the induction hypothesis

$$\langle L_\eta[D' \cap \eta, p_i/\eta^o], D' \cap \eta, p_i/\eta^o \rangle \not\models ZF .$$

We are left with the case that $\eta \in [\alpha_\lambda, \delta_\lambda]$. Set $\mathfrak{A}_\eta = \langle L_\eta[D' \cap \alpha_\lambda], D' \cap \alpha_\lambda \rangle$. We claim that $\mathfrak{A}_\eta \not\models ZF$. Assume by negation that $\mathfrak{A}_\eta \models ZF$ for $\eta \in [\alpha_\lambda, \delta_\lambda]$. Apply σ to $\mathfrak{A}_\eta$ and get $\langle L_{\eta^*}[D'], D' \rangle \models ZF$ where $\eta^* = \sigma(\eta)$. We can construct $A \cap \alpha^+$ and begin constructing s in $\langle L_{\eta^*}[D'], D' \rangle$ as in the proof of lemma 1.3 (since D' is generic and $\langle L_{\eta^*}[D'], D' \rangle \models ZF$). There cannot be a ξ s.t. $L_{\eta^*}[A \cap \alpha^+, s/\xi] \models "\xi = \alpha^{++}"$ since for the first ξ , $s/\xi \in L_{\eta^*}[D']$ and $L_{\eta^*}[A \cap \alpha^+, s/\xi] \models ZF$. But $\eta^* < \mu^*_\xi$ since $L_{\eta^*}[A \cap \alpha^+, s/\xi] \models \xi \geq \alpha^+$. This is a contradiction to $L_{\eta^*}[A \cap \alpha^+, s/\xi] \models ZF$ and $s/\xi \in \mathbf{S}_{\alpha^+}$. Then all of s and in turn μ_s can be defined in $\langle L_{\eta^*}[D'], D' \rangle$ which is impossible.

Q.E.D. (Claim 2)

So far we have dealt with theorems that deal directly with $\mathbf{S}_\alpha$. In the rest of the proof we have to worry about situations where there are $r_i \in \mathbf{S}_\alpha$ for $i < \lambda$, $\lim(\lambda)$ and a claim that $\bigcup_{i<\lambda} r_i \in \mathbf{S}_\alpha$. This occurs in Lemma 2.14.3, Claim 1. There the verification that $r^*_i \in \mathbf{S}_{\beta_i}$ by the new definition is simple since either $\mu^*_{r^*_i}$ does not change or it is equal to $|r^*_i| + \omega$.

The next place we examine is Lemma 3.22(a). It must be shown that $p_\gamma \in \mathbf{S}_\gamma$, where $p_\gamma = \bigcup_{i<\lambda} p_{i\gamma}$ and $\mu^*_{p\gamma} = \mu' + \omega$. We recall some properties of μ' .

$\sigma\colon \langle L_{\mu'}[A',s'],A',s'\rangle \cong X^\gamma_\lambda \prec \mathfrak{A}^1_s$, $s \in \mathbf{S}_{\alpha+}$, $\sigma(s') = s$,

$\sigma(A') = A$, $\sigma(\alpha') = \alpha$, $\sigma(\alpha^*) = \alpha^+$, $p = \bigcup_{i<\lambda} p_i$,

$\sigma(p'_i) = p_i$, $D'_\delta = \bigcup_{i<\lambda} \tilde{p}'_{i\delta}$ for $\delta \in (\mathrm{card})_b$ and

$$D' = \smile_{\delta\in(\mathrm{card})_b \cap [\gamma,\alpha']} D'_\delta .$$

Set $\mathfrak{A}_\eta = \langle L_\eta[A \cap \eta, p_\gamma {\restriction} \eta^o], A \cap \eta, p_\gamma {\restriction} \eta^o\rangle$.

We must show $\forall\eta \le \mu'\ (\mathfrak{A}_\eta \not\models \mathrm{ZF})$.

For $\eta < |p_\gamma| = \gamma^{+b}$, $\mathfrak{A}_\eta \not\models \mathrm{ZF}$ since there is an $i < \lambda$ s.t. $p_\gamma {\restriction} \eta = p_i {\restriction} \eta$.

If η is a cardinal in b then $\sigma(\eta)$ is a cardinal in $\langle L_\beta[A],A\rangle$, setting $\eta^* = \sigma(\eta)$, $\langle L_{\eta^*}[A \cap \eta^*], A \cap \eta^*\rangle \not\models \mathrm{ZF}$; thus $\mathfrak{A}_\eta \not\models \mathrm{ZF}$.

For $\eta \in (\gamma^{+b},\alpha')$, if $\mathfrak{A}_\eta \models \mathrm{ZF}$, the proof of Lemma 3.20, Claim 1 implies that $A' \cap \eta$ and $D' \cap \eta$ are definable in $L_\eta[A \cap \gamma, p_\gamma]$. Hence:

(*) $\langle L_\eta[A' \cap \eta, D' \cap \eta], A' \cap \eta, D' \cap \eta\rangle \models \mathrm{ZF}$.

Set $\eta^* = \sigma(\eta)$. Then there is a $\beta \in \mathrm{card} \cap \alpha$ s.t. $\eta^* \in (\beta,\beta^+)$ and $\sigma(D' \cap \eta) = \bigcup_{\delta<\beta} \tilde{p}_\delta \cup \tilde{p}_\beta {\restriction} \eta^*$. There is an $i < \lambda$ s.t. $\tilde{p}_\beta {\restriction} \eta^* = \tilde{p}_{i\beta} {\restriction} \eta^*$. Now apply σ to (*) and get:

$$\langle L_{\eta^*}[A \cap \eta^*, \bigcup_{\delta<\beta} \tilde{p}_\delta \cup \tilde{p}_{i\beta} {\restriction} \eta^*], A \cap \eta^*, \bigcup_{\delta<\beta} \tilde{p}_\delta \cup \tilde{p}_{i\beta} {\restriction} \eta^*\rangle \models \mathrm{ZF} .$$

This is impossible since

$\langle L_{\eta^*}[A \cap \eta^*, p_{i\beta} {\restriction} \eta^*], A \cap \eta^*, p_{i\beta} {\restriction} \eta^*\rangle \not\models \mathrm{ZF}$ by virtue of $p_{i\beta} \in \mathbf{S}_\beta$.

For $\eta \in (\alpha',\alpha^*)$ the situation is similar using the proof of Lemma 3.20, Claim 2 and getting $D'_{\alpha'}/\eta \in L_\eta[A \cap \gamma, p_\gamma]$. Applying σ the same kind of contradiction is reached. Finally for $\eta \in (\alpha^*,\mu']$ we reach a similar contradiction using the proof of Lemma 3.20, Claim 3 to get $s'/\eta, A' \in L_\eta[A \cap \gamma, p_\gamma]$.

For the version of Lemma 3.25 the same variations are required and in the case of Theorem 3.2 (as proved in §3.7) the variations are easier.

This completes the discussion that the coding theorem goes through with the new definition of $\mathbf{S}_\alpha$. Now apply the coding theorem and get an $a \subseteq \omega$ s.t. a codes A, $L_\beta[a] \models ZF$ and $L_\alpha[a] \not\models ZF$ for $\alpha < \beta$.

Q.E.D. (Theorem 5.8)

5.3 FORCING WITH $0^{\#}$

We can use the coding theorem (0.1) to show the consistency of a real, $a \subseteq \omega$, with a particular property by starting with a ground model, $\langle M,A \rangle \models ZF+GCH+ \neg 0^{\#}$ (usually L). Then force over $\langle M,A \rangle$ with a class of conditions to get a $\langle M',A' \rangle$ that satisfies the properties required in §2.1 and the additional desired property. Then code up $\langle M',A' \rangle$ by an $a \subseteq \omega$. As an example we can use the above procedure to get an $a \subseteq \omega$ s.t. $L[a] \models \neg 0^{\#}$ and in $L[a]$ an unbounded sequence of L-cardinals has been collapsed.

It is not possible to use Theorem 0.2 in a similar manner without restrictions. For the rest of this section assume that $0^{\#}$ exists and that I is the canonical class of indiscernibles for L. An attempt to use Theorem 0.2 would start with L, define a class of conditions $\mathbb{P}$ and construct a $\mathbb{P}$-generic, G, in $L[0^{\#}]$. This is not always possible; in fact, if $0^{\#}$ exists, there are conditions for which no generic exists.

If $\mathbb{P}$ is the class of conditions that collapse every even successor cardinal on the immediate odd cardinal below (i.e., collapse $\aleph_{\delta+2i+2}$ on $\aleph_{\delta+2i+1}$ for $\lim(\delta)$ and $i < \omega$ - a general formulation of such conditions can be found in [Z]), then, using $0^{\#}$, a $\mathbb{P}$-generic, G, can be constructed that leaves I intact. Then Theorem 0.2 can be applied to get an $a \subseteq \omega$ s.t. $a \notin L[0^{\#}]$ and $L[a] \models \neg 0^{\#}$ and an unbounded class of L-cardinals is collapsed.

On the other hand if $\mathbb{H}$ are the conditions for collapsing odd onto even cardinals (i.e., $\aleph_{\delta+2i+1}$ onto $\aleph_{\delta+2i}$ for $\lim(\delta)$ and $i < \omega$), then no $\mathbb{H}$-generic over L exists. Let $\kappa \in I$ s.t. $cf(\kappa) > \omega$. Then $\mathbb{H}$ collapses κ^{+L} onto κ. If G were $\mathbb{H}$-generic over L then $L[G] \models \neg 0^{\#}$ and by Marginalia [DJ],

$$L[G] \models cf(\kappa^{+L}) = card(\kappa^{+L}) = \kappa .$$

It is known that $L[0^{\#}] \models card(\kappa^{+L}) = \omega$, so there is an $f \in L[0^{\#}]$, $f: \omega \longrightarrow \kappa^{+L}$ cofinally. There is a $g \in L[G]$, $g: \kappa^{+L} \longrightarrow \kappa$ cofinally, hence $g \circ f: \omega \longrightarrow \kappa$ cofinally which is a contradiction to $cf(\kappa) > \omega$.

The above follows from a more general restriction. We will show that if $\kappa \in I$ and $cf(\kappa) > \omega$ then we cannot add a Cohen generic to κ. Set:

$$\mathbb{H}(\kappa) = \{p \mid p: \kappa \longrightarrow \kappa^{+}, card(p) < \kappa\} ,$$

the set of conditions which collapses κ^{+} onto κ. Then if G is $\mathbb{H}(\kappa)$ generic over L, G induces a Cohen subset of κ as follows. Define a mapping, F, from $\mathbb{H}(\kappa)$ to $\mathbb{C}(\kappa) = \{q \mid q: \kappa \longrightarrow 2, card(q) < \kappa\}$ (the standard Cohen conditions) by:

$$[F(p)](\alpha) = \begin{cases} 0 & \text{if } \lim(p(\alpha)) \\ 1 & \text{otherwise} \end{cases}$$

for $\alpha \in \mathrm{dom}(p)$. It is easy to verify that $F''(G)$ is $\mathbf{C}(\kappa)$-generic.

We can prove the above claim for a broader definition of "Cohen Generic".

Definition 1 G is a Cohen generic subset of κ in the broad sense if G is $\mathbf{C}$-generic over L , where $\mathbf{C}$ is a set of conditions s.t.

(a) $\mathbf{C} \subseteq L_\kappa$.

(b) $\mathbf{C}$ is $<\kappa$-distributive.

(c) $G \cap L_\tau \in L$ for $\tau < \kappa$.

(d) $p \in \mathbf{C} \longrightarrow \mathrm{card}(p) < \kappa$.

(e) $G \notin L$.

$\mathbf{C}$ is called a set of Cohen conditions (in a broader sense).

Theorem 5.9 *Assume* $0^{\#}$ *exists and let* $\kappa \in I$ *s.t.* $\mathrm{cf}(\kappa) > \omega$. *Let* $\mathbf{C}$ *be a set of Cohen conditions (in the broad sense) for* κ . *Then no* $\mathbf{C}$*-generic exists.*

Proof Let $\langle \alpha_i \mid i < \omega \rangle$ be a monotone sequence converging to κ^{+L} . Define $p_i \in \mathbf{C}$, $\kappa_i \in I$ and $X_i \prec L_{\kappa^+}$ by induction on $i < \omega$. κ is a limit point of I since $\mathrm{cf}(\kappa) > \omega$. Let $\kappa_0 < \kappa$, $\kappa_0 \in I$ be arbitrary. Set:

$$X_i = \text{the least } X \prec L_{\kappa^+} \text{ s.t. } X \supseteq \kappa_i \cup \{\alpha_0, \ldots, \alpha_i\} \quad \text{and}$$

$$\Delta_i = \bigcap \{\Delta \in X_i \mid \Delta \text{ is dense in } \mathbf{C}\} .$$

Δ_i is dense in $\mathbf{C}$ since $\mathrm{card}(X_i) < \kappa$ and $\mathbf{C}$ is $< \kappa$-distributive.

p_i = the least $p \in C$ s.t. $p \in \Delta_i \cap G$.

κ_{i+1} = the least $\tau \in I$ s.t. $\tau > \kappa_i$ and $p_i \in L_\tau$ (here we use $p \in \mathbf{C}$ implies $\mathrm{card}(p) < \kappa$).

Set $\bar{\kappa} = \sup_{i<\omega} \kappa_i$. $\bar{\kappa} < \kappa$ since $\mathrm{cf}(\kappa) > \omega$.

Set $X_{\bar{\kappa}} = \bigcup_{i<\omega} X_i$. $\bar{\kappa} \nless X_{\bar{\kappa}}$ since $\bar{\kappa}$ cannot be defined by ordinals less than $\bar{\kappa}$ ($\bar{\kappa} \in I$) . Hence $X_{\bar{\kappa}} \cap \kappa = \bar{\kappa}$. Let $\pi: L_\delta \cong X_{\bar{\kappa}}$, then $\pi(\bar{\kappa}) = \kappa$ and $\langle \pi^{-1}(\alpha_i) \mid i < \omega \rangle$ is cofinal in $\bar{\kappa}^{+L}$. Hence $\delta = \bar{\kappa}^{+L}$.

Assume by negation that G is a $\mathbf{C}$-generic over L .

<u>Claim</u> $G \cap L_{\bar{\kappa}}$ is $\mathbf{C} \cap L_{\bar{\kappa}}$-generic over L .

Proof Let Δ be dense in $\mathbf{C} \cap L_{\bar{\kappa}}$, then $\Delta \in L_{\bar{\kappa}^+}$ (since $\mathbf{C} \subseteq L_\kappa$ and $L_{\bar{\kappa}} \prec L_\kappa$). Set $\pi(\Delta) = \Delta'$, Δ' is dense in $\mathbf{C}$ and there is an $i < \omega$ s.t. $\Delta' \in X_i$. Hence $p_i \in \Delta$ and $p_i \in L_{\kappa_{i+1}} \subseteq L_{\bar{\kappa}}$. Thus $\pi^{-1}(p_i) = p_i$ and $p_i \in G \cap \Delta \cap L_{\bar{\kappa}}$.

Q.E.D. (Claim)

If $G \cap L_{\bar{\kappa}}$ is $\mathbf{C} \cap L_{\bar{\kappa}}$-generic over L then by part (e) of the definition above and $L_{\bar{\kappa}} \prec L_\kappa$ we have $G \cap L_{\bar{\kappa}} \nless L$ which contradicts part (c).

Q.E.D.

Corollary 5.9.1 *If* $\kappa \in I$ *and* $\mathrm{cf}(\kappa) = \omega$ *then there is a Cohen generic,* G *, for* κ *s.t.* $G \in L[0^{\#}]$.

Proof Almost the same as Theorem 5.9. Let $\mathbf{C}$ be the Cohen conditions for κ . Let $\langle \alpha_i \mid i < \omega \rangle$ be a monotone sequence converging to κ^{+L} and $\langle \tau_i \mid i < \omega \rangle$ a monotone sequence converging to κ s.t. $\tau_i \in I$.

Set X_i = the least $X \prec L_{\kappa^+}$ s.t. $\tau_i \cup \{\alpha_0, \ldots, \alpha_i\} \subseteq X$,

$\Delta_i = \cap\{\Delta \mid \Delta \in X_i$ is dense in $\mathbf{C}\}$ and

p_{i+1} = the least $p \in \mathbf{C}$ s.t. $p \leq p_i$ and $p \in \Delta_i$.

Set $\bigcup_{i<\omega} p_i = G$, $X = \bigcup_{i<\omega} X_i$.

<u>Claim</u> G is $\mathbf{C}$-generic over L .

Proof $X \cap \kappa^+$ is transitive since $\kappa \in X$ and $L_{\kappa^+} \models \forall\alpha\exists f(f\colon \kappa \xrightarrow{\text{On}} \alpha)$. Hence $X \cap \kappa^+ = \kappa^+$ since $\{\alpha_i \mid i<\omega\} \subseteq X_\kappa$. Hence $X = L_{\kappa^+}$. So if Δ is dense in $\mathbf{C}$, there is an $i<\omega$ s.t. $\Delta \in X_i$ and $p_{i+1} \in G \cap \Delta$.

Q.E.D.

Theorem 5.9 shows that we cannot construct, using $0^\#$, a generic class for a class of conditions that adds a Cohen generic to arbitrary indiscernibles. For the case of cardinal collapsing we can construct a generic class if no Cohen set is added to indiscernibles. For the general class of conditions little can be said since less is known. We will define some types of class forcing and give sufficient conditions for the existence of a generic class in $L[0^\#]$. The classes of conditions defined will be L-definable.

Definition 2 Let $\mathbb{P}$ be an L-definable class of conditions. $\mathbf{P}$ is Easton if for every $\alpha \in \text{On}$ there is a set $\mathbf{P}^\alpha$ and a class $\mathbf{P}_\alpha$ s.t.

(a) $\bigcup_{\alpha\in\text{On}} \mathbf{P}^\alpha = \mathbf{P}$ and $\mathbf{P}^\alpha \subseteq \mathbf{P}^\beta$ if $\alpha \leq \beta$.

(b) If $p \in \mathbf{P}^\alpha$ and $q \in \mathbf{P}_\alpha$ then p and q are compatible and we denote by $p \cup q$ their least upper bound.

(c) For $p \in \mathbb{P}$ and $\alpha \in \text{On}$, there is a unique decomposition $p = p^\alpha \cup p_\alpha$ s.t. $p^\alpha \in \mathbf{P}^\alpha$, $p_\alpha \in \mathbf{P}_\alpha$ and $\mathbf{P} \cong \mathbf{P}^\alpha \times \mathbf{P}_\alpha$ by the mapping $p \longrightarrow \langle p^\alpha, p_\alpha \rangle$.

(Zarach [Z] calls this a coherent class of conditions.)

Definition 3 Let $\mathbb{P}$ be Easton and S a class of limit ordinals. $\mathbb{P}$ is continuous on S if $\forall\alpha \in S(\mathbb{P}^{\alpha} = \bigcup_{\beta<\alpha} \mathbb{P}^{\beta})$.

Definition 4 Let $\mathbb{P}$ be Easton and S a class of ordinals. $\mathbb{P}$ is short on S if $\forall\alpha \in S(p \in \mathbb{P}^{\alpha} \longrightarrow \mathrm{card}(p) < \alpha)$. Otherwise $\mathbb{P}$ is long on S.

Let $\mathbb{P}$ be a class of Easton conditions s.t. $\mathbb{P}$ is continuous on I. For $\kappa \in I$ we have that $\mathbb{P} \cap L_{\kappa} = \mathbb{P}^{\kappa}$ since $\mathbb{P} = \bigcup_{\alpha\in On} \mathbb{P}^{\alpha}$ and $L_{\kappa} \prec L$. This implies that $\mathbb{P}$ is short on I since $p \in \mathbb{P}^{\kappa} \longrightarrow p \in L_{\kappa} \longrightarrow \mathrm{card}(p) < \kappa$.

In L we cannot define a class of conditions continuous exactly on I, but only on a class that includes I, such as inaccessibles, Mahlo etc.

Lemma 5.10 *Let $\mathbb{P}$ be Easton and continuous on "at least I". Then $\mathbb{P}^{\kappa}$ is κ-AC for $\kappa \in I$.*

Proof Let $\Delta \subseteq \mathbb{P}^{\kappa}$ be a set of mutually incompatible conditions. Let $P(\alpha,x)$ be the defining formula for $\mathbb{P}$, i.e., $x \in \mathbb{P}^{\alpha} \equiv P(\alpha,x)$ for all α. Let,

$\theta(\kappa,\Delta) \equiv \forall x(P(\kappa,x) \longleftrightarrow \exists\alpha < \kappa(P(\alpha,x))) \wedge$ "$\Delta \subseteq \mathbb{P}^{\kappa}$ is a set of mutually incompatible conditions".

By the discussion above we have $L_{\kappa} \models \theta(\kappa,\Delta)$. κ is Mahlo since $\kappa \in I$, hence there is a regular $\bar{\kappa} < \kappa$ s.t. $L_{\bar{\kappa}} \models \theta(\bar{\kappa},\Delta \cap L_{\bar{\kappa}})$. We have: $\mathbb{P}^{\bar{\kappa}} = \bigcup_{\alpha<\bar{\kappa}} \mathbb{P}^{\alpha} = \mathbb{P} \cap L_{\bar{\kappa}}$ and $\Delta_{\bar{\kappa}} = \Delta \cap \mathbb{P}^{\bar{\kappa}}$ is a set of mutually incompatible conditions. We claim $\Delta_{\bar{\kappa}} = \Delta$. If $q \in \Delta \setminus \Delta_{\bar{\kappa}}$ then q must be incompatible with all of $\Delta_{\bar{\kappa}}$, contradicting the maximality of $\Delta_{\bar{\kappa}}$.

Q.E.D.

Theorem 5.11 *Let $\mathbb{P}$ be Easton and*

(a) $\mathbf{P}$ *is continuous on at least* I .

(b) $\mathbf{P}_\kappa$ *is* κ*-distributive for* $\kappa \in I$.

Then we can construct a $\mathbf{P}$*-generic* G *in* $L[0^\#]$ *s.t.* I *is a class of indiscernibles for* L[G] .

Proof By Lemma 5.10, $\mathbb{P}^\kappa$ is κ-AC for $\kappa \in I$.

Notation *If* $\gamma \in I$ *then* $\gamma' = \min(I \setminus \gamma+1)$. *If we write* $\langle \vec{\gamma} \in I^n \rangle$ *it is assumed that* $\gamma_0 < \gamma_1 < \dots < \gamma_{n-1}$.

We note the following properties derived from indiscenibility and remarkability.

(1) If $p_\gamma \in \mathbb{P}$ is uniformly definable from $\gamma \in I$, then $p_\gamma \in \mathbb{P}^{\gamma'}$ for $\gamma < \gamma' \in I$.

(2) If $p_{\gamma,\vec{\gamma}}$ is uniformly definable from $\langle \gamma,\vec{\gamma} \rangle \in I^{n+1}$ and $p_{\gamma,\vec{\gamma}} \in P^{\min(\vec{\gamma})}$, then $p_{\gamma,\vec{\gamma}} = p_{\gamma,\vec{\delta}}$ for all $\langle \gamma,\vec{\gamma} \rangle, \langle \gamma,\vec{\delta} \rangle \in I^{n+1}$.

(3) Let $p_{\gamma,\vec{\gamma}}$ be as in (2). Since $p_{\gamma,\vec{\gamma}}$ is independent from $\vec{\gamma}$, we can set $p_\gamma = p_{\gamma,\vec{\gamma}}$. In such a case $(p_\gamma)^\gamma = (p_\delta)^\delta$ for $\gamma,\delta \in I$.

We are interested in enumerating all the L-definable dense classes in $\mathbf{P}$. Since every $x \in L$ is definable from parameters in I and there are denumerably many formulas, each dense class has the following form:

$$\Delta = \{x \mid \psi(\chi,\vec{\gamma}) , \vec{\gamma} \in I\}$$

and if we change the tuple $\vec{\gamma} \in I$ to $\vec{\delta} \in I$ we also get a dense class in $\mathbf{P}$. Hence it is possible to define a Gödel numbering of such ψ so that

$$\langle \Delta^i_{\gamma_0,\ldots,\gamma_i} \mid i<\omega,\ \gamma_0 \ldots \gamma_i,\ \vec{\gamma} \in I\rangle$$

enumerates all the dense classes of $\mathbb{P}$.

We construct a sequence $\langle p^i_\gamma \mid \gamma \in I\rangle$ for $i<\omega$ s.t.

(a) $p^{n+1}_\gamma \leq p^n_\gamma$.

(b) $p^i_\gamma \in \mathbf{P}^{\gamma'}$ is uniformly definable in $\langle \gamma,\vec{\gamma}\rangle \in I^i$.

(c) $p^i_{\gamma_0} \cup p^i_{\gamma_1} \cup \ldots \cup p^i_{\gamma_i} \in \Delta^i_{\gamma_0,\ldots,\gamma_i}$.

Assume that we have already constructed such a sequence. Then by facts (1), (2) and (3) above p^i_γ and p^i_δ are compatible (this is implicit in (c)) and by (a) we have that p^i_γ and p^{i+1}_γ are compatible. So we may set

$$G = \bigcup_{i\in\omega} \bigcup_{\gamma\in I} p^i_\gamma .$$

By (c) G is $\mathbb{P}$-generic over L . Hence it remains to define $\langle p^i_\gamma \mid \gamma \in I\rangle$.

We define $\langle p^i_\gamma \mid \gamma \in I\rangle$ by induction on $i<\omega$.

p^o_γ = the L-least $p \in \mathbb{P}$ s.t. $p \in \Delta^o_\gamma$.

p^o_γ is clearly uniformly definable in γ .

Assume we have $\langle p^n_\gamma \mid \gamma \in I\rangle$. As a preliminary step to defining p^{n+1}_γ we define p^{*i}_γ for $i \leq n$.

p^{*o}_γ = the L-least $q \leq (p^n_\gamma)_\gamma$, $q \in \mathbb{P}_\gamma$ s.t. $\Delta^q_{\vec{\nu},\gamma}$ is dense in $\{r \in \mathbf{P}^\gamma \mid r \leq (p^n_\gamma)^\gamma\}$ for all $\langle \vec{\nu},\gamma\rangle \in I^{n+1}$.

This can be done since $\mathbb{P}^\gamma$ is γ-AC and $\mathbf{P}_\gamma$ is γ-distributive.

p^{*1}_γ = the L-least $q \leq (p^n_\gamma)_\gamma$, $q \in \mathbb{P}_\gamma$ s.t.

$$(\Delta^{p^{*o}_\delta}_{\vec{\nu},\gamma,\delta})^q \text{ is dense in } \{r \in \mathbf{P}^\gamma \mid r \leq (p^n_\gamma)^\gamma\}$$

for all $\langle \vec{\nu},\gamma,\delta\rangle \in I^{n+1}$.

By fact (2) p_γ^{*1} is independent of δ .

If we already have p_γ^{*i} and $i+1 < n$ then set

p_γ^{*i+1} = the L-least $q \le (p_\gamma^n)_\gamma$, $q \in \mathbf{P}_\gamma$ s.t.

$$(\Delta_{\vec{\nu},\gamma,\vec{\delta}}^{p_{\delta_n}^{*o}})^{p_{\delta_{n-1}}^{*1}} \ldots)^{p_{\delta_{n-i}}^{*i}} q$$

is dense in $\{r \in \mathbb{P}^\gamma \mid r \le (p_\gamma^n)^\gamma\}$ where $\langle \vec{\nu},\gamma,\vec{\delta}\rangle \in I^{n+1}$.

Again p_γ^{*i+1} is independent of $\vec{\delta} \in I$. Set:

p_γ^{*n} = the least $q \le (p_\gamma^n)_\gamma$, $q \in \mathbf{P}_\gamma$ s.t.

$$q \in (\Delta_{\gamma,\vec{\delta}}^{p_{\delta_n}^{*o}}) \ldots)^{p_{\delta_1}^{*n-1}} \quad \text{for } \langle \gamma,\vec{\delta}\rangle \in I^{n+1} .$$

Finally set $p_\gamma^{n+1} = p_\gamma^{*o} \cup \ldots \cup p_\gamma^{*n}$.

From the definition it is clear that p_γ^{n+1} is uniformly definable from $\langle \gamma,\vec{\delta}\rangle \in I^{n+1}$ independently of $\vec{\delta} \in I^n$ and that $p_\gamma^{n+1} \in \mathbf{P}^{\gamma'}$ (since the whole definition can be carried out in $L_{\gamma'}$). By the definition of Δ^q and the construction of p_γ^n it is clear that $p_{\gamma_o}^n \cup p_{\gamma_1}^n \cup \ldots \cup p_{\gamma_n}^n \in \Delta_{\gamma_o,\ldots,\gamma_n}^n$ for $n < \omega$, $\vec{\gamma} \in I^{n+1}$. Hence $G = \bigcup_{i<\omega} \bigcup_{\gamma\in I} p_\gamma^i$ is $\mathbb{P}$-generic over L .

<u>Claim</u> I is a class of indiscernibles for L[G] .

Proof We must show that for any $\phi(x_o,\ldots,x_n)$ and $\vec{\gamma},\vec{\delta} \in I^n$, $L[G] \models \phi(\vec{\gamma}) \longleftrightarrow \phi(\vec{\delta})$. Choose an $\vec{\eta} \in I^n$ s.t. the maps $\pi(\gamma_i) = \eta_i$ and $\sigma(\delta_i) = \eta_i$ for $i \le n$ can be extended to order preserving maps, $\pi',\sigma': I \longrightarrow I$ which are the identity from some point on. Thus it is sufficient to show that $L[G] \models \phi(\vec{\gamma}) \longleftrightarrow \phi(\pi(\vec{\gamma}))$ for any order preserving map $\pi: I \longrightarrow I$ which is eventually the identity. Any order preserving $\pi': I \longrightarrow I$ induces a unique $\pi: L \xrightarrow[\Sigma_\omega]{} L$, s.t. $\pi \supseteq \pi'$.

Assume we have a $\pi: L \xrightarrow[\Sigma_\omega]{} L$ s.t. $\pi \restriction On\backslash\nu = id \restriction On\backslash\nu$ for some $\nu \in I$ and $\vec{\eta} \in I^n$ s.t. $\max(\vec{\eta}) < \nu$. For $p \in \mathbb{P}$ we have

$$p \Vdash \phi(\vec{\gamma}) \quad \text{iff} \quad \pi(p) \Vdash \phi(\pi(\vec{\gamma})) ,$$

since the forcing relation can be defined from the indiscernibles greater than ν. It is also true that

$$[L[G] \models \phi(\vec{\gamma})] \quad \text{iff} \quad \exists p \in G(p \Vdash \phi(\vec{\gamma})) .$$

Hence it is sufficient to show that $p \in G \longrightarrow \pi(p) \in G$. By the construction of G it is clear that any $p \in G$ has the form $p = f(\vec{\eta})$ for $\vec{\delta} \in I^n$ and $n < \omega$ and for any $\delta \in I^n$, $f(\vec{\delta}) \in G$.

Q.E.D.

Theorem 5.11 works since generic sets are not directly added to L_κ for $\kappa \in I$. $\mathbb{P}^\kappa = \bigcup_{\alpha<\kappa} \mathbb{P}^\alpha$ which adds sets to L_κ as $\mathbb{P} = \bigcup_{\alpha\in On} \mathbb{P}^\alpha$ adds classes to L and $\mathbb{P}_\kappa$ adds nothing to L_κ since $\mathbb{P}_\kappa$ is κ-distributive. For cardinal collapsing the situation is summed up by: a generic can be constructed in $L[0^\#]$ iff there is no collapsing on $\kappa \in I$. In the general case we do not have a necessary and sufficient condition.

Long conditions (as defined above) are not often used. Theorem 1.5 is an example; long conditions are used in [J4] and the conditions we use are long (though not fully Easton). Many forcing tasks that are usually done by short conditions can also be done by using long conditions though they are not equivalent and different properties are preserved by each type of forcing.

It is possible to construct a generic class for long cardinal collapsing conditions (not collapsing on indiscernibles) using techniques similar to that of Theorem 5.3. The following attempts to generalize the above by defining a class of conditions with enough properties so that the proof for cardinal collapsing can go through.

Lemma 5.12 *Let* $\mathbb{P}$ *be a class of Easton conditions. Let* α *be regular s.t.* $\mathbf{P}^{\beta}$ *is* α-AC *and* $\mathbf{P}_{\beta}$ *is* α*-distributive. Let* $\langle \Delta_{\nu} \mid \nu < \gamma < \alpha \rangle$ *be a sequence dense classes (sets) in* $\mathbb{P}$ $(\mathbf{P}^{\delta}_{\theta},\ \theta < \beta < \delta)$. *Set*

$$\Delta = \{p \in \mathbb{P}_{\beta}\ (\mathbf{P}^{\delta}_{\beta}) \mid \Delta^{p} \text{ is dense in } \mathbb{P}^{\beta}\ (\mathbf{P}^{\beta}_{\theta})\} .$$

Then Δ *is dense in* $\mathbf{P}_{\beta}$ $(\mathbf{P}^{\delta}_{\beta})$.

Proof Implicit in Theorem 3.4.

Definition 5 Let $\mathbb{P}$ be Easton. $\mathbf{P}$ has property W iff there is a definable $X \subseteq \mathrm{On}$ and a definable $F\colon X \longrightarrow \mathrm{On}$ s.t.

(a) X is unbounded and F is strictly increasing.

(b) $\forall\alpha \in X$ $(F(\alpha)$ is regular $\wedge\ \mathbb{P}^{\alpha}$ is $F(\alpha)$ - AC $\wedge\ \mathbb{P}_{\alpha}$ is $F(\alpha)$-distributive.

(c) X and $F''X$ are totally discontinuous, i.e., X and $F''X$ do not contain any of their limit points.

(d) Let $\alpha \in X$ be s.t. $X \cap \alpha$ has no maximal element and $\sup(X \cap \alpha) = \sup_{\beta<\alpha}(F(\beta))$. Set $\kappa = \sup(X \cap \alpha)$. Then:

(1) $\mathbf{P}_{\kappa}$ is κ-distributive.

(2) For every $\langle \Delta_{\nu} \mid \nu \in X\ \ \kappa\ (= X \cap \alpha)\rangle$, where Δ_{ν} is dense in $\mathbb{P}_{\nu} \cap L_{\kappa}$ we have,

$\{p \in \mathbf{P}^{\kappa} \mid \forall\nu \in X\ \ \kappa\ \exists q \in \mathbf{P}_{\nu} \cap L_{\kappa}\ (p \leq q \wedge q \in \Delta_{\nu})\}$ is dense in $\mathbb{P}^{\kappa}$.

(3) For every $\langle \Delta_\nu \mid \nu \in X \cap \kappa \rangle$, where Δ_ν is dense in $\mathbb{P}^\kappa_\nu$ we have:

$\{p \in \mathbf{P}^\kappa \mid \forall \nu \in X \cap \kappa \ ((p)_\nu \in \Delta_\nu)\}$ is dense in $\mathbb{P}^\kappa$.

Comments

(1) If $\alpha = \min(X \backslash \kappa)$ for $\kappa \in I$, then α satisfies the conditions of (d); in particular

$$\sup(X \cap \kappa) = \kappa = \sup_{\beta \in X \cap \kappa} (F(\beta)) .$$

Hence for $\kappa \in I$, $\mathbb{P}_\kappa$ is κ-distributive and $\mathbf{P}^\kappa$ satisfies the diagonal property - (d)(2) and (3).

(2) The long conditions for collapsing even successor cardinals to odd cardinals below satisfy property W by taking $X = \{\aleph_{\delta+2i+1} \mid \lim(\delta) \text{ and } i < \omega\}$ and F the identity function where

$\mathbb{P}^{\aleph_{\delta+2i+3}}_{\aleph_{\delta+2i+1}}$ are the conditions which collapse $\aleph_{\delta+2i+4}$ on $\aleph_{\delta+2i+3}$.

Theorem 5.13 *Let* $\mathbb{P}$ *be Easton and satisfy property* W. *Then there is a* $\mathbf{P}$*-generic,* G, *in* $L[0^{\#}]$.

Proof We follow the proof of Lemma 5.3. Let $\langle \Delta_\alpha \mid \alpha \in On \rangle$ be a uniform enumeration of all dense classes in $\mathbf{P}$, s.t. if we define the same sequence in L_κ for $\kappa \in I$ we get that $\langle \Delta_\alpha \mid \alpha < \kappa \rangle$ is the uniform enumeration of all dense classes in $\mathbf{P} \cap L_\kappa$.

We define $p_\kappa \in \mathbf{P}^\kappa$ uniformly for $\kappa \in I$ as follows: For $\alpha \in X \cap \kappa$ set:

$$\Delta^*_\alpha = \{p \in \mathbb{P}_\alpha \cap L_\kappa \mid \forall \beta < F(\alpha) \ (\Delta^p_\beta \text{ is dense in } \mathbf{P}^\alpha)\} .$$

By Lemma 5.12, Δ^*_α is dense in $\mathbb{P}_\alpha \cap L_\kappa$ since $\mathbf{P}^\alpha$ is $F(\alpha)$-AC, $\mathbb{P}_\alpha$ is $F(\alpha)$-distributive and $L_\kappa \prec L$. By the diagonal property ((d)(2)),

$\Delta^*_\kappa = \{p \in \mathbb{P}^\kappa \mid \forall\alpha \in X\ \exists q \in \mathbf{P}_\alpha \cap L_\kappa\ (p \le q \wedge q \in \Delta^*_\alpha)\}$ is dense in $\mathbb{P}^\kappa$. Δ^*_κ is uniformly L definable in κ; hence we can define $p_\kappa \in \Delta^*_\kappa$ s.t. $p_\kappa = (p_\tau)^\kappa$ for $\kappa,\tau \in I$, $\kappa < \tau$.

We also have, by the definition of Δ^*_κ:

$\forall\alpha < \kappa\ \exists\beta \in X \cap \kappa\ \exists q \in \mathbf{P}_\beta \cap L_\kappa\ (p \le q \wedge \Delta^q_\alpha$ is dense in $\mathbb{P}^\beta)$.

For $\alpha \in X$ set:

$$G(\alpha) = \begin{cases} F(\beta) & \text{if } X \cap \alpha \text{ has a maximal element and } \beta = \max(X\cap\alpha) \\ \sup\limits_{\beta\in X\cap\alpha} (F(\beta)) & \text{if } X \cap \alpha \text{ has no maximal element.} \end{cases}$$

Note that $G(\alpha) < F(\alpha)$.

For $\alpha \in X \cap \kappa$, $n < \omega$ define:

$X^\alpha_{n,\kappa}$ = the least $X \prec L$ s.t. $X \supseteq G(\alpha) \cup \{\kappa,\tau_1,\ldots,\tau_n\}$, where $\langle\tau_1,\ldots,\tau_n\rangle \in I^n$, $\kappa < \tau_1 < \ldots < \tau_n$.

$X^\alpha_{n,\kappa}$ is uniformly definable in $\langle\kappa,\vec{\tau}\rangle$.

<u>Claim 1</u> Let $\alpha \in X$ be s.t. $X \cap \alpha$ has no maximal element and $\sup(X\cap\alpha) = \sup\limits_{\beta\in X\cap\alpha} F(\beta))$. Let $\langle\Delta_\nu \mid \nu \in X \cap G(\alpha)\rangle$ be a sequence of dense sets in $\mathbf{P}^\alpha_\nu$.

Set $\Delta = \{p \in \mathbb{P}^\alpha \mid \forall\nu \in X \cap G(\alpha)\ ((p)_\nu \in \Delta_\nu)\}$.

Then Δ is dense in $\mathbf{P}^\alpha$.

Proof Set $\kappa = G(\alpha)$. Then $\mathbf{P}^\alpha_\nu = \mathbf{P}^\kappa_\nu \times \mathbf{P}^\alpha_\kappa$ for $\nu < \kappa$. $\mathbf{P}^\alpha_\kappa$ is κ-distributive by (d)(1) of property W.

Set $\Delta^*_\nu = \{p \in \mathbf{P}^\kappa \mid \exists q \in \mathbf{P}^\alpha_\kappa\ (p \cup q \in \Delta_\nu)\}$. Δ^*_ν is dense in $\mathbb{P}^\kappa$. By the diagonal property (d)(3),

$\Delta^* = \{p \in \mathbf{P}^\kappa \mid \forall\nu \in X \cap \kappa\ ((p)_\nu \in \Delta^*_\nu)\}$ is dense in $\mathbf{P}^\kappa$.

Set $\Delta^{**}_\nu = \{q \in \mathbb{P}^\alpha_\kappa \mid \exists p \in \mathbf{P}^\kappa_\nu\ (q \cup p \in \Delta_\nu)\}$. Δ^{**}_ν is dense in $\mathbb{P}^\alpha_\kappa$ and hence $\bigcap_{\nu \in X \cap \kappa} \Delta^{**}_\nu = \Delta^{**}$ is dense in $\mathbb{P}^\alpha_\kappa$. It is obvious that $\Delta = \{r \in \mathbb{P}^\alpha \mid \exists p \in \Delta^* \exists q \in \Delta^{**} (q \cup p = r)\}$ and Δ is dense in $\mathbb{P}^\alpha$.

Q.E.D. (Claim 1)

<u>Claim 2</u> Let $\alpha \in X$. Let $\langle \Delta_\nu \mid \nu < G(\alpha)\rangle$ be a sequence of dense sets in $\mathbb{P}^\alpha$. Set:

$$\Delta = \{p \in \mathbb{P}^\alpha \mid \forall \nu < G(\alpha)(p \in \Delta_\nu \vee \exists \beta \in X \cap \alpha (\Delta_\nu^{(p)_\beta} \text{ is dense in } \mathbb{P}^\alpha))\} .$$

Then Δ is dense in $\mathbb{P}^\alpha$.

Proof The proof follows Lemma 3.4 closely.

<u>Case 1</u> $X \cap \alpha$ has a maximal element. Set $\beta = \max(X \cap \alpha)$. Then $G(\alpha) = F(\beta)$ and Lemma 5.12 gives us the conclusion of Claim 2.

<u>Case 2</u> $X \cap \alpha$ has no maximal element and

$$\sup(X \cap \alpha) < \sup_{\beta \in X \cap \alpha} F(\beta)) = G(\alpha) .$$

Let $\langle \alpha_i \mid i < \delta\rangle$ be an enumeration of $\{\beta \in X \cap \alpha \mid \sup(X \cap \alpha) < F(\beta)\}$. Then $\delta < F(\alpha_0)$. For $0 < i < \delta$ set:

$$\Delta^*_i = \{q \in \mathbb{P}^\alpha_{F(\alpha_0)} \mid \forall \nu < F(\alpha_i) (\Delta_\nu^{(q)_{\alpha_i}} \text{ is dense in } \mathbb{P}_{F(\alpha_0)}^{\alpha_i})\} .$$

By Lemma 5.12, Δ^*_i is dense in $\mathbb{P}^\alpha_{F(\alpha_0)}$. Hence $\Delta^* = \bigcap_{i<\delta} \Delta^*_i$ is dense in $\mathbb{P}^\alpha_{F(\alpha_0)}$, since $\mathbb{P}^\alpha_{F(\alpha_0)}$ is $F(\alpha_0)$-distributive. Hence $\Delta' = \{p \in \mathbb{P}^\alpha \mid (p)_{F(\alpha_0)} \in \Delta^*\}$ is dense in $\mathbb{P}^\alpha$ and $\Delta' \subseteq \Delta$.

<u>Case 3</u> $X \cap \alpha$ has no maximal element and

$$\sup(X \cap \alpha) = \sup_{\beta \in X \cap \alpha} (F(\beta)) = G(\alpha) .$$

For $\nu \in X \cap \alpha$ set

$$\Delta_\nu^* = \{q \in \mathbf{P}_\nu^\alpha \mid \forall\beta < F(\alpha)\ (\Delta_\beta^q \text{ is dense in } \mathbf{P}^\nu)\}.$$

By Lemma 5.12, Δ_ν^* is dense in $\mathbf{P}_\nu^\alpha$. By Claim 1,

$$\Delta' = \{p \in \mathbf{P}^\alpha \mid \forall\nu \in X \cap \alpha\ ((p)_\nu \in \Delta_\nu^*\} \text{ is dense in } \mathbf{P}^\alpha.$$

But $\Delta' \subseteq \Delta$.

Q.E.D. (Claim 2)

Corollary (to Claim 2) *Let* $\alpha, \gamma \in X$, $\gamma < \alpha$. *Let* $\langle \Delta_\nu \mid \nu < G(\alpha) \rangle$ *be a sequence of dense sets in* $\mathbf{P}_\gamma^\alpha$. *Set*

$$\Delta = \{p \in \mathbf{P}_\gamma^\alpha \mid \forall \nu < G(\alpha)\ (p \in \Delta_\nu \vee \exists \beta \in X \cap [\gamma, \alpha)$$
$$(\Delta_\nu^{(p)_\beta} \textit{ is dense in } \mathbf{P}_\gamma^\beta))\}.$$

Then Δ *is dense in* $\mathbf{P}_\gamma^\alpha$.

Proof The same as Claim 2

Definition 6 Let $\alpha, \beta \in X$, $\kappa \in I$ s.t. $\beta < \alpha < \kappa$. Let $n < \omega$. $\Sigma_{n,\kappa}^{\alpha,\beta}$ is the set of $p \in \mathbf{P}_\beta^\alpha$ s.t.

(*) If $\gamma \in (\beta, \alpha] \cap X$ and $\Delta \in X_{n,\kappa}^\gamma$, where Δ is dense in $\mathbf{P}_\beta^\gamma$.

Then $(p)^\gamma \in \Delta$ or $\exists \delta \in [\beta, \gamma) \cap X$ s.t. $\Delta^{(p)_\delta^\gamma}$ is dense in $\mathbf{P}_\beta^\delta$.

It is clear that $\Sigma_{n,\kappa}^{\alpha,\beta}$ is independent of $\kappa \in I$.

<u>Claim 3</u> $\Sigma_{n,\kappa}^{\alpha,\beta}$ is dense in $\mathbf{P}_\beta^\alpha$.

Proof We closely follow the proof of Lemma 3.10. We prove the claim by induction on α. By Claim 2 the set of $p \in \mathbf{P}_\beta^\alpha$ s.t. for every $\Delta \in X_{n,\kappa}^\alpha$, where Δ is dense in $\mathbf{P}_\beta^\alpha$, $p \in \Delta$ or $\exists \delta \in [\beta, \alpha)$ $(\Delta^{(p)_\delta}$ is dense in $\mathbf{P}_\beta^\delta)$ is dense in $\mathbf{P}_\beta^\alpha$. Hence we only have to show (*) for $\gamma < \alpha$.

Case 1 $X \cap \alpha$ has a maximal element. Set $\eta = \max(X \cap \alpha)$. Since α is already taken care of the claim follows from the induction hypothesis on η.

Case 2 $X \cap \alpha$ has no maximal element and

$$\sup(X \cap \alpha) < \sup_{\delta \in X \cap \alpha} (F(\delta)) .$$

Let $\langle \alpha_i \mid i < \eta \rangle$ be the enumeration of $\delta \in X \cap \alpha$ s.t. $\sup(X \cap \alpha) < F(\delta)$. Clearly $\eta < F(\alpha_o)$. By induction each $\sum_{n,\kappa}^{\alpha_i,\alpha_o}$ is dense in $\mathbf{P}_{\alpha_o}^{\alpha_i}$ for $i > 0$.

For $0 < i < \eta$ set $\Delta_i = \{p \in \mathbf{P}_{\alpha_o}^{\alpha} \mid (p)_{\alpha_o}^{\alpha_i} \in \sum_{n,\kappa}^{\alpha_i,\alpha_o}\}$. Δ_i is dense in $\mathbb{P}_{\alpha_o}^{\alpha}$ and hence $\bigcap_{i<\eta} \Delta_i$ is dense in $\mathbf{P}_{\alpha_o}^{\alpha}$ by $F(\alpha_o)$-distributivity.

Set $\Delta^* = \{p \in \mathbf{P}_{\beta}^{\alpha} \mid (p)^{\alpha_o} \in \sum_{n,\kappa}^{\alpha_o,\beta} \wedge (p)_{\alpha_o} \in \bigcap_{i<\eta} \Delta_i\}$. Δ^* is dense in $\mathbb{P}_{\beta}^{\alpha}$. Note that if $\Delta \in X_{n,\kappa}^{\alpha_i}$ is dense in $\mathbf{P}_{\beta}^{\alpha_i}$ then $\Delta' = \{p \in \mathbf{P}_{\alpha_o}^{\alpha_i} \mid \Delta^p$ is dense in $\mathbb{P}_{\beta}^{\alpha_o}\}$ is dense in $\mathbf{P}_{\alpha_o}^{\alpha_i}$ and $\Delta' \in X_{n,\kappa}^{\alpha_i}$. Hence as in Lemma 3.10, $\Delta^* \subseteq \sum_{n,\kappa}^{\alpha,\beta}$.

Case 3 $X \cap \alpha$ has no maximal element and

$$\sup(X \cap \alpha) = \sup_{\delta \in X \cap \alpha} (F(\delta)) .$$

Let $\langle \alpha_i \mid i < G(\alpha) \rangle$ be an enumeration of $X \cap \alpha$. By induction $\sum_{n,\kappa}^{\alpha_{i+1},\alpha_i}$ is dense in $\mathbf{P}_{\alpha_i}^{\alpha_{i+1}}$. Hence

$$\Delta_i = \{p \in \mathbb{P}_{\alpha_i}^{\alpha} \mid (p)_{\alpha_i}^{\alpha_{i+1}} \in \sum_{n,\kappa}^{\alpha_{i+1},\alpha_i}\} \text{ is dense in } \mathbf{P}_{\alpha_i}^{\alpha} .$$

Set $\Delta^* = \{p \in \mathbf{P}_{\beta}^{\alpha} \mid \forall i < G(\alpha)\ (p)_{\alpha_i} \in \Delta_i\}$. Δ^* is dense in $\mathbf{P}_{\beta}^{\alpha}$ by Claim 2. As in Case 2,

$$\{p \in \mathbf{P}^{\alpha}_{\beta} \mid p \in \Delta^{*} \wedge (p)^{\alpha_o} \in \Sigma^{\alpha_o,\beta}_{n,\kappa}\} \subseteq \Sigma^{\alpha_o,\beta}_{n,\kappa} .$$

Q.E.D. (Claim 3)

Definition 7 Let $\alpha \in X$ s.t. $X \cap \alpha$ has no maximal element and

$$\sup(X \cap \alpha) = \sup_{\delta \in X \cap \alpha} (F(\delta)) = G(\alpha) .$$

Let $\kappa \in I$, $\kappa \geq G(\alpha)$, $n < \omega$.

$\Sigma^{*G(\alpha)}_{n,\kappa}$ is the set of $p \in \mathbf{P}^{G(\alpha)}$ s.t. if $\gamma \in X \cap \alpha$ and $\Delta \in X^{\gamma}_{n,\kappa}$ is dense in $\mathbf{P}^{\gamma}$ then $(p)^{\gamma} \in \Delta$ or $\exists \beta \in X \cap \gamma$ ($\Delta^{(p)^{\gamma}_{\beta}}$ is dense in $\mathbf{P}^{\beta}$).

Corollary (to Claim 3) *If* α *is as in the preceding definition, then* $\Sigma^{*G(\alpha)}_{n,\kappa}$ *is dense in* $\mathbf{P}^{G(\alpha)}$.

Proof Same as in Claim 3, Case 3, using part d(3) of the definition following Lemma 5.12, instead of Claim 1.

It is now possible to construct $p^{i}_{\kappa} \in \mathbf{P}^{\kappa}$ for $i < \omega$ and $\kappa \in I$ uniformly L-definable in $\langle \kappa, \tau_1, \ldots, \tau_i \rangle \in I^{n+1}$ as follows:

$p^{o}_{\kappa} = p_{\kappa}$, previously defined.

p^{i+1}_{κ} = the L-least $p \in \mathbf{P}^{\kappa}$ s.t. $p \leq p^{i}_{\kappa}$ and $p \in \Sigma^{*\kappa}_{\kappa,i+1}$.

Clearly $p^{i}_{\kappa} = (p^{i}_{\tau})^{\kappa}$ for $\kappa < \tau$, $\kappa, \tau \in I$.

Set $G = \bigcup_{i \in \omega} \bigcup_{\kappa \in I} p^{i}_{\kappa}$. By previous techniques, G is $\mathbf{P}$-generic over L and I is preserved in $L[G]$.

Q.E.D.

6 . The fine-structural lemmas

6.1 AN INTRODUCTION

Here we shall prove two of the three auxiliary lemmas that were quoted in 2.3. The first two, 2.3 and 2.4 are related, and the proof of the second is a reformulation of that of the first. The rather general, almost category like structure we shall define, will yield these two lemmas as well as the $\square_\beta$ and $\square$ principles of [J1] (Theorems 6.36 and 6.47).

The third lemma, 2.5, provides the web of Cohen generic sets, and is proved in Chapter 7 using the notion of a 'quasi-morass'. Previous knowledge of morasses, however, is unnecessary although we shall be assuming that the reader is familiar with the fine structure of L (see [J1] or [DEV]).

We shall require a relativised fine structure for L^a for various sets of ordinals a, and partly to introduce this and partly to fix notation we shall first present a list of the salient facts of [DEV] Chapters 6 and 7 stated for a relativised J-hierarchy which we shall now define. The proofs of all the following are straightforward adaptations of those of L.

Lemma 6.1 *There is a rudimentary-in-A* (rud_A) *function* s_A *such that whenever* U *is transitive* $U \cup \{U\} \subseteq s_A(U)$ *and* $\bigcup_{n>\omega} s_A^n(U)$ *is the* rud_A *closure of* U *(denoted by* $\mathrm{rud}_A(U)$ *)*.

Definition $|J_0^A| = \emptyset$

$$|J_{\alpha+1}^A| = \mathrm{rud}_A(|J_\alpha^A|)$$

$$|J^A_\lambda| = \bigcup_{\nu<\lambda} |J^A_\nu| \quad \text{for limit } \lambda$$

Lemma 6.2 *Each* $|J^A_\nu|$ *is transitive;* $\alpha \le \beta \longrightarrow |J^A_\alpha| \subseteq |J^A_\beta|$ *and* $\mathrm{rank}(|J^A_\alpha|) = \omega\alpha = \mathrm{On} \cap |J^A_\alpha|$.

We shall attempt to adhere to the notation that $|J^A_\alpha|$ is the domain of J^A_α which in turn equals the structure $\langle J^A_\alpha, A\rangle$. So, for example:

$$P(|J^A_\alpha|) \cap |J^A_{\alpha+1}| \supseteq \mathrm{Def}(J^A_\alpha)$$

This formulation of $|J^A_{\alpha+1}|$ ensures that each $|J^A_\alpha|$ is amenable.

We have the auxiliary S_A-hierarchy :

$$S^A_0 = \emptyset$$

$$S^A_{\alpha+1} = s_A(S^A_\alpha)$$

$$S^A_\lambda = \bigcup_{\nu<\lambda} S^A_\nu \quad \text{for } \mathrm{Lim}(\lambda)$$

In the sequel we shall take $\Sigma^M_n(N)$ to mean Σ_n in M with parameters from (w.p.f.) N

$\Sigma_n(M)$ to mean $\Sigma^M_n(M)$ and

Σ^M_n to mean $\Sigma^M_n(\emptyset)$, i.e. 'uniformly $\Sigma_n(M)$ ' .

We then have the following:

Lemma 6.3 $\alpha \le \beta \longrightarrow S^A_\alpha \subseteq S^A_\beta$. $|J^A_\nu| = S^A_{\omega\nu}$.

Lemma 6.4 *Each* J^A_α *is amenable and if* $\langle J^A_\alpha, B\rangle$ *is amenable then there is a* Σ_1 *Skolem function* $h_{\langle J^A_\alpha, B\rangle}$ *for* $\langle J^A_\alpha, B\rangle$ *,* *i.e. if* $P \in \Sigma_1^{\langle J^A_\alpha, B\rangle}(\{x\})$ *for some* $x \in |J^A_\alpha|$ *then* $\exists y P(y) \longrightarrow \exists i < \omega P(h(i,x))$.

Lemma 6.5

a) $\langle S^A_\xi | \xi < \omega\nu\rangle$, $\langle J^A_\xi | \xi < \nu\rangle$ *are* $\Sigma_1^{(\langle J^A_\nu, A\rangle)}$.

b) There exists a $\Delta_1^{J^A_\nu}$ *map of* $[\omega\nu]^{<\omega}$ *(= finite subsets of* $\omega\nu$ *) onto* J^A_ν .

c) There exists a $\Delta_1(J^A_\nu)$ *map of* $\omega\nu$ *onto* $[\omega\nu]^{<\omega}$.

Lemma 6.6 *There exists a well-ordering* $<_{J^A}$ *of* L^A *such that setting* $<_{J^A_\nu} = <_{J^A} \cap J^A_\nu$ *we have*

a) $\nu \le \eta \longrightarrow <_{J^A_\nu}$ *is an initial segment of* $<_{J^A_\eta}$.

b) $<_{J^A_\nu}$, $\langle <_{J^A_\xi} | \xi < \nu\rangle$ *are* $\Sigma_1^{J^A}$.

Lemma 6.7 *The relation* $\{\vec{x} \mid J^A_\nu \models \phi^m_i(\vec{x})\}$ *is* $\Sigma_m^{J^A_\nu}$ *for* $m \ge 1$ *, where* ϕ^m_i *is a standard enumeration of the* Σ_m *formulae.*

Lemma 6.8 *There is a* Π_2 *statement "I'm a* J^A_ν*" such that for all transitive* $\langle U,A\rangle$ $\langle U,A\rangle \models \phi \leftrightarrow U = J^A_\nu$.

ϕ *is* $\forall \nu \exists \tau\ \nu<\tau \wedge \forall x \exists s \exists v\ (T(s) \wedge x \in s(v))$ *where* $T(s) \longleftrightarrow \exists \xi\ s = \langle S^A_\nu : \nu < \xi\rangle$.

Now for the condensation lemma:

Lemma 6.9

a) If $X \prec_{\Sigma_1} J^A_\nu$ *, then* $\exists \bar{A} \exists \bar{\nu}\ \langle X, A \cap X\rangle \simeq \langle J^{\bar{A}}_{\bar{\nu}}, \bar{A}\rangle$.

b) *If* $A \subset J_\nu^A, \beta \geq \nu$ *and* $|J_\nu^A| \subset X \prec_{\Sigma_1} J_\beta^A$ *then* $\exists\, \bar\beta \geq \nu$ *such that* $X \simeq J_{\bar\beta}^A$. *We can do a bit better than this already:*

c) *If* $A \subset J_\nu^A$, $\tau < \nu \leq \beta$, $J_\nu^A \models \forall x, \bar{\bar{x}} \leq \tau$ *and* $\tau \subset X \prec_{\Sigma_1} J_\beta^A$ *then* $\exists\, \beta', \bar\beta \geq \tau$, $X \simeq J_{\bar\beta}^{A'}$, *where* $A' = A \cap J_{\beta'}^A$.

Lemma 6.10 *Let* $\langle J_{\nu'}^{A'}, A', B'\rangle$ *be amenable and suppose* $\pi: \langle J_{\nu'}^{A'}, A'\rangle \xrightarrow[\Sigma_0]{} \langle J_\nu^A, A\rangle$ *cofinally, i.e.* $\pi''\omega\nu'$ *is cofinal in* $\omega\nu$. *Set* $B = \bigcup_{\xi<\omega\nu'} \pi(B' \cap S_\xi^{A'})$ *then*

a) $\langle J_\nu^A, A, B\rangle$ *is amenable.*

b) $\pi: \langle J_{\nu'}^{A'}, A', B'\rangle \xrightarrow[\Sigma_1]{} \langle J_\nu^A, A, B\rangle$.

c) B *is the unique* B *such that* $\pi: \langle J_{\nu'}^{A'}, A', B'\rangle \xrightarrow[\Sigma_0]{} \langle J_\nu^A, A, B\rangle$.

All of the proofs of the above are in [Dev]. Now a lemma which will be of considerable use later but here seems an appropriate place to state it.

Lemma 6.11 (The Interpolation Lemma) *Suppose we have the following diagram:*

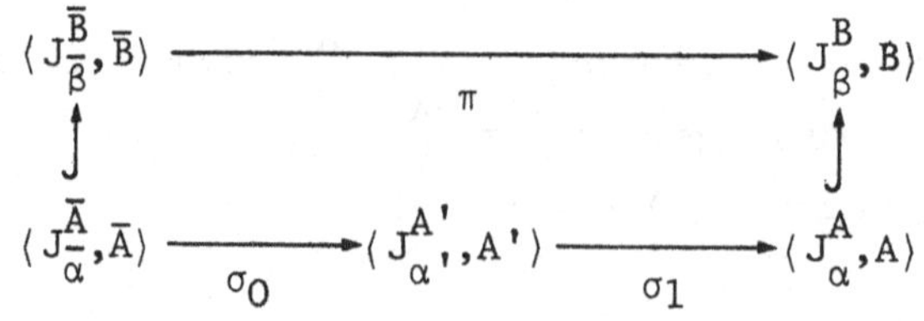

such that the whole diagram is commutative, all maps are Σ_0, σ_0 *is cofinal and* $\bar\alpha$ *is regular in* $J_{\bar\beta}^{\bar B}$ *and* $H_{\omega\bar\alpha}^{J_{\bar\beta}^{\bar B}} \subseteq J_{\bar\alpha}^{\bar A}$.
Then there exist an amenable $\langle J_{\beta'}^{B'}, B'\rangle$ *and maps* π_0, π_1 *such that the following diagram commutes,* π_0, π_1 *are* Σ_0, π_0 *is*

cofinal and $\alpha' = \pi_0(\bar\alpha)$ *if* $\bar\alpha < \bar\beta$.

$$\begin{array}{ccccc}
& & \overset{\pi}{\curvearrowright} & & \\
\langle J^{\bar B}_{\bar\beta}, \bar B\rangle & \xrightarrow[\pi_0]{} & \langle J^{B'}_{\beta'}, B'\rangle & \xrightarrow[\pi_1]{} & \langle J^{B}_{\beta}, B\rangle \\
\uparrow & & \uparrow & & \uparrow \\
\langle J^{\bar A}_{\bar\alpha}, A\rangle & \xrightarrow[\sigma_0]{} & \langle J^{A'}_{\alpha'}, A'\rangle & \xrightarrow[\sigma_1]{} & \langle J^{A}_{\alpha}, A\rangle
\end{array}$$

Proof If $\bar\alpha = \bar\beta$ there is nothing to prove. So we may take $\alpha = \pi(\bar\alpha)$. One can also take without loss of generality π to be cofinal in J^B_β since if this is not the case we may replace β by β^* where $\omega\beta^* = \sup \pi''\omega\bar\beta$. Then π is Σ_1 preserving. Set $X = \Sigma_1$ skolem hull in J^B_β of $\mathrm{ran}(\pi) \cup \mathrm{ran}(\sigma_1)$. Then $X \prec_{\Sigma_1} J^B_\beta$.

Claim $X \cap J^A_\alpha = \mathrm{ran}(\sigma_1)$.

Proof ($\supset$) is trivial. Consider ($\subset$). Let $x \in X \cap J^A_\alpha$. Then there are $z = \pi(\bar z)$, $y = \sigma_1(y')$ such that x = the unique $x \in J^B_\beta$ s.t.

$$J^B_\beta \models \phi[x,y,z] .$$

Since $\pi''\omega\bar\beta$ is cofinal in J_β there is $\nu = \pi(\bar\nu)$ s.t. $x,y,z \in S^B_\nu$ and

$$S^B_\nu \models \phi[x,y,z] .$$

Since σ_0 is cofinal in $J^{A'}_{\alpha'}$ there is $\eta' = \sigma_0(\bar\eta)$ such that $y' \in S^{A'}_{\eta'}$. Hence $y \in S^A_\eta$, where $\eta = \pi(\bar\eta)$. Define $\bar f : S^{\bar A}_{\bar\eta} \to J^{\bar A}_{\bar\alpha}$ by:

$$\bar f(u) = \text{the unique } v \in S^{\bar B}_{\bar\nu} \text{ s.t. } \langle S^{\bar B}_{\bar\nu}, \bar B \cap S^{\bar B}_{\bar\nu}\rangle \models \phi[v,u,\bar z] \quad \text{if such a } v \text{ exists}$$
$$= 0 \quad \text{otherwise.}$$

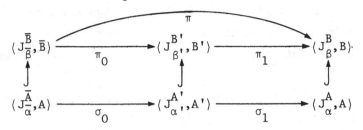

Then $\bar{f} \in J_{\bar{\beta}}^{\bar{B}}$ and by our assumption is thus in $J_{\bar{\alpha}}^{\bar{A}}$. Set $f' = \sigma_0(\bar{f})$, $f = \sigma_1(f') = \pi(\bar{f})$. Then f has the same definition in J_{β}^{B} and hence $x = f(y)$. Thus $x = \sigma_1 f'(y') \in \operatorname{ran}(\sigma_1)$. QED Claim.

Now let $\pi_1 : J_{\beta'}^{B'} \simeq \langle X, B \cap X \rangle$. Then $\pi_1 \restriction J_{\alpha'}^{A'} = \sigma_1$ by the above Claim. Hence $A' = B' \cap J_{\alpha'}^{B'}$. Since $\operatorname{ran}(\pi) \subset X = \operatorname{ran}(\pi_1)$ we may set: $\pi_0 = \pi_1^{-1}\pi$. Then $\pi_0 : J_{\bar{\beta}}^{\bar{B}} \xrightarrow[\Sigma_1]{} J_{\beta'}^{B'}$ cofinally and $\pi_1\pi_0 = \pi$. Clearly $\alpha = \pi(\bar{\alpha}) \in X$. But then $\pi_1(\omega\alpha') = \min \pi_1''(\omega\beta'/\omega\alpha') = \min(\mathrm{On} \cap (\operatorname{ran}(\pi_1)/\operatorname{ran}(\sigma_1)) = \omega\alpha$ by the above Claim. Hence $\pi_0(\bar{\alpha}) = \pi_1^{-1}\pi(\bar{\alpha}) = \alpha'$. QED

Definition Suppose that there is no Σ_n map of a bounded subset of some ρ less than ν onto ρ defined over J_ν^A then ρ is a Σ_n cardinal in J_ν^A.

Now given J_ν^a with $a \subset J_\nu^a$ we can proceed as in [J1] to carry out the normal fine structure theory using lemmas 6.1-6.9 above the supremum of a. If $\beta \geq \nu$ is such that ν is still a Σ_n-cardinal in J_β^a then we can define the notion of Σ_n-projectum and Σ_n-mastercode and Σ_n-parameter just as there, e.g.:

Definition $\rho_\beta^{n,a}$ = the largest $\rho \leq \beta$ such that $\langle J_\rho^a, A\rangle$ is amenable for all $A \in \Sigma_n(J_\beta^a) \cap P(\omega\rho)$. A Σ_n-mastercode for J_ν^a is a set $A \subset |J_{\rho_\beta^n}^a, a|$, $A \in \Sigma_n(J_\beta^a)$ such that for all $m \geq 1$

$$\Sigma_m(\langle J_{\rho_\beta^n}^a, a, A\rangle) = P(|J_{\rho_\beta^n}^a, a|) \cap \Sigma_{n+m}(J_\beta^a) .$$

Then we have:

Lemma 6.12 (The Uniformization Theorem) *Let* a, β, n *be as above. Then* *a)* J^a_β *is* Σ_{n+1} *uniformizable.*

b) There is a $\Sigma_n(J^a_\beta)$ *map of* $\omega\rho^{n,a}_\beta$ *onto* $|J^a_\beta|$ *. (Unless* $n = 0$ *in which case the map is* Σ_1*).*

c) J^a_β *has a* Σ_n*-mastercode.*

The other equivalent characterisations of ρ as the least ordinal such that there is a map of it onto the whole structure and such that it is the least ordinal, a subset of which is definable over the whole structure are as before. The definition of the canonical mastercodes etc. are the same with just one variation:

Definition Let a and β be as above. Set $p^{0,a}_\beta = \emptyset$. $A^{0,a}_\beta = a$.

$$p^{n+1,a}_\beta = \leq_{J^a_\beta}\text{-least } p \in J^a_{\rho^{n,a}_\beta} \text{ such that every } x \in J^a_{\rho^{n,a}_\beta} \text{ is } \Sigma_1\text{-definable in } \langle J^a_{\rho^{n,a}_\beta}, A^{n,a}_\beta\rangle \text{ wpf } \{p\} \cup J^a_{\rho^{n+1,a}_\beta} .$$

$$A^{n+1,a}_\beta = \{\langle i,x\rangle \mid i \in \omega,\ x \in J^a_{\rho^{n+1,a}_\beta} \wedge \langle J^a_{\rho^{n,a}_\beta}, A^{n,a}_\beta\rangle \models \phi_i[x, p^{n+1,a}_\beta]\}$$

Then $A^{n+1,a}_\beta$ is a Σ_{n+1} mastercode for J^a_β . $\langle J^a_{\rho^{n,a}_\beta}, A^{n,a}\rangle$ is then amenable and a is rud in $A^{n,a}_\beta$. and $|J^a_{\rho^{n,a}_\beta}| = |J^{A^{n,a}}_{\rho^{n,a}_\beta}|$.

Without going into acceptability considerations, this is almost as far as one may go with the straightforward relativisation. However if $J^A_\nu \models$ "$\forall x\ \bar{\bar{x}} \leq \tau$" $(\tau < \nu < \beta)$ we may extend the above definitions when the projectum drops to τ . In the following let m be least such that ν is no longer a Σ_m-cardinal (or, with the above hypothesis $\rho^m_\beta = \tau$) .

Lemma 6.13 (Extension of embeddings) *Let* $a, \nu, \beta, m\ (\tau)$ *be as above. Let* $\rho = \rho_\beta^{n,a}, A_\beta = A^{n,a}$ *for* $n < m\ (n \leq m)$. *Let* $i \geq 0$ *be such that*

$$\pi : J_{\bar{\rho}}^{\bar{A}} \xrightarrow[\Sigma_i]{} J_\rho^A$$

Set $\bar{a} = \pi^{-1}{}''a$. *Then:*

a) there is a unique $\bar{\beta}$ *such that* $\bar{\rho} = \rho_{\bar{\beta}}^{n,\bar{a}}$ *and* $\bar{A} = A_{\bar{\beta}}^{n,\bar{a}}$.
(so $|J_{\bar{\rho}}^{\bar{A}}| = |J_{\bar{\rho}}^{\bar{a}}|$ *).*

b) there is a unique $\tilde{\pi} \supset \pi$ *such that*

$$\tilde{\pi} : J_{\bar{\beta}}^{\bar{a}} \xrightarrow[\Sigma_{n+i}]{} J_\beta^a \quad \text{and} \quad \tilde{\pi}(p_{\bar{\beta}}^{j,\bar{a}}) = p_\beta^{j,a} \quad j \leq n\ (j \leq m)$$

$$\tilde{\pi} \restriction J^{\bar{a}}_{\rho_{\bar{\beta}}^{n-j}} : J^{\bar{A}^{n-j,\bar{a}}}_{\rho_{\bar{\beta}}^{n-j,\bar{a}}} \xrightarrow[\Sigma_{i+j}]{} J^{A^{n-j,a}}_{\rho_\beta^{n-j,a}}.$$

The proof is just as in [J1].

6.2 THE LEMMAS

We shall now proceed to describe the structure which will lead to a proof of 2.3 (and incidentally Theorem 6.36). This will be used to generate □-like sequences and yields quite an intuitive, although slightly longer, proof of e.g. $\square_\beta$. (For example the C_ν's will be readily definable.)

Definition S = the class of pairs $s = \langle \nu_s, a_s \rangle$ s.t.

I ν is p.r. closed or a limit of p.r. closed ordinals, with $\nu > \omega$

II $a_s \subset \nu$

III $\exists \alpha = \alpha_s$ s.t. $\alpha > \omega$ and $J_\nu^a \models$ "α's the largest cardinal"

IV ν is not a cardinal in L^a .

For $s \in S$ we shall write J_s for $J_{\nu_s}^{a_s}$ and J_β^s for $J_\beta^{a_s}$ (for $\beta \geq \nu$).

Now if $s \in S$, $s = \langle \nu, a \rangle$, $\alpha = \alpha_s$, then there exists a least $\mu \geq \nu$ s.t. ν is not a Σ_ω cardinal in J_μ^s (i.e. $\exists$a J_μ^s-definable map of α onto ν). Set μ_s equal to the least such μ . Then there will also be a least $n \geq 1$ s.t. ν is not a Σ_n-cardinal in J_μ^s .

Definition Set

$$n_s = n$$

$$\rho_s = \rho_{\mu_s}^{n-1,a}$$

$$A_s = A_\mu^{n-1,a}$$

p_s = the $<_{J_\rho^A}$-least p s.t.

$h''(\omega \times |J_\alpha^A| \times \{p\}) = |J_\rho^A|$, where $h = h_{J_\rho^A}$ is the canonical Σ_1 Skolem function for J_ρ^A . We shall often abuse notation by writing the last line as: $h(J_\alpha^A \cup \{p\}) = |J_\rho^A|$. By virtue of §1 Lemma 6.5(b) this J_α^A can be replaced by J_α or equivalently $[\omega\alpha]^{<\omega}$.

Definition Let $s = \langle \nu, a \rangle \in S$. Let $\alpha = \alpha_s$, $A = A_s$, $\rho = \rho_s$, $p = p_s$. Then

$\mathbb{F}(s) =_{\mathrm{Df}}$ the set of maps

$$f : J_\rho^A \xrightarrow[\Sigma_1]{} J_\rho^A \text{ s.t. } \langle \alpha, p \rangle \in \mathrm{ran}(f)$$

and $\nu \in \mathrm{ran}(f)$ if $\nu < \rho$.

Lemma 6.14 $f \in \mathbb{F}(s)$ *is uniquely determined by* $\alpha \cap \operatorname{ran}(f)$.

Proof Let $h = h_{J^A_\rho}$. Then $|J^A_\rho| = h(J^A_\rho \cup \{p\})$

$[\alpha = \alpha_s, \ p = p_s, \ \rho = \rho_s, \ A = A_s]$.

Let g be the $\Delta_1(J^A_\alpha)$ map of $\omega\alpha$ onto $|J^A_\alpha|$ guaranteed by Lemma 6.5(b) and (c) of §1. Then $|J^A_\alpha| = g''\alpha \quad (\alpha = \omega\alpha)$ and $|J^A_\rho| = h(g''\alpha \cup \{p\})$. Let $Y = \operatorname{ran}(f)$. As $\alpha, p \in Y$ and $Y \prec_{\Sigma_1} J^A_\rho$, we have

$$Y = h(g''(\alpha \cap Y) \cup \{p\}) .$$

So Y and hence f are determined by $Y \cap \alpha$. QED

Lemma 6.15 *Let* $f \in \mathbb{F}(s)$ s.t. $f : J^{\bar{A}}_{\bar{\rho}} \xrightarrow[\Sigma_1]{} J^A_\rho$.
Let $f(\langle\alpha,p\rangle) = \langle\alpha,p\rangle$ $[\rho, A, \alpha, p$ *as above*]. *Set* $\bar{\nu} = f^{-1}{}''\nu$, $\bar{a} = f^{-1}{}''a$, $\bar{s} = \langle\bar{\nu},\bar{a}\rangle$.

Then $\bar{s} \in S$, $\bar{\alpha} = \alpha_{\bar{s}}$, $\bar{\rho} = \bar{\rho}_s$, $\bar{A} = A_{\bar{s}}$, $\bar{p} = p_{\bar{s}}$ *and* $n_{\bar{s}} = n_s$.

Proof $a \subset \nu$ is rud in A . Thus $\bar{a} \subset \bar{\nu}$ is rud in $\bar{A}$ by the same definition. Clearly $\bar{\nu} \leq \bar{\rho}$ is regular in $J^{\bar{A}}_{\bar{\rho}}$ and $J^{\bar{A}}_{\bar{\nu}} = J^{\bar{a}}_{\bar{\nu}}$. Clearly also:

$$f \restriction J^{\bar{a}}_{\bar{\nu}} : J^{\bar{a}}_{\bar{\nu}} \xrightarrow[\Sigma_1]{} J^a_\nu .$$

But then ν is p.r. closed (or a limit of such) and

$$J^{\bar{a}}_{\bar{\nu}} \models \text{"}\bar{\alpha} \text{ is the largest cardinal"}.$$

Now if $\bar{\mu} \geq \bar{\rho}$ s.t. $\bar{\rho} = \rho^{n-1,\bar{a}}_{\bar{\mu}}$, $\bar{A} = A^{n-1,\bar{a}}_{\bar{\mu}}$ by §1 Lemma 6.13 and if $\bar{h} = h_{J^{\bar{A}}_{\bar{\rho}}}$ then

(1) $|J^{\bar{A}}_{\bar{\rho}}| = \bar{h}(J^{\bar{A}}_{\bar{\alpha}} \cup \{\bar{p}\})$, as $|J^{A}_{\rho}| = h(J^{A}_{\alpha} \cup \{p\})$.

But then $\bar{s} \in S$, $\bar{\alpha} = \alpha_{\bar{s}}$, $\bar{\mu} = \mu_{\bar{s}}$, $n = n_{\bar{s}}$, $\bar{\rho} = \rho_{\bar{s}}$, $\bar{A} = A_{\bar{s}}$. $\bar{p} \geq_{J^{\bar{A}}} p_{\bar{s}}$ by (1). Suppose $\bar{p} >_{J^{\bar{A}}} \bar{q} = p_{\bar{s}}$. Then there is $\bar{x} \in |J^{\bar{A}}_{\bar{\alpha}}|$ s.t. $\bar{p} = \bar{h}(i,\bar{x},\bar{q})$ for some $i < \omega$. Let $f(\langle\bar{x},\bar{q}\rangle) = \langle x,q\rangle$. Then $p = h(i,x,q)$ where $q <_{J^A} p$ and $x \in |J^{A}_{\alpha}|$. Hence $|J^{A}_{\rho}| = h(J^{A}_{\alpha} \cup \{q\})$; hence $p >_{J^A} q \geq_{J^A} p_s$. Contradiction! Hence $\bar{p} = p_{\bar{s}}$. QED

Definition $\mathbb{F}$ = the class of f where f is a triple $\langle\bar{s},|f|,s\rangle$ such that $s \in S$ and $\exists\, g \in \mathbb{F}(s)$ s.t. $\bar{\nu} = g^{-1}{}''\nu$ $\bar{a} = g^{-1}{}''a$, and $|f| = g \upharpoonright \bar{\nu}$.
We know by Lemma 6.1 that g is uniquely determined by $|f|$ and so we can set without ambiguity $\hat{f} = g$. By Lemma 6.2 we know $\bar{s} \in S$ and

$$\hat{f} : J^{A_{\bar{s}}}_{\rho_{\bar{s}}} \xrightarrow[\Sigma_1]{} J^{A_s}_{\rho_s} \text{ with } \hat{f}(\langle\alpha_{\bar{s}},p_{\bar{s}}\rangle) = \langle\alpha_s,p_s\rangle .$$

Also note that $\bar{\nu} < \rho_{\bar{s}} \longleftrightarrow \nu < \rho_s$, in which case $\hat{f}(\bar{\nu}) = \nu$ and $\hat{f}(\bar{a}) = a$.
For arbitrary $f = \langle\bar{s},|f|,s\rangle$, $g = \langle\bar{t},|g|,t\rangle$ we put:
$fg = \langle\bar{t},|f|.|g|,s\rangle$ if $t = \bar{s}$ and $\mathrm{ran}(|g|) \subset \mathrm{dom}(|f|)$.

$$f^{-1}g = \langle\bar{t},|f^{-1}|.|g|,\bar{s}\rangle \text{ if } t = s \text{ and } \mathrm{ran}(|g|) \subset \mathrm{ran}(|f|) .$$

We also set $d(f) = \bar{s}$, $r(f) = s$, but $\mathrm{dom}(f)$, $\mathrm{ran}(f)$ are defined as $\mathrm{dom}(|f|)$, $\mathrm{ran}(|f|)$ respectively. Similarly we put $f''X$ for $|f|''X$, $f \upharpoonright X$ for $|f| \upharpoonright X$, $f(x) = |f|(x)$ etc. The following couple of lemmas are immediate consequences of this definition.

Lemma 6.16

a) $f,g \in \mathbb{F}$, $r(g) = d(f)$, *and* $\mathrm{ran}(g) \subset \mathrm{dom}(f)$
$\longrightarrow fg \in \mathbb{F}$.

b) $f,g \in \mathbb{F}$, $r(f) = r(g)$, *and* $\mathrm{ran}(g) \subset \mathrm{ran}(f)$
$\longrightarrow f^{-1}g \in \mathbb{F}$.

Definition Set $id_s = \langle s, id\restriction\nu, s\rangle$. Then

Lemma 6.17

a) $s \in S \longrightarrow id_s \in \mathbb{F}$.

b) $f \in \mathbb{F} \longrightarrow f = id_{d(f)} \longleftrightarrow d(f) = r(f)$

$\longleftrightarrow \alpha_{d(f)} = \alpha_{r(f)}$.

Definition We write "$f:\bar{s} \Rightarrow s$" for "$f \in \mathbb{F}$, $\bar{s} = d(f)$, and $s = r(f)$" and we also write "$f \Rightarrow s$" for "$\exists\bar{s}(f:\bar{s} \Rightarrow s)$".

Def. $S_\alpha = \{s \in S \mid \alpha_s = \alpha\}$. Note S_α closed in $\sup S_\alpha$.

Definition Let $s \in S$, $\lambda \leq \nu_s$ then $s\restriction\lambda = \langle\lambda, a_\nu \cap \lambda\rangle$.

Definition Let $f:\bar{s} \Rightarrow s$ then $\lambda(f) = \sup f''\nu_s$.

Lemma 6.18 *Let* $f:\bar{s} \Rightarrow s$, $\lambda = \lambda(f)$. *Then* $s\restriction\lambda \in S_{\alpha_s}$ *and there is a unique* $f_o:\bar{s} \Rightarrow s\restriction\lambda$ s.t. $|f_o| = |f|$.

Proof Set $a' = a \cap \lambda$, $s' = \langle\lambda, a'\rangle = s\restriction\lambda$. Note first that λ is p.r. closed (or a limit of such) and $J_{s'} \models$ "α is the largest cardinal", since $\hat{f}\restriction J_{\bar{s}} : J_{\bar{s}} \xrightarrow[\Sigma_1]{} J_{s'}$.
Now let $\bar{\rho} = \rho_{\bar{s}}$, $\bar{A} = A_{\bar{s}}$, $\bar{p} = p_{\bar{s}}$, and similarly ρ, A, p with $n_{\bar{s}} = n_s = n$, using the Extension of Embeddings lemma.
By the Interpolation Lemma there are

$J^{A'}_{\rho'}, g_0, g_1$ s.t.

$$\begin{array}{ccccc} & & \hat{f} & & \\ \langle J^{\bar{A}}_{\bar{\rho}},\bar{A}\rangle & \xrightarrow{\quad g_0 \quad} & \langle J^{A'}_{\rho'},A'\rangle & \xrightarrow{\quad g_1 \quad} & \langle J^{A}_{\rho},A\rangle \\ \uparrow & & \uparrow & & \uparrow \\ \langle J^{\bar{A}}_{\bar{\nu}},\bar{A}\cap J^{\bar{A}}_{\bar{\nu}}\rangle & \xrightarrow{\hat{f}\restriction J^{\bar{A}}_{\bar{\nu}}} & \langle J^{A}_{\lambda},A\cap J^{A}_{\lambda}\rangle & \hookrightarrow & \langle J^{A}_{\nu},A\cap J^{A}_{\nu}\rangle \end{array}$$

where all maps are Σ_0 and g_0 is cofinal and hence Σ_1. By the existence of g_1 $\exists\, \mu'$ s.t.

$$\rho' = \rho^{n-1,a'}_{\mu'}, \quad A' = A^{n-1,a'}_{\mu'}. \quad \text{Let } \bar{\alpha} = \alpha_{\bar{s}},\ \alpha = \alpha_s .$$

Set $p' = g_0(\bar{p}) = g_1^{-1}(p)$. Let $\bar{h}, h', h$ be the uniform Σ_1 Skolem functions of the three structures at the top of the diagram. Then

$$\lambda \cap h'(J^{A'}_{\alpha} \cup \{p'\})$$

is cofinal in λ since $\bar{\nu} \subset \bar{h}(J^{\bar{A}}_{\bar{\alpha}} \cup \{\bar{p}\})$ and $g_0{}''\bar{\nu} = f''\bar{\nu}$ is cofinal in λ. Thus λ is not Σ_n-regular and hence not a Σ_n-cardinal in $J^{s'}_{\mu'}$. Hence $s' \in S$, $\mu' = \mu_{s'}$, $n = n_{s'}$, $\rho' = \rho_{s'}$, $A' = A_{s'}$ and $\alpha = \alpha_{s'}$. It only remains to prove:

Claim $p' = p_{s'}$.

$(p' \geq_{J^{A'}} p_{s'})$. Let $X = h'(J^{A'}_{\alpha} \cup \{p'\})$.

Set $\pi : \langle J^{A*}_{\rho*}, A*\rangle \xrightarrow{\sim} \langle X, A' \cap X\rangle$. We know that $|J^{A'}_{\lambda}| \subset X$ since $X \cap \lambda$ is cofinal in λ and α is the largest cardinal in $J^{A'}_{\lambda}$. But then $\pi \restriction |J^{A'}_{\lambda}| = \mathrm{id} \restriction |J^{A'}_{\lambda}|$, and $\pi(\lambda) = \lambda$ if $\lambda < \rho'$. Again by Ext. of Emb. $\exists\, \mu*$ s.t. $\rho* = \rho^{n-1,a'}_{\mu*}$, $A* = A^{n-1,a'}_{\mu*}$. But then $\rho* = \rho_{s'} = \rho'$, $A* = A_{s'} = A'$. Hence $h* = h'$ and

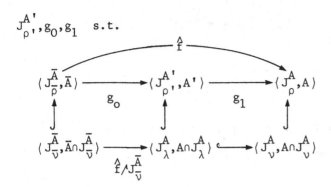

$p_{s'} \leq_{J^{A'}} p^* \leq_{J^{A'}} p'$, where $p^* = \pi^{-1}(p')$, since

$|J^{A'}_{\rho'}| = h'(J^{A'}_{\alpha} \cup \{p^*\})$.

$(p_{s'} \geq_{J^{A'}} p')$

If not then, letting $q' = p_{s'}$, we have

$q' <_{J^{A'}} p' \wedge p' = h'(i,x,q')$ for some $x \in J_\alpha$, $i < \omega$.

But then $p = h(i,x,q)$ where $q = g_1(q') < p$. Hence $|J^A_\rho| = h(J_\alpha \cup \{q\})$ and $p >_{J^A} q \geq_{J^A} p_s$. Contradiction!

QED (Claim)

But we have now shown that $g_0 \in \mathbb{F}(s)$. Hence $f_0 \in \mathbb{F}$ where $f_0 = \langle \bar{s}, g_0/\bar{\nu}, s' \rangle = \langle \bar{s}, |f|, s' \rangle$.

QED

Definition Such an f_0 will be called the *reduct* of f . Formally: Let $f : \bar{s} \Rightarrow s$, $\lambda = \lambda(f)$, then

$$f_0 = \langle \bar{s}, |f|, s/\lambda \rangle =_{df} \mathrm{red}(f) .$$

As a corollary to this proof we have two technical lemmas for later use.

Lemma 6.19 *Let* $f : \bar{s} \Rightarrow s$. *Then* $\lambda(f) < \nu \longleftrightarrow \sup \hat{f}''\omega\rho_{\bar{s}} < \omega\rho_s$.

Proof $(\leftarrow)$ Suppose not. Then in the diagram of Lemma 6.18 we can take $\omega\rho' = \sup \hat{f}''\omega\bar{\rho}$ and $g_1 = \mathrm{id}/J^{A'}_{\rho'}$. But then $\rho' = \rho_{s/\lambda} = \rho_s = \rho$ as $\lambda = \nu$. Contradiction.

$(\rightarrow)$ Again suppose not. Then g_1 is cofinal in the same diagram and so is Σ_1 . So $g_1 \in \mathbb{F}(s)$. But then $f_1 \in \mathbb{F}$ where $f_1 = \langle s', g_1/\lambda, s \rangle$. But $\alpha_{s'} = \alpha_s$; so $f_1 = \mathrm{id}_s$. Contradiction again.

QED

Lemma 6.20 *Let* $f : \bar{s} \Rightarrow s$ s.t. $\lambda(f) = \nu_s$. *Let* $s' = \langle \nu', a' \rangle$ *where* $a' \subset \nu'$. *Let*

$$g_0 : J_{\bar{s}} \xrightarrow[\Sigma_0]{} J_{s'} \,, \quad g_1 : J_{s'} \xrightarrow[\Sigma_1]{} J_s \quad \text{s.t. } g_1 g_0 = \hat{f} \restriction J_{\bar{s}} \,.$$

Then there are $k_0 : \bar{s} \Rightarrow s'$, $k_1 : s' \Rightarrow s$ s.t. $|k_0| = g_0 \restriction \nu_{\bar{s}}$, $|k_1| = g_1 \restriction \nu_s$.

Proof Both g_0 and g_1 are cofinal. So, setting ν, a, ρ, A, $\bar{\nu}, \bar{a}, \bar{\rho}, \bar{A}$ to their usual meanings, we have by the Interpolation Lemma the following diagram

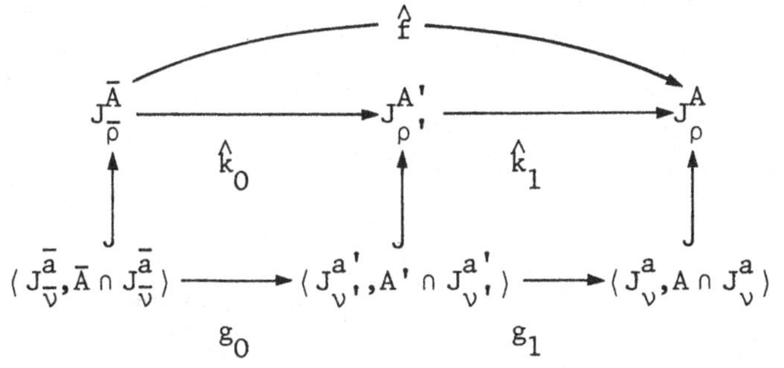

where $A' \cap J^{a'}_{\nu} = \bigcup_{\xi < \omega\bar{\nu}} g_0(\bar{A} \cap S^{\bar{a}}_{\xi})$. Then $\hat{k}_0$ and $\hat{k}_1$ are both cofinal by Lemma 6.19 and just as in the proof of Lemma 6.18 we show: $s' \in S$. $\rho' = \rho_{s'}$. $A' = A_{s'}$, $n_s = n_{s'}$ and $p' = p_{s'}$ where $p' = \hat{k}_0(p_{s'}) = \hat{k}_1^{-1}(p_s)$. Hence

$$k_0 = \langle \bar{s}, \hat{k}_0 \restriction \bar{\nu}, s' \rangle \in \mathbb{F}(s') \quad \text{and}$$

$$k_1 = \langle s', \hat{k}_1 \restriction \nu', s \rangle \in \mathbb{F}(s) \,.$$

The rest of the conclusion is immediate. QED

We may now define straight away what our $\square$ sequences will be.

Definition Let $s \in S$

$C_s^+ = \{\lambda(f) \mid f \Rightarrow s\}$

$C_s = C_s^+ \setminus \{\nu_s\}$.

The next theorem embodies the major part of the properties of C sequences that we shall require for 2.3

Theorem 6.21 *Let* $s \in S$, *then:*

a) $C_s \subset [\alpha_s, \nu_s)$ *is closed in* ν_s .

b) $\text{otp } C_s \leq \alpha_s$.

c) If $\lambda \in C_s$, *then* $s \upharpoonright \lambda \in S$ *and* $C_{s\upharpoonright\lambda} = \lambda \wedge C_s$.

d) If C_s *is bounded in* ν_s *then* $\text{cf}(\nu_s) = \omega$.

e) If $f : \bar{s} \Rightarrow s$, $\lambda(f) = \nu_s$, *then*

$$\hat{f} \upharpoonright J_{\bar{s}} : \langle J_{\bar{s}}, C_{\bar{s}} \rangle \xrightarrow[\Sigma_1]{} \langle J_s, C_s \rangle .$$

Def. Let $f : \bar{s} \Rightarrow s$. Then $\beta(f) =_{df}$ the largest $\beta \leq \alpha_{\bar{s}}$ s.t. $f \upharpoonright \beta = id \upharpoonright \beta$.

It easily follows that:

Lemma 6.22 *Let* $f : \bar{s} \Rightarrow s$, $\beta = \beta(f)$, *then*

a) $J_{\rho_{\bar{s}}}^{A_{\bar{s}}} \models$ " β *is a cardinal*"

b) $\beta = \alpha_s \longleftrightarrow f = id_s \longleftrightarrow f(\beta) \neq \beta$.

Definition Let $s \in S$, $q \in J_s$, $\gamma \leq \alpha_s$

$$f_{(\gamma,q,s)} =_{df} \langle \bar{s}, g \upharpoonright \bar{\nu}, s \rangle$$

where $g : J_{\bar{\rho}}^{\bar{A}} \tilde{\leftrightarrow} \langle X, A_s \cap X \rangle$ and $\bar{s} = \langle \bar{\nu}, \bar{a} \rangle$ where $\bar{\nu} = g^{-1}{}''\nu_s$, $\bar{a} = g^{-1}{}''a_s$, and X is the smallest $X \prec_{\Sigma_1} J_{\rho_s}^{A_s}$ s.t. $\gamma \cup \{r_s\} \subset X$ where

$$r_s = \langle p_s, \alpha_s, q \rangle \quad \text{if} \quad \nu_s = \rho_s$$
$$= \langle p_s, \alpha_s, q, \nu_s \rangle \quad \text{if} \quad \nu_s < \rho_s$$

i.e. $X = h_{J^{A_s}_{\rho_s}}(J_\gamma \cup \{r\})$ for $\gamma \geq 1$.

For $\gamma = 0$ we get $f_{(0,q,s)} = f_{(1,q,s)}$. It is clear that $f_{(\gamma,q,s)} \in \mathbb{F}$.

We now list some properties and consequences of these definitions, the proofs of which are mostly trivial.

Lemma 6.23

a) $f_{(\gamma,q,s)}$ *is the* <u>*least*</u> $f \Rightarrow s$ *s.t.* $f \upharpoonright \gamma = id \upharpoonright \gamma$ *and* $q \in ran(f)$ *in the sense that for any such* f

$$f^{-1} f_{(\gamma,q,s)} \in \mathbb{F}$$

(this is the <u>*minimality property*</u> *of* $f_{(\gamma,q,s)}$ *).*

b) *Let* $f = f_{(\gamma,q,s)}$. *Let* $\beta = \beta(f)$. *Then*

$$f = f_{(\beta,q,s)} .$$

c) $f_{(\alpha_s,0,s)} = id_s$.

d) *Let* $f : \bar{s} \Rightarrow s$, $\bar{\gamma} \leq \alpha_{\bar{s}}$, $f''\bar{\gamma} \subset \gamma \leq \alpha_s$, $\bar{q} \in |J_{\bar{s}}|$, $\hat{f}(\bar{q}) = q$ *then*

$$ran(\hat{f}\hat{f}_{(\bar{\gamma},\bar{q},\bar{s})}) \subset ran(\hat{f}_{(\gamma,q,s)}) .$$

From this and a) we get: if $\beta(f) \geq \gamma$ *then*

$$ff_{(\gamma,\bar{q},\bar{s})} = f_{(\gamma,q,s)}$$

e) *Let* $g = f_{(\gamma,q,s)}$, $\lambda = \lambda(g)$. *Set* $g_0 = red(g)$, *then*

$q \in |J_{s/\lambda}|$ *and* $g_0 = f_{(\gamma,q,s/\lambda)}$.

Proof of e) Let $s_0 = s/\lambda$. We know that $g_0 \Rightarrow s_0$, and $J_\gamma^{A_s} \cup \{q\} \subset \mathrm{ran}(\hat{g}_0)$. Also $J_{\nu_{s_0}}^{A_{s_0}} \cap A_{s_0} = J_{\nu_{s_0}}^{A_s} \cap A_s$ by the proof of Lemma 6.18. Hence $J_\gamma^{A_{s_0}} = J_\gamma^{A_s}$. Set:

$$g_0' = f_{(\gamma,q,s/\lambda)} \ ; \ k = g_0^{-1} g_0' .$$

(The latter exists by the minimality of g_0' .) Then $r(k) = d(g_0) = d(g)$. Hence $gk \in \mathbb{F}$. But then $J_\gamma^{A_s} \cup \{q\} \subset \mathrm{ran}(gk)$. Hence $(gk)^{-1}g = k^{-1} \in \mathbb{F}$ by a). Hence $k = \mathrm{id}_{d(g_0')}$ and $g_0 = g_0'$. QED

The next lemma tells us, reasonably enough, that our definitions are preserved through "=>".

Lemma 6.24 *Let* $f : \bar{s} \Rightarrow s$ *cofinally (i.e.* $\lambda(f) = \nu_s$*). Let* $\bar{\gamma} < \alpha_{\bar{s}}$, $\gamma = f(\bar{\gamma})$, $\bar{q} \in J_{\bar{s}}$, $q = \hat{f}(\bar{q})$. *Set:*

$$\bar{g} = f_{(\bar{\gamma},\bar{q},\bar{s})} \ ; \ g = f_{(\gamma,q,s)} .$$

Then

a) $\lambda(\bar{g}) < \nu_{\bar{s}} \longleftrightarrow \lambda(g) < \nu_s$

b) If $\lambda(\bar{g}) < \nu_{\bar{s}}$ *, then*

$$f(\lambda(\bar{g})) = \lambda(g) \text{ and } f(\beta(\bar{g})) = \beta(g) .$$

Proof Set $X = \mathrm{ran}(\hat{g})$, $\bar{X} = \mathrm{ran}(\hat{\bar{g}})$. Set r_s as in Lemma 6.22. Then $X = h_{J_{\rho_s}^{A_s}}(J_\gamma^{A_s} \cup \{r_s\})$. Hence $X \subset J_{\rho_s}^{A_s}$ is $\Sigma_1^{J_{\rho_s}^{A_s}}(\{\gamma,r_s\})$. Likewise $\bar{X}$ is $\Sigma_1^{J_{\rho_{\bar{s}}}^{A_{\bar{s}}}}(\bar{\gamma},r_{\bar{s}})$ by the same definition. We first prove a).

($\rightarrow$) Let $\lambda(\bar{g}) < \nu_{\bar{s}}$. Then by Lemma 6.19, $\bar{\eta} = \sup(\bar{X} \cap On) < \rho_{\bar{s}}$. Set $\eta = \hat{f}(\bar{\eta})$. The statement

$$\forall \tau \in \bar{X}\ (\tau < \bar{\eta}) \text{ is } \Pi_1^{J_{\rho_{\bar{s}}}^{A_{\bar{s}}}}(\{\bar{\eta}, \bar{\gamma}, r_{\bar{s}}\}) .$$

But $\hat{f}$ is Π_1-preserving, hence the corresponding statement holds for $J_{\rho_s}^{A_s}$ in $\eta, \gamma, r_{\bar{s}}$. Hence $\sup(X \cap On) \leq \eta < \rho_s$. Hence $\lambda(g) < \nu_s$. QED ($\rightarrow$)

($\leftarrow$) Let $\lambda(g) < \nu_s$. Since $\mathrm{ran}(\hat{f}/\rho_{\bar{s}})$ is cofinal in ρ_s there is $\eta \in \mathrm{ran}(\hat{f})$ such that $On \cap X \subset \eta$. Let $\eta = \hat{f}(\bar{\eta})$. Repeating the argument we get: $\bar{X} \cap On \subset \bar{\eta} < \rho_{\bar{s}}$. Hence $\lambda(\bar{g}) < \nu_{\bar{s}}$. QED a)

Now for b). Assume $\lambda(\bar{g}) < \nu_{\bar{s}}$. Clearly it suffices to prove:

Claim $\bar{X} \in |J_{\rho_{\bar{s}}}^{A_{\bar{s}}}|$ *and* $\hat{f}(\bar{X}) = X$, since then we have:

$$f(\beta(\bar{g})) = \hat{f}(\sup\{\gamma | \gamma \subset \bar{X}\}) = \sup\{\gamma | \gamma \subset X\} = \beta(g)$$

$$f(\lambda(\bar{g})) = \hat{f}(\sup(\bar{X} \cap \bar{\nu})) = \sup(X \cap \nu) = \lambda(g)$$

So set

$$\omega\bar{\eta} = \sup(On \cap \bar{X}) \ ; \quad \bar{A} = A_{\bar{s}} \cap |J_{\bar{\eta}}^{A_{\bar{s}}}| .$$

$$\omega\eta = \hat{f}(\omega\bar{\eta}) \ ; \quad A = \hat{f}(\bar{A}) .$$

Then

$$A = A_s \cap |J_\eta^{A_s}| , \quad \text{since } \hat{f}(|J_{\bar{\eta}}^{A_{\bar{s}}}|) = |J_\eta^{A_s}| .$$

Set:

$$X_{\bar{\eta}} = h_{J_{\bar{\eta}}^{\bar{A}}}(J_{\bar{\gamma}}^{\bar{A}} \cup \{r_{\bar{s}}\})$$

$$X_\eta = h_{J_\eta^A}(J_\gamma^A \cup \{r_s\}) .$$

Then clearly

(i) $\hat{f}(X_{\bar{\eta}}) = X_\eta$

(ii) $X_{\bar{\eta}} \subset \bar{X}$; $X_\eta \subset X$

But $\langle \bar{X}, A^{\bar{s}} \cap \bar{X} \rangle \prec_{\Sigma_1} J_{\rho_{\bar{s}}}^{A_{\bar{s}}}$ and

$$\langle \bar{X}, A^{\bar{s}} \cap X \rangle \prec_{\Sigma_0} \langle |J_{\bar{\eta}}^{\bar{A}}|, \bar{A} \rangle \prec_{\Sigma_0} \langle |J_{\rho_{\bar{s}}}^{A_{\bar{s}}}|, A_{\bar{s}} \rangle .$$

Hence a Σ_1 statement w.p.f. $\bar{X}$ holds in $J_{\bar{\eta}}^{\bar{A}}$ iff it holds in $J_{\rho_{\bar{s}}}^{A_{\bar{s}}}$. By the definition of $\bar{X}$, this tells us that:

(iii) $\bar{X} \subset X_{\bar{\eta}}$

Now the statement of (iii) is $\Pi_1^{J_{\rho_{\bar{s}}}^{A_{\bar{s}}}}(\{\bar{\gamma}, X_{\bar{\eta}}, r_{\bar{s}}\})$. Hence the corresponding statement holds in $J_{\rho_s}^{A_s}$ and we have:

(iv) $X \subset X_\eta$

(i)-(iv) together give us $\hat{f}(\bar{X}) = \hat{f}(X_{\bar{\eta}}) = X_\eta = X$.

QED

Definition Let $s \in S$, $q \in J_s$.

$B^+(q,s) = \{\beta(f_{(\gamma,q,s)}) \mid \gamma \leq \alpha_s\}$

$B(q,s) = B^+(q,s) \setminus \{\alpha_s\}$.

Then $B(q,s)$ is the set of $\beta = \beta(f)$ s.t. $\beta < \alpha_s$ and $f = f_{(\beta,q,s)}$. The following are easily verified:

Lemma 6.25

a) Let γ *be a limit point of* $B(q,s)$. *Then*

$$\operatorname{ran}(f_{(\gamma,q,s)}) = \bigcup_{\beta \in B(q,s) \cap \gamma} \operatorname{ran}(f_{(\beta,q,s)}) .$$

b) *Let* $\gamma \leq \alpha_s$, $q \in J_s$. *Let* $\bar{s} = d(f_{(\gamma,q,s)})$, $f_{(\gamma,q,s)}(\bar{q}) = q$. *Then* $\gamma \cap B(q,s) = B(\bar{q},\bar{s})$.

c) *Let* $g = f_{(\gamma,q,s)}$, $\lambda = \lambda(g)$, $g_0 = \text{red}(g)$. *Then* $\gamma \cap B(q,s{\restriction}\lambda) = \gamma \cap B(q,s)$.

Proof a) is trivial. b) follows from 6.23d). c) by 6.23e) and b) above. QED

Definition Let $s \in S$, $q \in J_s$.

$$\Lambda^+(q,s) = \{\lambda(f_{(\gamma,q,s)}) \mid \gamma \leq \alpha_s\}$$

$$\Lambda(q,s) = \Lambda^+(q,s) \setminus \{\nu_s\} .$$

Notice that $\Lambda(q,s) \subset C_s$ for all $q \in J_s$ and we can regard them as "first approximations" to the sets C_s . The next group of lemmas in fact show how good an approximation they are, by establishing similar properties that we want for the C_s's .

Lemma 6.26(i) *Let* $s \in S$, $q \in J_s$.

a) $\Lambda(q,s) \subset (\alpha_s, \nu_s)$ *and is closed in* ν_s .

b) $\text{otp}(\Lambda(q,s)) \leq \alpha_s$.

c) *If* $\lambda \in \Lambda(q,s)$ *then* $q \in J_{s{\restriction}\lambda}$ *and* $\Lambda(q,s{\restriction}\lambda) = \lambda \cap \Lambda(q,s)$.

Proof Firstly a). The first half of the conjunct is trivial. To show closure suppose η to be a limit point of $\Lambda(q,s)$. We claim $\eta \in \Lambda^+(q,s)$. For $\lambda \in \Lambda(q,s) \cap \eta$ pick $\beta_\lambda \in B(q,s)$ s.t. $\lambda(f_{(\beta_\lambda,q,s)}) = \lambda$. Clearly $\beta_\lambda < \beta_{\lambda'}$ for $\lambda < \lambda'$. Let $\gamma = \sup_\lambda \beta_\lambda$. Then

$$\lambda(f_{(\gamma,q,s)}) = \sup_\lambda \lambda(f_{(\beta_\lambda,q,s)}) = \eta \quad \text{by Lemma 6.25a).}$$

QED a)

b) follows from $\Lambda(q,s) \subset \{\lambda(f_{(\gamma,q,s)}) \mid \gamma \leq \alpha_s\}$.

c) Let $\lambda \in \Lambda(q,s)$. Let $\lambda = \lambda(g)$ for $g = f_{(\beta,q,s)}$, $\beta = \beta(g)$. Let $g : \bar{s} \Rightarrow s$, $g(\bar{q}) = q$. Set $g_0 = \text{red } g$. Then $g_0 = f_{(\beta,q,s\wedge\lambda)}$ by 6.23e).
For $\gamma \geq \beta$ we have that $\lambda = \lambda(f_{(\gamma,q,s\wedge\lambda)}) \leq \lambda(f_{(\gamma,q,s)})$
For $\gamma \leq \beta$ we have

$$|f_{(\gamma,q,s\wedge\lambda)}| = |g_0||f_{(\gamma,\bar{q},\bar{s})}| \quad \text{(by 6.23d)}$$
$$= |g||f_{(\gamma,\bar{q},\bar{s})}|$$
$$= |f_{(\gamma,q,s)}| .$$

Thus $\lambda(f_{(\gamma,q,s\wedge\lambda)}) = \lambda(f_{(\gamma,q,s)})$ and the conclusion follows. QED

Lemma 6.26(ii) *Let* $f : \bar{s} \Rightarrow s$ *cofinally. Let* $\bar{q} \in |J_{\bar{s}}|$, $\hat{f}(\bar{q}) = q$. *Then*

a) $\Lambda(\bar{q},\bar{s}) = \emptyset \rightarrow \Lambda(q,s) = \emptyset$.

b) $f''\Lambda(\bar{q},\bar{s}) \subset \Lambda(q,s)$.

c) If $\bar{\lambda} = \max \Lambda(\bar{q},\bar{s})$ *and* $\lambda = f(\bar{\lambda})$ *, then* $\lambda = \max \Lambda(q,s)$.

Proof

a) If $\Lambda(\bar{q},\bar{s}) = \emptyset$, then $f_{(0,\bar{q},\bar{s})}$ is cofinal in $\nu_{\bar{s}}$. Hence $ff_{(0,\bar{q},\bar{s})}$ is cofinal in ν_s . But $\text{ran}(ff_{(0,\bar{q},\bar{s})}) \subset \text{ran}(f_{(0,q,s)})$ by 6.23d). Hence $\lambda(f_{(0,q,s)}) = \nu_s$ and $\Lambda(q,s) = \emptyset$ also. QED a)

b) is trivial by Lemma 6.24.

c) Let $\bar{\lambda} = \max \Lambda(\bar{q},\bar{s})$. Set $\bar{\beta} = \max\{\beta \mid \lambda(f_{(\beta,\bar{q},\bar{s})}) \leq \bar{\lambda}\}$. Then $\lambda(f_{(\bar{\beta},\bar{q},\bar{s})}) = \bar{\lambda}$ and $\lambda(f_{(\bar{\beta}+1,\bar{q},\bar{s})}) = \nu_{\bar{s}}$. By Lemma 6.24

we have $\lambda(f_{(\beta,q,s)}) = \lambda$ for $f(\langle\bar\beta,\bar\lambda,\bar q\rangle) = \langle\beta,\lambda,q\rangle$. On the other hand by the argument for a), $\lambda(f_{(\beta+1,q,s)}) = \nu_s$. Hence $\lambda = \max \Lambda(q,s)$. QED

Before stating 6.26(iii) notice the following facts:
Let $s \in S$ then

(1) "$\mu = \mu_s$" is uniformly p.r. in s (i.e. there is a p.r. relation R s.t.

$$\forall\mu\,\forall s\in S\ (\mu=\mu_s \longleftrightarrow R(\mu,s)))\ .$$

(2) $f(s) = \rho_s$ is uniformly p.r. in s,μ_s (i.e. there is a p.r. function f s.t. $\forall s\in S\ (\rho_s = f(s,\mu_s)))$.

(3) $f(s) = A_s$ is uniformly p.r. in s,μ_s as is $f(s) = \rho_s$.

(4) $f_{(\beta,q,s)} = f(\beta,q,s)$ is uniformly p.r. in s,μ_s , β,q for $\beta < \alpha_s$, $q \in |J_s|$.

(5) $f(q,s) = \Lambda(q,s)$ is uniformly p.r. in s,μ_s and q .

Lemma 6.26(iii) *Let* $f,\bar s,s,\bar q,q$ *be as in 6.26(ii). Then*

$$\hat f\restriction J_{\bar s} : \langle J_{\bar s},\Lambda(\bar q,\bar s)\rangle \xrightarrow[\Sigma_1]{} \langle J_s,\Lambda(q,s)\rangle$$

Proof It is enough to show that $\hat f(\Lambda(\bar q,\bar s)\cap\tau) = \Lambda(q,s)\cap f(\tau)$ for arbitrarily large $\tau < \nu_{\bar s}$. But by 6.26(ii) this follows from

(*) $\hat f(\Lambda(\bar q,\bar s)\cap\bar\lambda) = \Lambda(q,s)\cap\lambda$, for $\bar\lambda\in\Lambda(\bar q,\bar s)$, $f(\bar\lambda) = \lambda$.

But (*) is immediate from (5) above since

$$\hat f(\Lambda(\bar q,\bar s)\cap\bar\lambda) = \hat f(\Lambda(\bar q,s\restriction\bar\lambda)) = \Lambda(q,s\restriction\lambda)$$

$$= \lambda\cap\Lambda(q,s) \quad \text{by } 12(\text{i}).$$ QED

Lemmas 6.18 and 6.26(i) show easily that 6.26(iii) is equivalent to:

Corollary 6.26(iv) Let $f : \bar{s} \Rightarrow s$, $\bar{q} \in |J_{\bar{s}}|$, $\hat{f}(\bar{q}) = q$. *Then*

$$\hat{f}/J_{\bar{s}} : \langle J_{\bar{s}}, \Lambda(\bar{q},\bar{s}) \rangle \xrightarrow[\Sigma_0]{} \langle J_s, \Lambda(q,s) \rangle .$$

It turns out that we can analyse the sets C_s in terms of these $\Lambda(q,s)$. We start by defining a canonical sequence ℓ^i_s $(1 \leq i \leq m)$ of elements of C_s. We shall then find that C_s can be "generated" from just the $\Lambda(\ell^i_s, s)$ and using this we shall be able to extend the results of Lemma 6.26 to read "C_s" for "$\Lambda(q,s)$" which is just what we need.

Initially we shall define $\ell^i_{\lambda s} \in \lambda \cap C_s$ for <u>any</u> $\lambda \leq \nu_s$. This will later turn out to be redundant!

Definition Let $s \in S$, $\lambda \leq \nu_s$.

$\ell^i_{\lambda s}$ is defined by:

$\ell^0_{\lambda s} = 0$

$\ell^{i+1}_{\lambda s} = \max(\Lambda(\ell^i_{\lambda s}, s) \cap \lambda)$ if this maximum exists

undefined otherwise.

We also set $\ell^i_s = \ell^i_{\nu_s s}$.

Lemma 6.27(i) *If* $\ell^i_{\lambda s}$ *is defined and* $\ell^i_{\lambda s} < \rho \leq \lambda$, *then* $\ell^i_{\lambda s} = \ell^i_{\rho s}$.

Proof By induction on i.

Lemma 6.27(ii)

a) $\langle \ell^i_{\lambda s} \mid i < \omega \rangle$ *is monotone.*

b) $i > 0 \longrightarrow \ell^i_{\lambda s} \in \lambda \cap C_s$.

Proof Trivial.

Lemma 6.27(iii) *Let* $f : \bar{s} \Rightarrow s$ *cofinally. Then* $f(\ell^i_{\bar{s}}) = \ell^i_s$.

Proof By induction on i just using 6.26(ii).

Lemma 6.27(iv) *Let* $f : \bar{s} \Rightarrow s$, $\lambda = \lambda(f)$. *Then* $f(\ell^i_{\bar{s}}) = \ell^i_{s\upharpoonright\lambda} = \ell^i_{\lambda s}$.

Proof By induction. Assume true for i . Then we have $\lambda \in \Lambda(\ell^i_{\lambda s}, s) \rightarrow \lambda \cap \Lambda(\ell^i_{\lambda s}, s) = \Lambda(\ell^i_{\lambda s}, s\upharpoonright\lambda)$ by 6.26(i)c). So

$$
\begin{aligned}
f(\ell^{i+1}_{\bar{s}}) &= f_0(\ell^{i+1}_{\bar{s}}) = \ell^{i+1}_{s\upharpoonright\lambda} \quad \text{by 6.27(iii)} (f_0 = \text{red } f) \\
&= \max \Lambda(\ell^i_{s\upharpoonright\lambda}, s\upharpoonright\lambda) \\
&= \max \Lambda(\ell^i_{\lambda s}, s\upharpoonright\lambda) \quad \text{by induction hypothesis} \\
&= \max \Lambda(\ell^i_{\lambda s}, s) \cap \lambda \quad \text{by the remark above} \\
&= \ell^{i+1}_{\lambda s} .
\end{aligned}
$$

QED

Lemma 6.27(v) *Let* $f : \bar{s} \Rightarrow s$, $\bar{\tau} < \nu_{\bar{s}}$, $\tau = f(\bar{\tau})$, *then* $\ell^i_{\tau s} \simeq f(\ell^i_{\bar{\tau}\bar{s}})$.

Proof As above.

Lemma 6.27(vi) *Let* $\tau \leq \nu_s$. $\lambda = \min(C^+_s/\tau)$. *Then* $\ell^i_{\tau s} \simeq \ell^i_{\lambda s} \simeq \ell^i_{s\upharpoonright\lambda}$.

Proof The last "$\simeq$" is Lemma 6.27(iv). We prove $\ell^i_{\tau s} \simeq \ell^i_{\lambda s}$ by induction on i using $\Lambda(q,s) \cap \tau = \Lambda(q,s) \cap \lambda$, since $\Lambda(q,s) \subset C_s$ and $C_s \cap (\lambda \setminus \tau) = \emptyset$. QED

6.27(vi) gives us the redundancy mentioned earlier: for all τ, ν_s there exists $\lambda \in C^+_s$ such that $\ell^i_{\tau s} \simeq \ell^i_{s\upharpoonright\lambda}$ for all i . So the greater generality of the definition of $\ell^i_{\tau s}$ is only apparent.

Lemma 6.27(vii) $\ell_s^j \in \mathrm{ran}(f_{(0,\ell_s^i,s)})$ *for* $j \le i$.

Proof Let $f = f_{(0,\ell_s^i,s)}$, $f : \bar{s} \Rightarrow s$, $\lambda = \lambda(f)$. Then

$$f(\ell_{\bar{s}}^j) = \ell_{\lambda s}^j = \ell_s^j \quad \text{by 6.27(iv) and (i).}$$ QED

Lemma 6.28 ℓ_s^i *is undefined for some* $i < \omega$.

Proof Suppose not. Set

$$\beta^i = \max\{\beta < \alpha_s \mid \lambda(f_{(\beta,\ell_s^i,s)}) < \nu_s\} .$$

Then β^i is uniquely characterised by the conditions:

$$\lambda(f_{(\beta^i,\ell_s^i,s)}) = \ell_s^{i+1} , \quad \lambda(f_{(\beta^i+1,\ell_s^i,s)}) = \nu_s .$$

Claim $\beta^{i+1} < \beta^i \ \forall\, i < \omega$.

Proof Set $f = f_{(\beta^{i+1},\ell^{i+1},s)}$. Let $f : \bar{s} \Rightarrow s$. Set $f_0 = \mathrm{red}(f)$. Then $f_0 : \bar{s} \Rightarrow s \wedge \ell_s^{i+2}$. Now suppose $\beta^i < \beta^{i+1}$. Then $(\beta^i+1) \cup \{\ell_s^i\} \subset \mathrm{ran}(f)$, and hence $\mathrm{ran}(f_{(\beta^{i+1},\ell_s^i,s)}) \subset$ $\subset \mathrm{ran}(f)$. But then $\lambda(f) = \nu_s > \ell_s^{i+2}$, a contradiction. So suppose $\beta^i = \beta^{i+1}$, then β^i is the first thing moved by f and hence $\beta^i \notin \mathrm{ran}(f)$. Set $g = f_{(\beta^i,\ell_s^i,s)}$, $g_0 = f_{(\beta^i,\ell_s^i,s\wedge\ell_s^{i+2})}$, $\bar{g} = f_{(\beta^i,\ell_{\bar{s}}^i,\bar{s})}$. Then $g = f\bar{g}$ and $g_0 = f_0\bar{g}$ by Lemma 6.23d). Since $\lambda(g) = \ell_s^{i+1} < \ell_s^{i+2}$ Lemma 6.24 gives us

$$f(\beta(\bar{g})) = f_0(\beta(\bar{g})) = \beta(g_0) = \beta(g) = \beta^i .$$

Hence $\beta^i \in \mathrm{ran}(f)$. Another contradiction. Thus $\beta^{i+1} < \beta^i$.

Thus there must exist an i such that ℓ^i_s is undefined.

QED

It follows by Lemma 6.27(iv) that $\ell^i_{\lambda s}$ is undefined for some i.

Definition $m_{\lambda s}$ = that maximal $m < \omega$ s.t. $\ell^m_{\lambda s}$ is defined. Set

$$\ell_{\lambda s} =_{df} \ell^{m_{\lambda s}}_{\lambda s}$$

$$m_s =_{df} m_{\nu_s s}$$

$$\ell_s =_{df} \ell_{\nu_s s}$$

By our own definitions, $\Lambda(\ell_{\lambda s}, s) \cap \lambda$ is either empty or unbounded in λ. The following gives us some information about the former.

Lemma 6.29 *Let* $\Lambda(\ell_{\lambda s}, s) \cap \lambda = \emptyset$. *Set* $\ell = \ell_{\lambda s}$. *Then*

a) $\ell = 0 \longrightarrow C_s \cap \lambda = \emptyset$

b) $\ell > 0 \longrightarrow \ell = \max(C_s \cap \lambda)$

c) $\lambda \in C^+_s \longrightarrow \lambda = \lambda(f_{(0,\ell,s)})$

Proof Set $\rho = \min(C^+_s \setminus (\ell+1))$.

Claim 1) $\ell_{\lambda s} = \ell_{\rho s}$.

Proof $\ell_{\lambda s} = \ell_{\ell+1,s}$ by Lemma 6.27(i), and $\ell_{\ell+1,s} = \ell_{\rho s}$ by 6.27(vi).

Claim 2) $\lambda(f_{(0,\rho,s)}) = \rho$.

Proof Let $f : \bar{s} \Rightarrow s$, $\lambda(f) = \rho$. Then $f(\ell_{\bar{s}}) = \ell_{\rho s} = \ell_{\lambda s} = \ell$. But $\Lambda(\ell, s) \cap \rho = \emptyset$. Hence $\Lambda(\ell_{\bar{s}}, \bar{s}) = \emptyset$ since

$f''\Lambda(\ell_{\bar{s}},\bar{s}) \subset \Lambda(\ell, s/\rho) = \Lambda(\ell,s) \cap \rho = \emptyset$ by 6.26(i) and (ii).

QED 2)

By 2) we conclude $\rho \geq \lambda$, since otherwise $\Lambda(\ell,s) \cap \lambda \neq \emptyset$. a) and b) are then immediate. c) follows from 2) by:

$$\lambda \in C_s^+ \longrightarrow \lambda = \rho .$$

QED

Lemma 6.30 *Let* λ *be an element or fixed point of* C_s^+. *Then* $\lambda(f_{(\beta,\ell_{\lambda s},s)}) = \lambda$ *for some* $\beta \in B(\ell_{\ell s}, s)$.

<u>Case 1)</u> $\lambda \cap \Lambda(\ell_{\lambda s}, s) = \emptyset$.

Then λ is not a limit point of C_s, since $\lambda \cap (C_s^+ \setminus (\ell_{\lambda s}+1)) = \emptyset$, by Lemma 6.29. Hence $\lambda \in C_s^+$ and $\lambda = \lambda(f_{(0,\ell_{\lambda s},s)})$ by 6.29 again.

<u>Case 2)</u> $\lambda \cap \Lambda(\ell_{\lambda s}, s)$ is unbounded in λ.

For $\eta \in \lambda \cap \Lambda(\ell_{\lambda s}, s)$, let $\beta_\eta \in B(\ell_{\lambda s}, s)$ such that $\lambda(f_{(\beta_\eta,\ell_{\lambda s},s)}) = \eta$. Then $\beta_\eta < \beta_{\eta'}$ for $\eta < \eta'$. Set $\gamma = \sup_\eta \beta_\eta$. Then $\lambda(f_{(\gamma,\ell_{\lambda s},s)}) = \sup_\eta \lambda(f_{(\beta_\eta,\ell_{\lambda s},s)}) = \lambda$.

QED Case 2)

By the remark before Lemma 15 these two cases exhaust the possibilities.

QED

Corollary 6.31

a) C_s *is closed in* ν_s.

b) $C_s^+ = \{\lambda(f_{(\beta,q,s)}) \mid \beta \leq \alpha_s \wedge q \in J_s\}$

We can now polish off the remaining properties of the C sequences that we need. Note first that by b) above

(*) $f(s) = C_s^+$ is uniformly p.r. in s, μ_s for $s \in S$

(cf. the remarks before 6.26(iii)).

Lemma 6.32 $\lambda \in C_s \longrightarrow \lambda \cap C_s = C_{s\upharpoonright\lambda}$.

Proof By induction on λ . Set $\rho = \ell_{\lambda s} = \ell_{s\upharpoonright\lambda}$. Then there is β s.t. $\lambda = \lambda(f_{(\beta,\rho,s)}) \in \Lambda(\rho,s)$. Hence $\Lambda(\rho,s\upharpoonright\lambda) = \lambda \cap \Lambda(\rho,s)$. Set $\Lambda = \Lambda(\rho,s\upharpoonright\lambda)$.

Case 1) $\Lambda = \emptyset$.

If $\rho = 0$ then $C_{s\upharpoonright\lambda} = C_s \cap \lambda = \emptyset$ by Lemma 6.29. If $\rho > 0$ then $\rho = \max C_{s\upharpoonright\lambda} = \max(C_s \cap \lambda)$ by Lemma 6.29 again. Hence $C_{s\upharpoonright\rho} = C_{s\upharpoonright\lambda} \cap \rho = C_s \cap \rho$ by the induction hypothesis. Hence $C_{s\upharpoonright\lambda} = C_s \cap \lambda = C_{s\upharpoonright\rho} \cup \{\rho\}$.

Case 2) Λ unbounded in λ .

$C_{s\upharpoonright\lambda} \cap \tau = C_s \cap \tau = C_{s\upharpoonright\tau}$ for $\tau \in \Lambda$ by the induction hypothesis. Hence $C_{s\upharpoonright\lambda} = C_s \cap \lambda = \bigcup_{\tau\in\Lambda} C_{s\upharpoonright\tau}$.

QED

Combining (*) above with this last lemma we get.

Lemma 6.33 *Let* $f : \bar{s} \Rightarrow s$ *be cofinal. Then*

$$\hat{f} \upharpoonright J_{\bar{s}} : \langle J_{\bar{s}}, C_{\bar{s}} \rangle \xrightarrow[\Sigma_1]{} \langle J_s, C_s \rangle$$

Proof We must show

$$\hat{f}(C_s \cap \tau) = C_\tau \cap f(\tau) \quad \text{for arbitrarily large } \tau < \nu_{\bar{s}} .$$

Case 1) $\Lambda(\ell_{\bar{s}},\bar{s})$ is unbounded in $\nu_{\bar{s}}$.

Let $\bar{\tau} \in \Lambda(\ell_{\bar{s}},\bar{s})$, $\tau = f(\bar{\tau})$. Then $\tau \in \Lambda(\ell_s,s) \subset C_s$. Applying (*) above we get: $\hat{f}(C_{\bar{s}} \cap \bar{\tau}) = \hat{f}(C_{\bar{s}\upharpoonright\bar{\tau}}) = C_{s\upharpoonright\tau} = C_s \cap \tau$.

Case 2) $\Lambda(\ell_{\bar{s}},\bar{s}) = \emptyset$.

Set $\bar{\ell} = \ell_{\bar{s}}$, $\ell = \ell_s = f(\bar{\ell})$. Applying Lemma 6.29, we see that if $\bar{\ell} = 0$, then $\ell = 0$ and $C_{\bar{s}} = C_s = \emptyset$. If $\bar{\ell} > 0$ then $\ell > 0$ and $\bar{\ell} = \max C_{\bar{s}}$, $\ell = \max C_s$.

Using (*) we get:

$$\hat{f}(C_{\bar{s}}) = \hat{f}(C_{\bar{s}\wedge\bar{\ell}} \cup \{\bar{\ell}\}) = C_{s\wedge\ell} \cup \{\ell\} = C_s . \qquad \text{QED}$$

Lemma 6.34 $\sup C_s < \nu_s \longrightarrow cf(\nu_s) = \omega$.

Proof Let $\ell = \max(C_s)$. Then $\lambda(f_{(0,\ell,s)}) = \nu_s$. Set $X = \nu_s \cap \operatorname{ran}(f_{(0,\ell,s)})$. Then X is countable and cofinal in ν_s. QED

And finally:

Lemma 6.35 $\operatorname{otp}(C_s) \le \alpha_s$.

Proof For $p \in J_s$ set: $\beta(p,s) = 1 + \sup\{\beta | \lambda(f_{(\beta,p,s)}) < \nu_s\}$. Then $1 \le \beta(p,s) \le \alpha_s$ and $\operatorname{otp} \Lambda(p,s) \le \beta(p,s)$. Set $\tilde{\beta}(s) = \beta(\ell_s^o, s) + \ldots + \beta(\ell_s^m, s)$ where $m = m_s$. By definition of the ℓ_s^i it is clear that if $m > 0$ then for $0 \le i \le m$, $\beta(\ell_s^i, s) < \alpha_s$ and in fact by Lemma 6.28 that $\beta(\ell_s^o, s) > \ldots > \beta(\ell_s^m, s)$. And also if $m = 0$ then $\beta(\ell_s^o, s)$ is either 1 or α_s. Hence $\tilde{\beta}(s) \le \alpha_s$. The result follows from the following claim.

Claim $\operatorname{otp} C_s \le \tilde{\beta}(s)$.

Proof By induction on ν_s. Let it hold below ν_s. Let $\ell = \ell_s$. Then $\tilde{\beta}(s) = \tilde{\beta}(s\wedge\ell) + \beta(\ell, s)$ (putting $\tilde{\beta}(\emptyset) = 0$). Set $\Lambda = \Lambda(\ell, s)$. If $\Lambda = \emptyset$ then

$$\operatorname{otp} C_s = \operatorname{otp} C_{s\wedge\ell} + 1 \le \tilde{\beta}(s\wedge\ell) + 1 = \tilde{\beta}(s)$$

(putting $C_\emptyset = \emptyset$). Now let Λ be unbounded in ν_s.

For $\lambda \in \Lambda$ we have $\ell_{s\restriction\lambda} = \ell$ and hence

$$\tilde{\beta}(s\restriction\lambda) = \tilde{\beta}(s\restriction\ell) + \beta(\ell, s\restriction\lambda) .$$

But

$$\beta(\ell, s) = \sup_{\lambda\in\Lambda} \beta(\ell, s\restriction\lambda) .$$

Hence

$$\text{otp } C_s \leq \sup_{\lambda\in\Lambda} \text{otp } C_{s\restriction\lambda} \leq \sup_{\lambda\in\Lambda} \tilde{\beta}(s\restriction\lambda) = \tilde{\beta}(s)$$

QED

Lemmas 6.31-.35 constitute the five clauses of Theorem 6.21.

QED (Theorem 6.21)

As an immediate corollary to this proof we can obtain the following result stated but not proved in [J1] (Remark (3) on p286).

Theorem 6.36 *Let* β *be a cardinal. Let* $A \subset \beta^+$ s.t. $\bar{\bar{\nu}}^{L[A\cap\nu]} \leq \beta$ *for all p.r. closed* $\nu < \beta^+$. *Then* $\Box_\beta$ *holds.*

Proof Let S' = set of all p.r. closed $\nu \in (\beta, \beta^+)$ s.t. $L_\nu[A\cap\nu] \models$ "β is the largest cardinal". It is easily seen that S' is cub in β^+ . Then $\nu \in S' \longrightarrow \langle\nu, A\cap\nu\rangle \in S$. Set $C_\nu = C_{\langle\nu, A\cap\nu\rangle}$ for $\nu \in S'$. Then the clauses of Theorem 6.21 show that $\langle C_\nu \mid \nu \in S'\rangle$ satisfies a well-known equivalent form of $\Box_\beta$. QED

We shall use the proof of Theorem 6.21 to prove 2.3. A little more work will be required however and as a first step we show:

Lemma 6.37 *Let* $\lambda \in C_s^+$. *There is a unique* $k = k_{\lambda s}$ s.t. $k : J_{\rho_{s\restriction\lambda}}^{A_{s\restriction\lambda}} \longrightarrow_{\Sigma_0} J_{\rho_s}^{A_s}$ *and* $k\restriction\lambda = \text{id}\restriction\lambda$, $k(p_{s\restriction\lambda}) = p_s$. *Moreover,* *if* $g : \bar{s} \Rightarrow s$ s.t. $\lambda(g) = \lambda$ *and* $g_0 = \text{red}(g)$ *then* $k\hat{g}_0 = \hat{g}$.

Proof Let g be as above. By the proof of Lemma 6.18 there is a k satisfying the above so we are only required to prove uniqueness. Let $y \in |J_{\rho_{s/\lambda}}^{A_{s/\lambda}}|$. Let h_λ, h be the canonical Σ_1 Skolem functions for $J_{\rho_{s/\lambda}}^{A_{s/\lambda}}$ and $J_{\rho_s}^{A_s}$ respectively. Then $\exists x \in |J_{s/\lambda}|$ s.t. $y = h_\lambda(i, x, p_{s/\lambda})$ for some $i < \omega$. Hence $k(y) = h(i, x, p_s)$. QED

Lemma 6.38 *Let* C_s *be unbounded in* ν_s . *Then*

$$|J_{\rho_s}^{A_s}| = \bigcup_{\lambda \in C_s} \text{ran}(k_{\lambda s}) .$$

Proof We first note that $\text{ran}(k_{\lambda s}) \subset \text{ran}(k_{\lambda' s})$ for $\lambda \leq \lambda'$, since $k_{\lambda s} = k_{\lambda' s} k_{\lambda s/\lambda'}$. Set $X = \bigcup_{\lambda \in C_s} \text{ran}(k_{\lambda s})$. It follows easily that $X \prec_{\Sigma_1} \langle |J_{\rho^*}^{A_s}|, A_s \cap |J_{\rho^*}^{A_s}| \rangle$ where $\omega\rho^* = \sup(\text{On} \cap X)$. Hence it suffices to show:

Claim $\rho^* = \rho_s$.

Proof Let $\ell = \ell_s$. Then $\Lambda(\ell, s)$ is unbounded in ν_s . Let $\gamma = \sup\{\beta \mid \lambda(f_{(\beta,\ell,s)}) < \nu_s\}$. Then $\lambda(f_{(\gamma,\ell,s)}) = \nu_s$ and hence $\sup(\text{On} \cap \text{ran}(\hat{f}_{(\gamma,\ell,s)})) = \rho_s$ by Lemma 6.19. But $\text{ran}(\hat{f}_{(\gamma,\ell,s)}) = \bigcup_{\beta<\gamma} \text{ran}(\hat{f}_{(\beta,\ell,s)})$. For $\beta < \gamma$ set: $\lambda_\beta = \lambda(f_{(\beta,\ell,s)})$. Then $\text{ran}(\hat{f}_{(\beta,\ell,s)}) \subset \text{ran}(k_{\lambda_\beta s})$ by Lemma 6.37. Hence $\text{ran}(\hat{f}_{(\gamma,\ell,s)}) \subset X$ and so $\sup(\text{On} \cap X) = \rho_s$.
QED

Lemma 6.39 *Let* C_s *be unbounded in* ν_s . *Let* $f : \langle J_{\bar{s}}, \bar{C} \rangle \xrightarrow{\Sigma_1} \langle J_s, C_s \rangle$. *Then* $\bar{s} \in S$, $\bar{C} = C_{\bar{s}}$ *and* $\langle \bar{s}, f \restriction \nu_{\bar{s}}, s \rangle \in \mathbb{F}$.

Proof $\bar{s} \in S$ is obvious. Set $X = \text{ran}(f)$.

Set $|Y| = h_{J^{A_s}_{\rho_s}}(X \cup \{r_s\})$ where $r_s = \langle p_s, \nu_s \rangle$ if $\nu_s < \rho_s$

$= p_s$ otherwise.

Then $Y \prec_{\Sigma_1} J^{A_s}_{\rho_s}$ s.t. $p_s, \alpha_s \in |Y|$ and $\nu_s \in |Y|$ if $\nu_s < \rho_s$.

Hence setting $g : \langle J^A_\rho, A \rangle \overset{\sim}{\longleftrightarrow} \langle |Y|, A_s \cap |Y| \rangle$ we have $g \in \mathbb{F}(s)$

It suffices to prove:

Claim $X = |J_s| \cap |Y|$ (since from this we get $g \supset f$, $\nu_{\bar{s}} = g^{-1}{}''\nu_s, a_{\bar{s}} = g^{-1}{}''\nu_s$ and using (*) as in 6.33, $\bar{C} = C_{\bar{s}}$).

Proof Let $y \in |Y| \cap |J_s|$. Set $h_s = h_{J^{A_s}_{\rho_s}}$. Then

$$y = h_s(i, x, r_s) \quad \text{for some} \quad x \in X,\ i < \omega .$$

By Lemma 6.38 there is some $\lambda \in C_s$ s.t. "$y = h_s(i,x,r_s)$" holds relativised to $\text{ran}(k_{\lambda s})$; hence $y = h_{s/\lambda}(i,x,r_{s/\lambda})$. Hence $\exists \lambda \in C_s\ \langle i,x,r_{s/\lambda} \rangle \in \text{dom}(h_{s/\lambda})$. Since $X \prec_{\Sigma_1} \langle J_s, C_s \rangle$, there is such a $\lambda \in X$. Then:

$$k_{\lambda s} h_{s/\lambda}(i,x,r_{s/\lambda}) = h_s(i,x,r_s) = y .$$

Hence $y = h_{s/\lambda}(i,x,r_{s/\lambda}) \in X$. QED

The above result was obtained on the premiss that C_s was unbounded in ν_s . To cater for cases in which it is not we shall define a sequence $\langle \xi^s_i \mid i < \omega \rangle$ such that $\xi^s_i \in (\alpha_s, \nu_s)$, $\inf C_s < \xi^s_0$, $\langle \xi^s_i \mid i < \omega \rangle$ is monotone and converges to ν_s . By ensuring that whenever $f : J_{\bar{s}} \xrightarrow[\Sigma_0]{} J_s$ s.t. $\{\xi^s_i \mid i < \omega\} \subset \text{ran}(f)$ then $\langle \bar{s}, f/\nu_{\bar{s}}, s \rangle \in \mathbb{F}$ we can imitate the above result. So to this end:

Definition Let $s \in S$ s.t. $\lambda(f_{(0,\ell,s)}) = \nu_s$, where $\ell = \ell_s$ (i.e. $\sup C_s = \ell_s < \nu_s$). Let h_s be the usual Σ_1 Skolem function for $J_{\rho_s}^{A_s}$. Clearly $\nu_s \cap h_s''(\omega \times \{\ell, p_s\})$ is cofinal in ν_s. For $n < \omega$ set

$$u_n = \nu_s \cap h_s''(n \times \{\ell, p_s\}) .$$

Then $u_n \in |J_s|$ and is finite. And $\bigcup_n u_n$ is cofinal in ν_s. Set

$$\xi_n^s = \prec n, \ell, \alpha_s, u_0, \ldots, u_n \succ .$$

Then ξ_n^s is monotone with $\xi_0^s > \ell$ and $\sup_n \xi_n^s = \nu_s$.

Lemma 6.40 *Let* $f : \bar{s} \Rightarrow s$ s.t. $\sup C_s < \nu_s$. *Let* $\xi_n' = f(\xi_n^{\bar{s}})$. *Then* $\xi_n' = \xi_n^s$ *for* $n < \omega$.

Proof Trivial.

Lemma 6.41 *Let* $\sup C_s < \nu_s$. *Let* $f : J_{\bar{s}} \xrightarrow{\Sigma_0} J_s$ s.t. $\forall n\ \xi_n^s \in \mathrm{ran}(f)$. *Then* $\bar{s} \in S$, $\langle \bar{s}, f \upharpoonright \nu_{\bar{s}}, s\rangle \in \mathbb{F}$.

Proof First notice that in fact $f : J_{\bar{s}} \xrightarrow{\Sigma_1} J_s$ since $\sup_n \xi_n^s = \nu_s$, so $\bar{s} \in S$ is immediate. Now set $g = f_{(0,\ell,s)}$, $g : s' \Rightarrow s$ where $\ell = \ell_s$. Then $\mathrm{ran}(g \upharpoonright |J_{s'}|)$ is the smallest $X \prec_{\Sigma_1} J_s$ s.t. $\xi_n^s \in X$ for $n < \omega$. But now we can use Lemma 6.20 substituting $s, \bar{s}, s', g, f, f^{-1} \circ \hat{g} \upharpoonright J_{s'}$ for $s, s', \bar{s}, f, g_1, g_0$ there and the conclusion follows.

QED

We now substitute $\{\xi_n^s \mid n < \omega\}$ for C_s of Theorem 6.21 in case $\sup C_s < \nu_s$. Formally:

Definition $C_s' = \begin{cases} C_s & \text{if } \sup C_s = \nu_s \\ \{\xi_n^s \mid n < \omega\} & \text{if not.} \end{cases}$

We may collect all the properties of C_s' in the following theorem.

Theorem 6.42

a) C'_s *is uniformly p.r. in* s,μ_s *for* $s \in S$.

b) $C'_s \subset (\alpha_s,\nu_s)$ *is cub in* ν_s *and* otp $C'_s \leq \alpha_s$.

c) If λ *is a limit point of* C_s *, then* $s/\!\!\wedge\lambda \in S$ *and* and $C_{s/\wedge\lambda} = \lambda \cap C_s$.

d) If $f : \bar{s} \Rightarrow s$ *then* $\hat{f}/\!\!\wedge J_{\bar{s}} : \langle J_{\bar{s}}, C_{\bar{s}}\rangle \xrightarrow[\Sigma_0]{} \langle J_s, C_s\rangle$.

e) If $f : \langle J_{\bar{s}}, \bar{C}\rangle \xrightarrow[\Sigma_1]{} \langle J_s, C_s\rangle$ *then* $\bar{s} \in S$*,* $\langle \bar{s}, f/\!\!\wedge\nu_{\bar{s}}, s\rangle \in \mathbb{F}$ *,* *and* $\bar{C} = C_{\bar{s}}$.

We can now at last prove 2.3.

Theorem 6.43 2.3 *holds.*

Proof Suppose that $p \in \$_\alpha$, $p \neq \emptyset$. Then $\mathring{p} \in S$ where $\mathring{p} = \langle \mu^o_p, (A \cap \alpha) \cup p^*\rangle$. [Recall that $p^* = \{\mu^o_{p/\wedge\xi} \mid p(\xi) = 1\}$.] Thus $J_{\mathring{p}} = \mathfrak{A}^o_p$. We claim that 2.3 holds with $C_p = C'_{\mathring{p}}$. Theorem 6.42 a) and b) provide 2.3 (i), (ii) and (iv) and e) provides (v). We are left with (iii). Let $p \in \mathbf{S}_\alpha$ and suppose λ is a limit point of $C_p = C'_{\mathring{p}}$. If $\lambda = \mu^o_p$ there is nothing to prove, so assume $\lambda < \mu^o_p$. By 2.2 $\langle \mu^i_{p/\wedge\xi} \mid i < \omega, \xi < |p|\rangle$ is $\Sigma_1(\mathfrak{A}^o_s)$, so if $|p|$ is a limit then $\{\mu^o_{p/\wedge\xi} \mid \xi < |p|\}$ is cub in μ^o_p , or if $|p|$ is a successor then $\{\mu^i_{p/\wedge\eta} \mid i < \omega\}$ is unbounded in μ^o_p if $|p| = \eta + 1$. In either case $J_{\mathring{p}} \models$ "ψ" where ψ is of the form $\forall\, \xi \exists \zeta > \xi \phi(\zeta,\alpha)$ and $\phi(\zeta,\alpha)$ is Σ_1 and says "$(\exists\tau$ s.t. $\zeta = \mu^o_{p/\wedge\tau}) \vee (\exists n < \omega\, \zeta = \mu^n_{p/\wedge\eta})$" . Now let $f : \bar{p} \Rightarrow \mathring{p}$ s.t. $\lambda(f) = \lambda$. Then $\hat{f}/\!\!\wedge J_{\bar{p}} : J_{\bar{p}} \xrightarrow[\Sigma_1]{} J_{\mathring{p}}$, and ψ is Π_2 thus $J_{\bar{p}} \models$ "ψ" . But $\hat{f}/\!\!\wedge J_{\bar{p}} : J_{\bar{p}} \xrightarrow[\Sigma_1]{} J_{\mathring{p}/\wedge\lambda}$ cofinally. By the special form of ψ this means $J_{\mathring{p}/\wedge\lambda} \models$ "ψ" . In either case this implies $\exists\, \sigma$ s.t. $\lambda = \mu^o_{p/\wedge\sigma}$. This together with Theorem 6.42 c) yields (iii). QED

The proof of 2.4 will be an adaptation of this proof. We start by enlarging our class S to an S^* where ν_s for $s \in S^*$ need not be a "pseudo" successor cardinal, but only must not be regular in L^{a_s}. We then proceed to prove an analogue of Theorem 6.21, but only in outline since once the basic definitions have been laid down, the proof will be identical in almost all respects.

Definition S^* is the class of all pairs $\langle \nu_s, a_s \rangle$ s.t. ν_s is p.r. closed or a limit of such, $\nu_s > \omega$, $a_s \subset \nu_s$ such that ν_s is not a regular cardinal of L^a. As remarked earlier $S \subset S^*$. By μ_s we denote the least $\mu \geq \nu_s$ s.t. ν_s is not Σ_ω-regular in J_μ^a. Let $n = n_s$ be the least $n \geq 1$ s.t. $\exists f \in \Sigma_n(J_\mu^a)$, a map of some ordinal less than ν unboundedly into ν. Obviously $\rho_{\mu_s}^n \leq \nu \leq \rho_{\mu_s}^{n-1}$.

Definition $\rho_s = \rho_{\mu_s}^{n-1}$. $A_s = A_{\mu_s}^{n-1,a}$ [Note that these define the same objects if $s \in S$.]

$p_s^* = <_{J_\rho^A}$-least p s.t.

$$|J_\rho^A| = h(J_\nu \cup \{p\}) .$$

(Note that we do not necessarily have $p_s^* = p_s$ for $s \in S$.) Clearly there will be some $\tau < \nu$ s.t.

$$\sup(h(J_\tau \cup \{p^*\}) \cap \nu) = \nu \quad [h = h_{J_\rho^A}] .$$

Hence there is a maximal $\alpha^* = \alpha_s^*$ s.t. $\alpha^* < \nu$ and $h(J_{\alpha^*} \cup \{p^*\}) \cap \omega\nu \subset \omega\alpha^*$. α_s^* will behave as an analogue of α_s but we may have $\alpha_s^* = 0$. For $s \in S$ any of these possibilities may be true: $\alpha_s^* < \alpha_s$, $\alpha_s^* > \alpha_s$, $\alpha_s^* = \alpha_s$. But $\omega\alpha_s^*$ must be strongly admissible if $\alpha_s^* \neq 0$.

Definition $\mathbb{F}^*(s)$ is the set of maps f s.t.

$$f : J^{\bar{A}}_{\bar{\rho}} \xrightarrow[\Sigma_1]{} J^A_\rho \quad \text{s.t.} \quad \langle p^*,\alpha^*\rangle \in \text{ran } f$$

and $\nu \in \text{ran}(f)$ if $\nu < \rho$. We then get the slightly weaker version of Lemma 6.14.

Lemma 6.44 $f \in \mathbb{F}^*(s)$ *is uniquely determined by* $\nu \cap \text{ran}(f)$.

But notice that in Lemma 6.15 (and subsequently) this was all that was ever used. As an analogue of Theorem 6.36 we get:

Lemma 6.45 *Let* $f \in \mathbb{F}^*(s)$ s.t. $f : J^{\bar{A}}_{\bar{\rho}} \xrightarrow[\Sigma_1]{} J^A_\rho$. *Let* $f(\langle\bar{p},\bar{\alpha}\rangle) = \langle p^*,\alpha^*\rangle$. *Set* $\bar{\nu} = f^{-1}{}''\nu, \bar{a} = f^{-1}{}''a$ *and* $\bar{s} = \langle\bar{\nu},\bar{a}\rangle$. *Then* $s \in S^*$, $\bar{\rho} = \rho_{\bar{s}}$, $\bar{A} = A_{\bar{s}}$, $\bar{p} = p^*_{\bar{s}}$, $\bar{\alpha} = \alpha^*_{\bar{s}}$ *and* $n_{\bar{s}} = n$.

Proof $\exists$ unique $\bar{\mu}$, by Ext. of Embeddings s.t. $\bar{\rho} = \rho^{n-1,\bar{a}}_{\bar{\mu}}$, $\bar{A} = A^{n-1,\bar{a}}_{\bar{\mu}}$ and as before $|J^{\bar{a}}_{\bar{\nu}}| = |J^{\bar{A}}_{\bar{\nu}}|$ and $f{\restriction}J_{\bar{s}} : J_{\bar{s}} \xrightarrow[\Sigma_1]{} J_s$. Thus $\bar{\nu}$ is p.r. closed or a limit of such. Set $\bar{h} = h_{J^{\bar{A}}_{\bar{\rho}}}$, $h = h_{J^A_\rho}$.

(1) $\bar{p}$ is defined from $J^{\bar{A}}_{\bar{\rho}}$ as p^* was from J^A_ρ.

Proof Just as before.

(2) $\bar{\alpha}$ is defined from $J^{\bar{A}}_{\bar{\rho}}$ as α^* is defined from J^A_ρ.

Proof Set $H(\xi,\zeta) \longleftrightarrow h(J_\xi \cup \{p^*\}) \cap \nu \subset \zeta$

$\bar{H}(\xi,\zeta) \longleftrightarrow \bar{h}(J_\xi \cup \{p\}) \cap \nu \subset \zeta$

Then H is $\Pi_1^{J^A_\rho}(\{r\})$ where $r = p^*$ if $\nu = \rho$

$= \langle p^*,\nu\rangle$ otherwise.

Also $\bar{H}$ is $\Pi_1^{J_{\bar{\rho}}^{\bar{A}}}(\{\bar{r}\})$ (where $\bar{r}$ is defined similarly) by the same definition. Then

$$J_{\bar{\rho}}^{\bar{A}} \models \text{"}\bar{H}(\bar{\alpha},\bar{\alpha})\text{"} \quad \text{since} \quad J_{\rho}^{A} \models \text{"}H(\alpha^*,\alpha^*) \quad \text{but}$$

$$J_{\bar{\rho}}^{\bar{A}} \models \text{"}\neg\bar{H}(\xi,\xi)\text{"} \quad \text{for} \quad \bar{\alpha} < \xi < \bar{\nu} \quad \text{since}$$

$$J_{\rho}^{A} \models \text{"}\neg H(f(\xi),f(\xi)\text{"} \quad \text{as} \quad \alpha^* < f(\xi) < \nu \text{ .}$$

(3) $\bar{\nu}$ is Σ_{n-1}-regular in $J_{\bar{\mu}}^{\bar{a}}$.

Proof Suppose not. Then $n > 1$, since $\bar{\nu}$ is regular in $J_{\bar{\mu}}^{\bar{a}}$, and $\bar{\rho}$ must equal $\bar{\nu}$. [Since if $\bar{\nu}$ is not Σ_{n-1}-regular in $J_{\bar{\mu}}^{\bar{a}}$ $\exists$ a new $\Sigma_{n-1}(J_{\bar{\mu}}^{\bar{a}})$ subset of $\bar{\nu}$, so $\rho_{\bar{\mu}}^{n-1} \leq \nu$.] Set:

$$\delta = \rho_{\mu}^{n-2,a} \qquad q = p_{\mu}^{n-1,a}$$

$$B = A_{\mu}^{n-2,a} \qquad k = h_{J_{\delta}^{B}}$$

Define $\bar{\delta},\bar{q},\bar{B},\bar{k}$ similarly.

Let $\hat{f} : J_{\bar{\delta}}^{\bar{B}} \xrightarrow[\Sigma_2]{} J_{\delta}^{B}$ be the unique such extension of f with $\hat{f}(\bar{q}) = q$. By our assumption there must be $\bar{\gamma} < \bar{\nu}$ s.t. $\sup(\bar{k}(J_{\bar{\gamma}} \cup \{\bar{q}\}) \cap \bar{\nu}) = \bar{\nu}$. Define $\bar{g} : J_{\bar{\gamma}} \longrightarrow \bar{\nu}$ by:
$\bar{g}(\langle i,x\rangle) = \bar{k}(i,x,\bar{q})$ if $\bar{k}(i,x,\bar{q})$ is defined and less than $\bar{\nu}$.

$= 0$ otherwise.

Then $\sup \bar{g}''J_{\bar{\gamma}} = \bar{\nu}$. Now define $g : J_{\gamma} \longrightarrow \nu$ from k,q,γ,ν as $\bar{g}$ was defined from $\bar{k},\bar{q},\bar{\gamma},\bar{\nu}$. It is clear that $\bar{g}$ is rudimentary over $J_{\bar{\rho}}^{\bar{A}}$ in the parameter $\bar{\gamma}$, and similarly g using γ , and the same definition over J_{ρ}^{A} . But $\text{ran}(g)$ must be bounded in ν . Hence:

$$\exists \lambda < \nu \forall x \in J_{\gamma} \exists \xi < \lambda \;\; g(x) = \xi \text{ .}$$

But this is $\Sigma_1(J^A_\rho)$ so the corresponding statement

$$\exists\lambda < \bar{\nu}\forall x \in J_{\bar{\gamma}}\exists\xi < \lambda \quad \bar{g}(x) = \xi$$

holds in $J^{\bar{A}}_{\bar{\rho}}$. Hence $\sup \bar{g}''J_{\bar{\gamma}} < \bar{\nu}$. Contradiction. QED (3)

(4) ν is not Σ_n-regular in $J^{\bar{a}}_{\bar{\mu}}$.

Proof Suppose not. As in (2) regularity of $\bar{\nu}$ can be expressed as

(*) $\forall\xi < \bar{\nu}\exists\zeta < \bar{\nu} \quad \bar{H}(\xi,\zeta)$.

Let $\lambda = \sup f''\bar{\nu}$. Then

(**) $\forall\xi < \lambda\exists\zeta < \lambda \quad H(\xi,\zeta)$.

Hence $\lambda < \nu$, otherwise ν would be Σ_1-regular in J^A_ρ . Thus $\alpha^* < \lambda < \nu$ and $H(\lambda,\lambda)$ holds in J^A_ρ by (**), contradicting the definition of α^* . QED

Definition $\mathbb{F}^*$ is the class of triples $\langle\bar{s}, f\upharpoonright\nu_{\bar{s}}, s\rangle$ with $s \in S^*$, $f \in \mathbb{F}^*(s)$.

$f : \bar{s} \Rightarrow_* s$ is again defined to mean $f = \langle\bar{s}, |f|, s\rangle \in \mathbb{F}^*$.

$f \Rightarrow_* s \longleftrightarrow \exists\bar{s} \in S^*$ s.t. $f : \bar{s} \Rightarrow_* s$.

$\hat{f}$ is defined as before using $\mathbb{F}^*(s)$.

Lemma 6.17 is proven as before with the obvious modifications and the definition of $\lambda(f)$ remains unchanged. Lemma 6.18 may be proven incorporating some changes as in Lemma 6.45. Note that (in the notation of Lemma 6.18) $\exists\bar{\tau} < \bar{\nu}$ s.t. $\sup(\bar{h}(J_{\bar{\tau}} \cup \{p\})\cap\bar{\nu}) = \bar{\nu}$, thus $\lambda \cap h'(J_{g_0(\bar{\tau})} \cup \{p'\})$ is cofinal in λ since $g_0''\bar{\nu}$ is. Thus λ is not Σ_n-regular. Proving λ Σ_{n-1}-regular is done as in (3) above; that $p' \geq_{J^{A'}} p_{s'}$ is now easier and the reverse is the same.

Lemmas 6.19 and 6.20 are as before. We set:

Def. $C^*_s = \{\lambda(f) \mid f \Rightarrow_* s \wedge \lambda(f) < \nu_s\}$ for $s \in S^*$.

We then prove:

Theorem 6.46 *Let* $s \in S^*$.

a) $C^*_s \subset \nu_s$ *is closed in* ν_s .

b) $\text{otp}\ C^*_s < \nu_s$.

c) If $\lambda \in C^*_s$ *then* $s \wedge \lambda \in S^*$ *and* $C^*_{s\wedge\lambda} = \lambda \cap C^*_s$.

d) If C^*_s *is bounded in* ν_s *, then* $\text{cf}(\nu_s) = \omega$.

e) If $f : \bar{s} \Rightarrow_* s$, $\lambda(f) = \nu_s$ *then*

$$\hat{f} \wedge J_{\bar{s}} : \langle J_{\bar{s}}, C^*_{\bar{s}} \rangle \xrightarrow[\Sigma_1]{} \langle J_s, C^*_s \rangle .$$

The proof of this is just a repetition of the proof of Theorem 6.21: we define $\beta(f)$, $B^*(q,s)$, $\Lambda^*(q,s)$, $f^*_{(\gamma,q,s)}$ as before although we now obtain $\text{otp}\ B^*(q,s) < \nu_s$ not $\leq \alpha^*_s$. This in turn gives us $\text{otp}\ \Lambda^*(q,s) < \nu_s$ and ultimately that $\text{otp}\ C^*_s < \nu_s$.

Notice that we have effectively obtained the following Theorem of [J1]:

Theorem 6.47 (V = L) *There is* $\langle C_\nu \mid \nu$ *p.r. closed and singular*$\rangle$ *s.t.*

a) $C_\nu \subset \nu$ *is closed in* ν .

b) $\text{otp}\ C_\nu < \nu$.

c) $\lambda \in C_\nu \longrightarrow \lambda$ *is p.r. closed and singular and* $C_\lambda = \lambda \cap C_\nu$.

d) $\sup C_\nu < \nu \longrightarrow \text{cf}(\nu) = \omega$.

Proof Set $C_\nu = C^*_{\langle\nu,\emptyset\rangle}$ in Theorem 6.46. QED

Also in [J1] was the following.

Theorem 6.48 (V = L) *Let* $E = \{\nu \mid C_\nu = \emptyset\}$. *Then* $E \cap \kappa$ *is stationary in any regular* $\kappa > \omega$.

Proof Just as in [J1], $E' \subset E$ where E' is the class of p.r. closed ν s.t. $n_{\langle\nu,\emptyset\rangle} = 1$ and $\mu_{\langle\nu,\emptyset\rangle} = \beta + 1$ for some β (cf. Lemma 5.2 of [J1]). But $E'' \subset E'$ where E'' is the class of ν s.t.

a) $J_\nu \models ZF^-$

b) $\exists\gamma < \nu, \exists q \in J_\nu$ s.t. each $x \in J_\nu$ is J_ν-definable w.p.f. $J_\gamma \cup \{q\}$.

Lastly 5.1 of [J1] shows that $E'' \cap \kappa$ is stationary in any regular κ . QED

We can now prove the simplest case of 2.4.

Lemma 6.49 2.4 *holds for the case* $p = \emptyset$.

Proof Note that $p = \emptyset \rightarrow \mathfrak{A}^o_p = \mathfrak{A}^o_\alpha$.

Set $s = \langle\alpha, A \cap \alpha\rangle$

$$D_p = C^*_s \cap \text{Card}. \qquad \left.\begin{array}{l} p_\beta = \emptyset \\ \pi^p_\beta = \text{id} \restriction \mathfrak{A}^o_\beta \end{array}\right\} \beta \in D_p$$

(i)-(v) of 2.4 follow easily. QED

We now establish an analogue of Theorem 6.46 in order to complete 2.4 in case $|p| > \alpha$.

Suppose that $a \subset \nu$, $\tau < \nu$ and $J^a_\nu \models$ "τ is the largest cardinal" then, supposing that τ is a Σ_m-cardinal in J^a_β for $\beta \geq \nu$, we may use §1 Lemma 6.9c) together with §1 Lemma 6.16

to define a canonical Σ_m map from a subset of $J^{A^m_\beta}_{\rho^m_\beta}$ onto J^a_β as follows:

Def. Let ν,τ,a,β be as above and let $\rho^{m,a}_\beta$ be defined (i.e. τ is a Σ_m-cardinal in J^a_β). Then define $F^{m,a}_\beta$ by induction on m as follows:

$$F^o = id \upharpoonright J^a_\beta$$

$$F^{m+1} = F^m \tilde{h} \quad \text{where} \quad \tilde{h}(\langle i,x\rangle) \simeq h_{J^{A^m_\beta}_{\rho^m_\beta}}(i,x,p^{m+1}) .$$

[Note every $y \in J^{A^m_\beta}_{\rho^m_\beta}$ is of the form $h(i,x,p^{m+1})$ for some $x \in J_{\rho^{m+1}_\beta}$. So F^{m+1} takes a subset of $J_{\rho^{m+1}_\beta}$ onto J^a_β as required.]

Now define $S^\dagger$ to be the class of $s \in S$ s.t. α_s is regular in J_s but not in L^{a_s}. Let $s = \langle \nu_s, a_s\rangle \in S^\dagger$; $\alpha = \alpha_s$. Define $\mu^\dagger = \mu^\dagger_s$ as the least $\mu \geq \nu$ s.t. α is not Σ_ω-regular in J^a_μ. Let $n^\dagger = n^\dagger_s$ be the least $n \geq 1$ s.t. α is not Σ_n-regular in J^a_μ.

Define $\rho^\dagger = \rho^\dagger_s = \rho^{n^\dagger-1,a}_{\mu^\dagger}$; $A^\dagger = A^\dagger_s = A^{n^\dagger-1,a}_{\mu^\dagger}$

$p^\dagger = p^\dagger_s$ = the $<_{J^{A^\dagger}_{\rho^\dagger}}$-least p s.t.

$$|J^{A^\dagger}_{\rho^\dagger}| = h_{J^{A^\dagger}_{\rho^\dagger}}(J_{\alpha_s} \cup \{p\})$$

$\alpha^\dagger = \alpha^\dagger_s$ is defined from α_s just as α^* was defined from ν. Note that possibly $\rho^\dagger = \alpha_s$.

Definition $t = t_s$ = the $<_{J^{A^\dagger}_{\rho^\dagger}}$-least t s.t.

$$F^{n^\dagger-1,a}_{\mu^\dagger}(t) = \nu \quad \text{if} \quad \nu < \mu^\dagger$$
$$= \alpha_s \quad \text{otherwise.}$$

Definition $\mathbb{F}^{\dagger}(s)$ is the set of f s.t. $\langle p^{\dagger},\alpha^{\dagger},t_s \in \mathrm{ran}(f)$ and

$$f : J_{\bar{\rho}}^{\bar{A}} \underset{\Sigma_1}{\longrightarrow} J_{\rho^{\dagger}}^{A^{\dagger}} .$$

As before we get these analogues of Lemmas 6.1 and 6.2.

Lemma 6.50 $f \in \mathbb{F}^{\dagger}(s)$ *is uniquely determined by* $\alpha_s \cap \mathrm{ran}(f)$.

Lemma 6.51 *Let* $f \in \mathbb{F}^{\dagger}(s)$ s.t. $f : J_{\bar{\rho}}^{\bar{A}} \underset{\Sigma_1}{\longrightarrow} J_{\rho^{\dagger}}^{A^{\dagger}}$ *with* $f(\langle\bar{p},\bar{t},\bar{\alpha}\rangle) = \langle p^{\dagger},t_s,\alpha^{\dagger}\rangle$. *Then there is a unique* $s \in S^{\dagger}$ s.t. $\langle\bar{\rho},\bar{A},\bar{p},\bar{\alpha},\bar{t}\rangle = \langle\rho_{\bar{s}}^{\dagger},A_{\bar{s}}^{\dagger},p_{\bar{s}}^{\dagger},\alpha_{\bar{s}}^{\dagger},t_{\bar{s}}\rangle$ *and* $n^{\dagger} = n_{\bar{s}}^{\dagger}$.

Note: to cater for the possibility that $\rho^{\dagger} < \nu$, $\bar{s}$ may be defined as follows: Let $\bar{a},\bar{\mu}$ be the unique pair s.t. $\bar{\rho} = \rho_{\bar{\mu}}^{n^{\dagger}-1,\bar{a}}$, $\bar{A} = A_{\bar{\mu}}^{n^{\dagger}-1,\bar{a}}$. Let $\tilde{f} : J_{\bar{\mu}}^{\bar{a}} \to J_{\mu^{\dagger}}^{a}$ be the usual canonical extension of f . Set $\bar{\nu} = \tilde{f}^{-1}{}''\nu$. Then $\bar{s} = \langle\bar{\nu},\bar{a}\rangle$ and $\tilde{f}(\alpha_{\bar{s}}) = \alpha$.

An extra lemma is needed to provide the correct analogue of Lemma 6.18. It shows that for $\lambda \le \alpha_s$ we only have at most one way of defining a reasonable notion of $s/^{\!\prime}\lambda$ (analogous to s/λ for $s \in S$). Compare Lemma 6.37.

Lemma 6.52 *Let* $\lambda \le \alpha_s$. *Then there exists at most one pair,* $\bar{s},\bar{k}$ *s.t.* $\bar{s} = \langle\bar{\nu},\bar{a}\rangle \in S^{\dagger}$, $\lambda = \alpha_{\bar{s}}$ *and, setting* $\langle\bar{A},\bar{\rho},\bar{p},\bar{t}\rangle = \langle A_{\bar{s}}^{\dagger},\rho_{\bar{s}}^{\dagger},p_{\bar{s}}^{\dagger},t_{\bar{s}}\rangle$ *, we have:*

$$\bar{k} : J_{\bar{\rho}}^{\bar{A}} \underset{\Sigma_0}{\longrightarrow} J_{\rho^{\dagger}}^{A^{\dagger}}, \ \bar{k}/^{\!\prime}\lambda = \mathrm{id}/^{\!\prime}\lambda, \ \bar{k}(<\bar{p},\bar{t}>) = <p^{\dagger},t> .$$

Proof Let $s' = \langle\nu',a'\rangle,k'$ be a second such pair. Set $\langle A',\rho',t',p'\rangle = \langle A_{s'}^{\dagger},\rho_{s'}^{\dagger},t_{s'},p_{s'}^{\dagger}\rangle$. For $\langle i,x\rangle \in J_{\lambda}$ set

$$\bar{h}(\langle i,x\rangle) \simeq h_{J_{\bar{\rho}}^{\bar{A}}}(i,x,\bar{p})$$

$$h'(\langle i,x\rangle) \simeq h_{J_{\rho'}^{A'}}(i,x,p') .$$

By the assumption that $\lambda = \alpha_{\bar{s}} = \alpha_{s'}$ we clearly have $\bar{h}''|J_\lambda| = |J_{\bar{\rho}}^{\bar{A}}|$, $h'''|J_\lambda| = |J_{\rho'}^{A'}|$. Now let

$$\omega\bar{\eta} = \sup \bar{k}''\omega\bar{\rho}$$

$$\omega\eta' = \sup k'''\omega\rho' .$$

We then have:

1) $\bar{k}\bar{h}(\langle i,x\rangle) \simeq h_{\langle J_{\bar{\eta}}^{A^\dagger}, A^\dagger \cap J_{\bar{\eta}}^{A^\dagger}\rangle}(i,x,p^\dagger)$

$k'h'(\langle i,x\rangle) \simeq h_{\langle J_{\eta}^{A^\dagger}, A^\dagger \cap J_{\eta'}^{A^\dagger}\rangle}(i,x,p^\dagger)$.

If $\eta = \eta'$ the conclusion is immediate. Assume w.l.o.g. that $\bar{\eta} < \eta'$. Then $\mathrm{ran}(\bar{k}) \subset \mathrm{ran}(k')$ by 1). Define a map: $\ell : J_{\bar{\rho}}^{\bar{A}} \underset{\Sigma_0}{\longrightarrow} J_{\rho'}^{A'}$ by

$$\ell\bar{h}(\langle i,x\rangle) = h'(\langle i,x\rangle) .$$

So we then shall have:

2) $\ell \restriction |J_\lambda^{\bar{A}}| = \mathrm{id} \restriction |J_\lambda^{\bar{A}}|$ and $\ell(\bar{t}) = t'$.

Canonically extend ℓ to $\tilde{\ell}$ with $n = n^\dagger$

$$\tilde{\ell} : J_{\bar{\mu}}^{\bar{a}} \underset{\Sigma_{n-1}}{\longrightarrow} J_{\mu'}^{a'} .$$

Then $\tilde{\ell}(\lambda) = \lambda$ and $\tilde{\ell}(\bar{\nu}) = \tilde{\ell}(F_{\bar{\mu}}^{n-1,\bar{a}}(\bar{t})) = F_{\mu'}^{n-1,a'}(t') = \nu'$, if $\nu < \mu$; if $\nu = \mu$ then $\bar{\nu} = \bar{\mu}$ and $\nu' = \mu'$. Hence

3) $\tilde{\ell}\restriction\bar{\nu} = \mathrm{id}\restriction\bar{\nu}$; $\tilde{\ell}(\bar{\nu}) = \nu'$ if $\nu' < \nu'$

$\bar{\nu} = \bar{\mu}$ if $\nu' = \mu'$.

But then $\bar{a} = a' \cap \bar{\nu}$. This implies that $J_\beta^{\bar{a}} \subset J_\beta^{a'}$ for

$\beta \geq \nu$. If $\bar{\mu} < \mu'$ we would therefore have $J^{\bar{a}}_{\bar{\mu}+1} \subset J^{a'}_{\mu}$ and λ would not be regular in $J^{a'}_{\mu}$, a contradiction. Hence $\bar{\mu} = \mu'$ and $\tilde{\ell}\restriction\bar{\mu} = \mathrm{id}\restriction\bar{\mu}$. But then $\bar{\nu} = \nu'$ by 3) and $\bar{s} = s'$. $\bar{k} = k'$ follows by 1). QED

We incorporate this result in the following definition.

Definition If $\lambda \leq \alpha_s, \bar{s}, \bar{k}$ are as in Lemma 6.52, set

$$s\restriction\lambda = \bar{s}\ ,\quad k_{\lambda s} = \bar{k}\ .$$

By Lemma 6.51 we may define $\mathbb{F}^{\dagger}$ to be the class of triples $\langle \bar{s}, f\restriction\alpha_{\bar{s}}, s\rangle$ s.t. $s \in S^{\dagger}$, $f \in \mathbb{F}^{\dagger}(s)$ and

$$f : J^{A^{\dagger}_{\bar{s}}}_{\rho^{\dagger}_{\bar{s}}} \xrightarrow[\Sigma_1]{} J^{A^{\dagger}_{s}}_{\rho^{\dagger}_{s}}\ .$$

If $g = \langle \bar{s}, f\restriction\alpha_{\bar{s}}, s\rangle \in \mathbb{F}^{\dagger}$ with f as above, set $\hat{g} = f$ as before.

Define $d(g)$, $r(g)$, as before and let the notations $g : \bar{s} \Rightarrow_{\dagger} s$, $g \Rightarrow_{\dagger} s$ be the obvious variants.

Define $\lambda(g)$ as $\sup f''\alpha_{\bar{s}}$, so $\lambda(g) \leq \alpha_s$ if $g \Rightarrow_{\dagger} s$.

Using Lemma 6.52 we can formulate the analogue of Lemma 6.18.

Lemma 6.53 *Let* $f : s \Rightarrow_{\dagger} s$, $\lambda = \lambda(f)$. *Then* $s\restriction\lambda$ *exists and* $f_0 \in \mathbb{F}^{\dagger}$ *where* $f_0 = \langle \bar{s}, |f|, s\restriction\lambda\rangle$. *Moreover* $\hat{f} = k_{\lambda s}\hat{f}_0$.

The analogue of Lemma 6.19 is as before.

Define $C^{\dagger}_s = \{\lambda(f) \mid f \Rightarrow_{\dagger} s \cap \lambda(f) < \alpha_s\}$.

We then shall have:

Theorem 6.54 *Let* $s \in S^{\dagger}$.

a) $C_s^{\dagger} \subset \alpha_s$ *is closed in* α_s .

b) $\mathrm{otp}\ C_s^{\dagger} < \alpha_s$.

c) If $\lambda \in C_s^{\dagger}$ *, then* s/λ *exists and* $C_{s/\lambda}^{\dagger} = C_s^{\dagger} \cap \lambda$ *;* *moreover* $k_{\lambda s} k_{\eta s/\lambda} = k_{\eta s}$ *for* $\eta \in C_{s/\lambda}^{\dagger}$.

d) If $C_s^{\dagger}$ *is bounded in* α_s *then* $\mathrm{cf}(\alpha_s) = \omega$.

e) If $f : \bar{s} \Rightarrow_{\dagger} s$, $\lambda(f) = \alpha_s$ *then*

$$f/J_{\alpha_{\bar{s}}}^{a_{\bar{s}}} : \langle J_{\alpha_{\bar{s}}}^{a_{\bar{s}}}, C_{\bar{s}}^{\dagger} \rangle \xrightarrow[\Sigma_1]{} \langle J_{\alpha_s}^{a_s}, C_s^{\dagger} \rangle .$$

We are now almost home. However we must establish just one more lemma - which provides a connection between $\mathbb{F}$ and $\mathbb{F}^{\dagger}$.

Definition Let $s \in S^{\dagger}$. Let $\lambda \le \alpha_s$ s.t, $s/\lambda = \bar{s}$ exists. Define

$$\tilde{k}_{\lambda s} : J_{\mu_{\bar{s}}^{\dagger}}^{a_{\bar{s}}} \xrightarrow[\Sigma_{n^{\dagger}-1}]{} J_{\mu}^{a}$$

to be the canonical extension of

$$k_{\lambda s} : J_{\rho_{\bar{s}}^{\dagger}}^{A_{\bar{s}}^{\dagger}} \xrightarrow[\Sigma_0]{} J_{\rho_s^{\dagger}}^{A_s^{\dagger}} .$$

Lemma 6.55 *Let* $s \in S^{\dagger}$, $\lambda \in C_s^{\dagger}$. $\bar{s} = s/\lambda$. *and* $\tilde{k}_{\lambda s}$ *be as in the previous definition. Set* $\eta = \sup \tilde{k}_{\lambda s}{}''\nu_{\bar{s}}$. *Then* $\eta \in C_s^{+}$ *and* $\langle \bar{s}, \tilde{k}_{\lambda s} / \nu_{\bar{s}}, s/\eta \rangle \in \mathbb{F}$. *Thus there is the natural connection between* $\mathbb{F}$ *and* $\mathbb{F}^{\dagger}$ *that one would hope for.*

Proof Set $k = k_{\lambda s}$, $\tilde{k} = \tilde{k}_{\lambda s}$.

<u>Case 1</u> $\mu_s^{\dagger} < \mu_s$.

This is impossible, since α_s singular in $J_{\mu_{s+1}^{\dagger}}^{a_s}$ means there is a new subset of α_s in $J_{\mu_{s+1}^{\dagger}}^{a_s}$ and hence also a map of $\omega\alpha_s$ onto $\mu_s^{\dagger} \geq \nu$. Thus ν is not a cardinal in $J_{\mu_s}^{a_s}$.

<u>Case 2</u> $\mu_s < \mu_s^{\dagger}$.

It is easily seen that $\tilde{k}(\mu_{\bar{s}}) = \mu_s$ and

$$\tilde{k}\upharpoonright J_{\mu_{\bar{s}}}^{a_s} : J_{\mu_{\bar{s}}}^{a_{\bar{s}}} \xrightarrow[\Sigma_\omega]{} J_{\mu_s}^{a_s} .$$

Hence $\tilde{k}\upharpoonright J_{\rho_{\bar{s}}}^{A_{\bar{s}}} \in \mathbb{F}(s)$ and $k' \in \mathbb{F}$ where $k' = \langle \bar{s}, \tilde{k}\upharpoonright\nu_{\bar{s}}, s\rangle$.

Hence $\langle \bar{s}, \tilde{k}\upharpoonright\nu_{\bar{s}}, s\upharpoonright\eta\rangle = \text{red}(k') \in \mathbb{F}$ also.

<u>Case 3</u> $\mu_s = \mu_s^{\dagger}$ but $n_s < n_s^{\dagger}$.

We again show that $k' \in \mathbb{F}$. Let $\bar{s} = \langle \bar{\nu}, \bar{a}\rangle$. Set $\bar{\mu}, \bar{\rho}, \bar{A}, \rho, A = \mu_{\bar{s}}^{\dagger}, \rho_{\bar{\mu}}^{n-1,\bar{a}}, A_{\bar{\mu}}^{n-1,\bar{a}}, \rho_{\mu_s}^{n-1,a}, A_{\mu_s}^{n-1,a}$ respectively and $\bar{\alpha} = \alpha_{\bar{s}} = \lambda$, where $n = n_s$.

We wish to show that $\bar{\rho} = \rho_{\bar{s}}, \bar{A} = A_{\bar{s}}$. It is easy to see that:

(1) $\tilde{k}\upharpoonright J_{\bar{\rho}}^{\bar{A}} : J_{\bar{\rho}}^{\bar{A}} \xrightarrow[\Sigma_1]{} J_{\rho}^{A}$.

(2) $\rho_{\bar{\mu}}^{n,\bar{a}} = \bar{\alpha}$ (since $\rho_{\mu}^{n,a} = \alpha$) .

(3) $\tilde{k}(p_{\bar{\mu}}^{n,\bar{a}}) = p_{\mu}^{n,a}$.

(4) $\tilde{k}(\bar{\alpha}) = \alpha$.

(5) $\tilde{k}(\bar{\nu}) = \nu$ if $\nu < \mu_s$.

(6) $\bar{\rho} < \bar{\mu} \to \tilde{k}(\bar{\rho}) \geq \rho$.

Proof of 6 Suppose not i.e. that $\tilde{k}(\bar{\rho}) = \rho' < \rho$. Then

$J^{\bar{a}}_{\bar{\mu}} \models "\psi(\bar{\rho})"$, where $\psi(\bar{\rho})$ is the Σ_{n+1}-statement

$\exists q(\{\langle x,y\rangle \mid \phi(x,y,q)\}$ is a function $\wedge$

$\wedge \forall y \exists x \in J_{\bar{\rho}}(\phi(x,y,q)))$ where ϕ is Σ_{n-1} .

But $\tilde{k} : J^{\bar{a}}_{\bar{\mu}} \xrightarrow[\Sigma_n]{} J^{a}_{\mu}$ (as $n_s < n^{\dagger}_s$) . Hence

$J^{a}_{\mu} \models \psi(\rho')$.

So $\rho' \geq \rho^{n-1,a}_{\mu} = \rho$, a contradiction! QED (6)

By (6) and (4) we have: $\bar{\rho} > \bar{\alpha}$ since $\tilde{k}(\bar{\alpha}) = \alpha < \rho$. Hence by (2):

(7) $\bar{\rho} = \rho_{\bar{s}}$, $\bar{A} = A_{\bar{s}}$.

Clearly $\bar{\rho} \geq \bar{\nu}$. If $\nu < \rho$, then $\bar{\nu} < \bar{\rho}$, since otherwise $\tilde{k}(\bar{\rho}) = \nu < \rho$. Hence

(8) $\nu < \rho \rightarrow \bar{\nu} < \bar{\rho} \wedge \tilde{k}(\bar{\nu}) = \nu$.

Finally, we note that $p = p_s$ is definable as the $<_{J^A_\rho}$-least p s.t.

$\exists i \exists x \in J_\lambda \; p^{n,a}_{\mu} = h_{J^A_\rho}(i,x,p)$. $\bar{p} = p_{\bar{s}}$

can be defined similarly in terms of $p^{n,\bar{a}}_{\bar{\mu}}$. With (3) this yields:

(9) $\tilde{k}(\bar{p}) = p$.

But this shows that $\tilde{k} \upharpoonright J^{\bar{A}}_{\bar{\rho}} \in \mathbb{F}(s)$. QED Case 3

Case 4 $\mu^{\dagger}_s = \mu_s$ and $n_s = n^{\dagger}_s$.
Then $\langle \rho_s, A_s, p_s\rangle = \langle \rho^{\dagger}_s, A^{\dagger}_s, p^{\dagger}_s\rangle$. Let f witness that $\lambda \in C^{\dagger}_s$,

i.e. let $f : u \Rightarrow_{\dagger} s$ s.t. $\lambda(f) = \lambda$. Then

$$\hat{f} : J^{A^{\dagger}_u}_{\rho^{\dagger}_u} \xrightarrow[\Sigma_1]{} J^{A}_{\rho} \text{ s.t. } \hat{f}(\alpha_u) = \alpha,\ \hat{f}(p^{\dagger}_u) = p ,$$

and $\hat{f}(\nu_u) = \nu$ if $\nu < \rho$. Hence $\hat{f} \in \mathbb{F}(s)$ and

$$\langle \mu^{\dagger}_u, \rho^{\dagger}_u, A^{\dagger}_u, p^{\dagger}_u \rangle = \langle \mu_u, \rho_u, A_u, p_u \rangle .$$

Set $f' = \langle u, \hat{f}\restriction\nu_u, s \rangle$ then $f' \in \mathbb{F}$ and $\hat{f}' = \hat{f}$.
Now let $g = \text{red}(f)$. Then $g : u \Rightarrow_{\dagger} \bar{s}$.
Hence $\hat{f} = k_{\lambda s}\hat{g}$ (cf. Lemma 6.53) where

$$\hat{g} : J^{A_u}_{\rho_u} \xrightarrow[\Sigma_1]{} J^{A^{\dagger}_{\bar{s}}}_{\rho^{\dagger}_{\bar{s}}} ,\quad n^{\dagger}_{\bar{s}} = n_u = n ,\quad \hat{g}(\alpha_u) = \bar{\alpha} = \lambda$$

and $g(\nu_u) = \bar{\nu}$ if $\bar{\nu} < \rho^{\dagger}_{\bar{s}}$. Clearly $\rho^{\dagger}_{\bar{s}} > \bar{\alpha}$; hence $\rho^{\dagger}_{\bar{s}} \geq \bar{\nu}$. ν_u is regular in $J^{A_u}_{\rho_u}$ if $\nu_u < \rho_u$; hence either $\bar{\nu} = \rho^{\dagger}_{\bar{s}}$ or $\bar{\nu} < \rho^{\dagger}_{\bar{s}}$ and $\bar{\nu}$ is regular in $J^{A^{\dagger}_{\bar{s}}}_{\rho^{\dagger}_{\bar{s}}}$.
In either case $\bar{\nu}$ is Σ_{n-1}-regular in $J^{\bar{a}}_{\mu^{\dagger}_{\bar{s}}}$; hence
$\mu^{\dagger}_{\bar{s}} = \mu_{\bar{s}} = \mu$; $\rho^{\dagger}_{\bar{s}} = \rho_{\bar{s}} = \bar{\rho}$; $A^{\dagger}_{\bar{s}} = A_{\bar{s}} = \bar{A}$ and $p^{\dagger}_{\bar{s}} = p_{\bar{s}} = \bar{p}$.
Since $\hat{g}(p_u) = \bar{p}$, we conclude that $\hat{g} \in \mathbb{F}(\bar{s})$. Set $g' = \langle u, \hat{g}\restriction\nu_u, \bar{s} \rangle$. Then $g' \in \mathbb{F}$ and $\hat{g}' = \hat{g}$.

Claim $\lambda(g') = \bar{\nu}$.
Let $\bar{\eta} = \lambda(g')$. Set $g_0 = \text{red}(g')$, $\bar{s}_0 = \bar{s}\restriction\bar{\eta}$. Then $g_0 : u \Rightarrow \bar{s}_0$. Thus $\hat{g}_0 : J^{A_u}_{\rho_u} \xrightarrow[\Sigma_1]{} J^{A_{\bar{s}_0}}_{\rho_{\bar{s}_0}}$, $\hat{g}_0(\alpha_u) = \bar{\alpha} = \lambda$.
Since α_u is not Σ_1-regular in $J^{A_u}_{\rho_u}$ and $g_0''\alpha_u$ is cofinal in $\bar{\alpha}$, $\bar{\alpha}$ is not Σ_1-regular in $J^{A_{\bar{s}_0}}_{\rho_{\bar{s}_0}}$. Hence $\bar{s}_0 \in S^{\dagger}$. It is clear that $\hat{g}_0(\langle t_u, p^{\dagger}_u \rangle) = \langle t_{\bar{s}_0}, p_{\bar{s}_0} \rangle$. By Lemma 6.37

$$\exists\ \ell \text{ s.t. } \ell : J^{A_{\bar{s}_0}}_{\rho_{\bar{s}_0}} \xrightarrow[\Sigma_0]{} J^{\bar{A}}_{\bar{\rho}} , \text{ s.t. } \ell\hat{g}_0 = \hat{g}' .$$

But then $\ell(\bar{\alpha}) = \bar{\alpha} = \lambda$, $\ell(\langle t_{\bar{s}_0}, p^{\dagger}_{\bar{s}_0} \rangle) = \langle t_{\bar{s}}, p^{\dagger}_{\bar{s}} \rangle$. Hence the pair $\bar{s}_0, k\ell$ satisfy the defining conditions for $s/\!\!\wedge\lambda, k_{\lambda s}$. Hence $\bar{s}_0 = s/\!\!\wedge\lambda = \bar{s}$. Thus $\bar{\eta} = \nu_{\bar{s}_0} = \nu_{\bar{s}} = \bar{\nu}$.

QED Claim

Hence $\eta = \sup \tilde{k}/\!\!\wedge\bar{\nu} = \sup \tilde{k}\circ\hat{g}/\!\!\wedge\nu_u = \sup \hat{f}/\!\!\wedge\nu_u = \lambda(f')$ and so $\eta \in C^{\dagger}_s$. Let $\mathring{f} = \text{red}(f')$, $s' = s/\!\!\wedge\eta$. Then $\mathring{f} : u \Rightarrow s'$. We may now use Lemma 6.20 with $u, \bar{s}, s', \mathring{f}, \hat{g}/\!\!\wedge J_u, \tilde{k}/\!\!\wedge J_{\bar{s}}$ (which is Σ_0 and cofinal - hence Σ_1) replacing $\bar{s}, s', s, f, g_0, g_1$ respectively. By the conclusion of 6.20, $\langle \bar{s}, \tilde{k}/\!\!\wedge\nu_{\bar{s}}, s' \rangle \in \mathbb{F}$ as required.

QED

Theorem 6.56 2.4 *holds.*

Proof Let $p \in \mathbf{S}_{\alpha}$. The case $|p| = \alpha$ has already been dealt with in Lemma 6.49. Again define $\mathring{p}$ as $\langle \mu^{o}_{p}, (A\cap\alpha) \cup p^* \rangle$, then $\mathring{p} \in S^{\dagger}$ so we may set $D_p = C^{\dagger}_{\mathring{p}} \cap \text{Card}$, and for $\beta \in D_p$ let p_{β} be that $p' \in \mathbf{S}^*_{\beta}$ s.t. $\mathring{p}' = \mathring{p}/\!\!\wedge\beta$. Finally set $\pi^p_{\beta} = \tilde{k}_{\beta\mathring{p}}/\!\!\wedge \mathfrak{A}^o_{p_{\beta}}$. Then Theorem 6.54 a) and b) provide (i) and (ii). c) to e) together with the definition of π^p_{β} yield (iii). (iv) and (v) are straightforward.

QED

7 . The Cohen-generic sets

In this chapter we shall prove 2.5. It is here that we shall use our assumption of the existence of ◇-sequences (3) of 2.1). Replace "α" by "β^+" in the statement of 2.5. β will remain fixed for the rest of the proof. We aim, for any $p \in \mathbf{S}_{\beta^+}$ and for any $\xi < |s|$, to provide a sequence of subsets of β^+, $\langle b_{p\restriction\zeta} \mid \xi \leq \zeta \leq |s| \rangle$, so that any particular subsequence of them, are Cohen generic over $\mathfrak{A}^1_{p\restriction\xi}$. We shall not have a direct inductive definition of these b_p for $p \in \mathbf{S}_{\beta^+}$, but we shall approximate each b_p in β^+-steps, with bounded subsets of β^+, from below in a manner which will be familiar to those who are acquainted with morass-type arguments. We shall do this in such a way that the final b_p sets "cohere" correctly with the desired properties.

To this end define the following:

Definition T_{β^+} = the set of pairs $s = \langle \nu, a \rangle$ s.t. $\nu \in (\beta^+, \beta^{++})$, $a \subset \nu$, ν is p.r. closed or a limit of such, and setting $\hat{a} = (A \cap \beta^+) \cup a$ we have

(i) $L_\nu[\hat{a} \cap \xi] \models \bar{\bar{\xi}} = \beta^+$ for $\xi \in (\beta^+, \nu)$

(ii) $L^{\hat{a}} \models \bar{\bar{\nu}} = \beta^+$

For $s = \langle \nu, a \rangle \in T_{\beta^+}$, set as usual $\nu_s = \nu$, $a_s = a$, $\hat{a}_s = \hat{a}$ and $b_s = \langle L^{\hat{a}}_\nu, \hat{a} \rangle$.
For $\lambda \leq \nu$ set $s\restriction\lambda = \langle \lambda, a \cap \lambda \rangle$.
Define $\bar{s} \leq s \longleftrightarrow \bar{s}, s \in T_{\beta^+}$ and $\exists \lambda$ s.t. $\bar{s} = s\restriction\lambda$.
Clearly then "$\leq$" is a tree and $\bar{s} \leq s$ iff $\bar{s} = s\restriction\lambda$ for a p.r. closed (or ...) $\lambda \in (\beta^+, \nu]$ for $s \in T_{\beta^+}$.

Definition For $\delta < \beta^+$ let $\mathbf{Q}^\delta$ be the set of Cohen conditions of size $<\beta^+$ for adding a δ-sequence of subsets of β^+, i.e. $p \in \mathbf{Q}^\delta \longleftrightarrow p \in \bigcup_{\alpha<\beta^+} 2^{\delta\times\alpha}$, with the usual ordering. Then $\mathbf{Q}^\delta \subset H_{\beta^+}$ is uniformly H_{β^+}-definable in the parameter δ. The following Theorem gives the desired result.

Theorem 7.1 *There is a sequence* $\langle b_s \mid s \in T_{\beta^+}\rangle$ s.t.

a) $b_s \subset [\beta,\beta^+)$.

b) If $s \le t \in T_{\beta^+}$, $\delta < \beta^+$ *and* g *injects* δ *into* $\{r \mid s \le r \le t\}$ *then* $<b_{g(i)} \mid i < \delta>$ *meets every* $\Delta \in b_s$ s.t. $\Delta \subset \mathbf{Q}^\delta$ *is dense in* $\mathbf{Q}^\delta$, *i.e. is* $\mathbb{Q}^\delta$*-generic over* b_s.

c) Let $\eta > \nu_s$ *be p.r. closed* s.t. $L_\eta^{\hat{a}_s} \models "\overset{=}{\nu}_s = \beta^+"$. *Then* $b_s \in |L_\eta^{\hat{a}_s}|$ *is uniformly* $L_\eta^{\hat{a}_s}$*-definable in* ν_s.

Corollary 7.2 *Theorem 7.1 proves 2.5*

Proof This consists of just noticing that for $p \in \mathbf{S}_{\beta^+}$, $\mathfrak{A}^1_p$ is of the form b_s where $s = <\mu^1_p, A \cap \beta^+ \cup p^*>$. Note that μ^2_p is p.r. closed and bigger than μ^1_p, thus c) above provides (ii) of 2.5.

QED (Corollary)

We shall approximate T_{β^+} by a "quasi-morass". We shall define T_α, for α a member of a certain cub in β^+ set U, in an analogous manner to T_{β^+}.

Definition U = the set of $\alpha \in (\beta,\beta^+]$ s.t. β is the largest cardinal of L^A_α.

For $\alpha \in U$, T_α = the set of $s = \langle\nu,a\rangle$ s.t. $\nu > \alpha$, $a \subset \nu$, ν is p.r. closed and (i) and (ii) hold in the definition of T_{β^+} with $\hat{a} = (A \cap \alpha) \cup a$. For $s \in T_\alpha$ define $\nu_s, a_s, \hat{a}_s, b_s$ as before. Set $T = \bigcup_\alpha T_\alpha$. For $s \in T$

let α_s = the unique α s.t. $s \in T_\alpha$. Define s/λ as before and extend the definition of "$\le$" to "$\exists \alpha, \bar{s}, s \in T_\alpha$ and $\exists \lambda \bar{s} = s/\lambda$" for $\bar{s} \le s$. Clearly $\le / T_\alpha$ is a tree for each α. For $s \in T$ set $\hat{s} = \langle \nu_s, \hat{a}_s \rangle$. Then $\hat{s} \in S$ (cf. 6. §2). Set C_s to be $C'_{\hat{s}}$ of 6. §2. Theorem 6.42 may be couched in our present terms as:

Lemma 7.3 *There is* $\langle C_s \mid s \in T \rangle$ s.t.

a) $C_s \subset (\alpha_s, \nu_s)$ *is cub in* ν_s *and* $\langle b_s, C_s \rangle$ *is amenable.*

b) *If* λ *is a limit point of* C_s, *then* $s/\lambda \le s$ *and* $C_{s/\lambda} = \lambda \cap C_s$.

c) *If* $\eta > \nu_s$ *is p.r. closed s.t.* $L_\eta^{\hat{a}_s} \models "\bar{\bar{\nu}}_s \le \alpha_s"$ *then* $C_s \in |L_\eta^{\hat{a}_s}|$ *is uniformly* $L_\eta^{\hat{a}_s}$*-definable.*

d) *Let* $\bar{s} = \langle \bar{\nu}, \bar{a} \rangle$, $\bar{\alpha} < \bar{\nu}$, $\hat{a} = (A \cap \bar{\alpha}) \cup \bar{a}$, $\bar{b} = L_{\bar{\nu}}^{\hat{a}}$. *If* $f : \langle \bar{b}, \bar{C} \rangle \xrightarrow[\Sigma_1]{} \langle b_s, C_s \rangle$ s.t. $f/\bar{\alpha} = id/\bar{\alpha}$ *and* $f(\bar{\alpha}) = \alpha_s$ *then* $\bar{s} \in T$ *and* $\bar{C} = C_{\bar{s}}$.

Definition $f : \bar{\mathfrak{A}} \xrightarrow[Q]{} \mathfrak{A}$ means that $f : |\bar{\mathfrak{A}}| \longrightarrow |\mathfrak{A}|$ and $\bar{\mathfrak{A}} \models \forall \nu \exists \tau > \nu \phi(\nu, \tau, \vec{x}) \longleftrightarrow \mathfrak{A} \models \forall \nu \exists \tau > \nu \phi(\nu, \tau, f(\vec{x}))$ for all $\phi \in \Sigma_1$, where $\bar{\mathfrak{A}}$ and $\mathfrak{A}$ are transitive p.r. closed models.

Definition Let $\bar{s}, s \in T$, $\bar{s} \preccurlyeq s$ iff $\exists$ f: $\langle b_{\bar{s}}, C_{\bar{s}} \rangle \xrightarrow[Q]{} \langle b_s, C_s \rangle$ s.t. $f/\alpha_{\bar{s}} = id/\alpha_{\bar{s}}$ and $f(\alpha_{\bar{s}}) = \alpha_s$. If $\bar{s} \le s$, the map f is uniquely determined and we set $\pi_{\bar{s}s} = f$. The structure $M = \langle T, \le, \prec, \langle \pi_{\bar{s}s} : \bar{s}, s \in T, \bar{s} \prec s \rangle \rangle$ is then a "quasi-morass" in that it satisfies the following axioms:

I a) $\prec$ is a tree and $\pi_{\bar{s}s}/\nu_{\bar{s}}$ for $\bar{s} \prec s$ satisfies the requirements of the above definition.

b) $U = \{\alpha \mid \exists s \in T$ with $\alpha = \alpha_s\}$ is closed.

c) Each T_α is closed in "$\le$" for $\alpha \in U$.

d) $\{\alpha_{\bar{s}} \mid \bar{s} \preccurlyeq s\}$ is closed in α_s.

II a) $\bar{r} \leq \bar{s} \prec s$, $r = \pi_{\bar{s}s}(\bar{r}) \to \bar{r} \prec r$ and $\pi_{\bar{r}r} = \pi_{\bar{s}s} \upharpoonright b_{\bar{s}}$.

b) If s is a limit in $\prec$, then $b_s = \bigcup_{\bar{s} \prec s} \operatorname{ran}(\pi_{\bar{s}s})$.

c) If s is not maximal in $\leq$, then s is a limit point in $\prec$.

d) If $\lambda = \sup \pi_{\bar{s}s}{}''\nu_{\bar{s}}$, $s_0 = s \upharpoonright \lambda$, then $\bar{s} \preccurlyeq s_0$ and $\pi_{\bar{s}s_0} = \pi_{\bar{s}s}$.

e) Let $\bar{s}$ be a limit in $\leq$, $\bar{s} \preccurlyeq s$ and $\sup \pi_{\bar{s}s}{}''\nu_{\bar{s}} = \nu_s$. If $\alpha \in \bigcap_{r < \bar{s}} \{\alpha_t \mid r \preccurlyeq t \preccurlyeq \pi_{\bar{s}s}(r)\}$ then

$$\exists s' (\alpha = \alpha_{s'} \wedge \bar{s} \preccurlyeq s' \preccurlyeq s) .$$

The only difference between the quasi-morass here and a gap 1 morass at β^+ , is that in the full morass we would require $\leq \upharpoonright T_\alpha$ (T_α usually being written S_α) to be a well-ordering rather than a tree and we would further insist $\bar{\bar{T}}_\alpha < \beta^+$.

The following verification that M is a quasi-morass, is in fact part of the usual morass proof.

Lemma 7.4 M *is a quasi-morass.*

Proof I and II a) are trivial.

b) If s is a limit in $\preccurlyeq$ then $\alpha_s = \sup\{\alpha_{\bar{s}} \mid \bar{s} \preccurlyeq s\}$. But every $x \in b_s$ is b_s-definable with parameters from α_s , and thus these parameters will be in $\operatorname{ran}(\pi_{\bar{s}s})$ for some $\bar{s} \preccurlyeq s$.

c) Let $s' > s$. As α_s is regular in $b_{s'}$, then in $b_{s'}$, we may construct a chain of elementary submodels of $b_s : \langle b_\xi : \xi \in B \rangle$ with $|b_\xi| \cap \alpha_s = \xi$ and B cub in α_s . Each b_ξ will be isomorphic to some b_t with $t \preccurlyeq s$ and the result follows.

d) This is just Lemma 6.18, with $b_s, b_{\bar{s}}$ replacing $J_s, J_{\bar{s}}$ respectively and $\tilde{\pi}_{\bar{s}s}$ (the canonical extension of $\pi_{\bar{s}s}$)

replacing $\hat{f}$ etc.

e) Let $\bar{t} < \bar{s}$. Let $X_{\bar{t}} = \mathrm{ran}(\pi_{t't}{}''\nu_{t'})$ where t' is such that $\bar{t} \preccurlyeq t' \preccurlyeq \pi_{\bar{s}s}(\bar{t})$. Let $X = \bigcup_{\bar{t}<\bar{s}} X_{\bar{t}}$. We then use the Interpolation Lemma in the following diagram to obtain $J^{A'}_{\rho'}$, verifying that $\rho' = \rho_{s'}$, $A' = A_{s'}$ etc. as before.

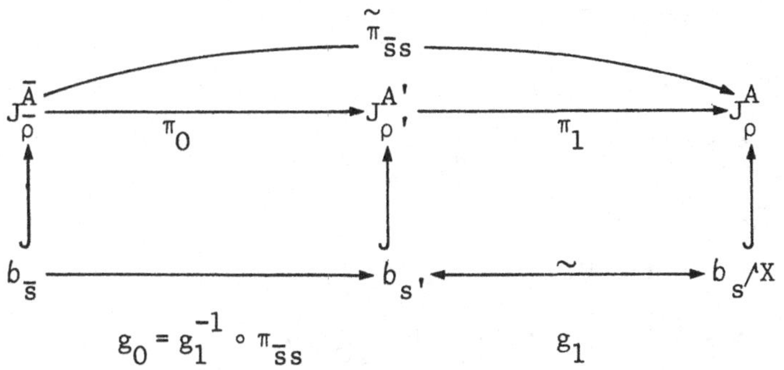

Note that each $b_t/X_{\bar{t}} \underset{Q}{\prec} b_t \underset{\Sigma_0}{\prec} b_s$. Then $b_t/X_{\bar{t}} \underset{\Sigma_0}{\prec} b_s$. Now $X \cap \nu_s$ is cofinal in ν_s and by the above b_s/X is a union of a chain of Σ_0-elementary substructures of b_s . Thus $b_s/X \underset{g_1^{-1}}{\cong} b_{s'} \underset{Q}{\longrightarrow} b_s$. So $s' \preccurlyeq s$ and $\alpha_{s'} = \alpha$.

QED

Definition $\bar{s} \prec_* s$ iff $\bar{s}$ immediately precedes s in $\prec$.

$$r \dashv s \text{ iff } \exists\, \bar{s},\bar{r}(\bar{s} \prec_* s, \bar{r} < \bar{s} \wedge \bar{r} \prec_* r \prec \pi_{\bar{s}s}(\bar{r})) .$$

Notice that $\bar{r}$ and $\bar{s}$ are uniquely determined by r,s . The following is readily verified.

III a) $\dashv$ is a tree.

b) $r \dashv s \longrightarrow \alpha_r < \alpha_s$.

c) If $\bar{s}$ is a limit in $<$, $\bar{s} \prec_* s$ and $\sup \pi_{\bar{s}s}{}''\nu_s = \nu_s$, then s is a limit in $\dashv$ and $\alpha_s = \sup_{r \dashv s} \alpha_r$.

[c) is immediate from II e).]

Definition $W(s) = \{t \mid \exists \bar{s} \preccurlyeq s,\ t \leq \bar{s} \vee \bar{s} \leq t\}$

$W^*(s) = \{t \mid \exists \bar{s} \prec_* s,\ t \leq \bar{s} \vee \bar{s} \leq t\}$

[Note that $r \dashv s \longrightarrow W^*(s) \subset W^*(r)$].

We now construct by induction on ν_s, sets $b_s (s \in T)$ and simultaneously $b_{st} (s \in T, t \in W(s))$, satisfying the following list of conditions:

a) $b_s \subset [\beta, \alpha_s)$.

b) $\bar{s} \preccurlyeq s \longrightarrow b_{\bar{s}} = \alpha_{\bar{s}} \cap b_s$.

c) $b_{st} \subset [\beta, \alpha_s)$.

d) $\bar{s} \preccurlyeq s \wedge t \in W(\bar{s}) \longrightarrow b_{\bar{s}t} = \alpha_{\bar{s}} \cap b_{st}$.

e) $\bar{s} \preccurlyeq s \wedge t \leq \bar{s} \longrightarrow b_{st} = b_{\pi_{\bar{s}s}(t)}$.

f) $s \leq t \rightarrow b_{st} = b_t$.

g) $r \dashv s \wedge t \in W^*(s) \longrightarrow b_{rt} = \alpha_r \cap b_{st}$.

There are in fact two further conditions to be met which we shall formulate presently. The intention is of course that $\langle b_s \mid s \in T_{\beta+} \rangle$ shall satisfy Theorem 7.1. For this to work a wilderness of dense sets of conditions in $\mathbf{Q}^\delta$ must be met. To ensure this we shall require that the approximating sequence $\langle b_s \mid s \in T \setminus T_{\beta+} \rangle$ satisfy a diagonalisation requirement, which will require a few preliminary remarks for its formulation.

Lemma 7.5 *There is a uniform canonical* $\Sigma_1^{\langle b_t, C_t \rangle}$*-definable map of* α_t *onto* b_t *for* $t \in T$.

Proof <u>Case 1</u> C_t (which is of the form $C'_{\hat{t}}$ remember) was the ω-sequence $\{\xi_n^{\hat{t}} \mid n < \omega\}$. Thus the map $f : \omega \longrightarrow \nu_t$ defined by $f(n) = \xi_n$ is clearly $\Sigma_1^{\langle b_t, C_t \rangle}$. Suppose $t \in T_{\hat{\alpha}_t}$ then since $\xi_n < \nu_t$ there is a b_t-least map $g : \alpha_t \twoheadrightarrow J_{\xi_n}^{\hat{a}_t}$ which is Σ_1-definable in b_t from α_t and ξ_n. Hence putting these two facts together, and using the fact that α_t is b_t-definable (without parameters) as "the largest cardinal" we obtain a map $h' : \alpha_t \twoheadrightarrow b_t$ which is $\Sigma_1^{\langle b_t, C_t\rangle}$.

<u>Case 2</u> Otherwise. In which case $C_{\hat{t}}$ was unbounded in ν_t. In the notation of 6.2 let $h = h_{J_{\rho_{\hat{t}}}^{A_{\hat{t}}}}$, $p = p_{\hat{t}}$. Then $|J_{\rho_{\hat{t}}}^{A_{\hat{t}}}| = h(J_{\alpha_t} \cup \{p\})$. In particular $|b_t| \subset |J_{\rho_{\hat{t}}}^{A_{\hat{t}}}|$. So if $x \in |b_t|$ $\exists y \in J_{\alpha_t}$, $i < \omega$ s.t. $x = h(i,y,p)$. But $C_{\hat{t}}$ is unbounded in ν_t. Thus

$$x = h(i,y,p) \longleftrightarrow \exists \lambda \in C_{\hat{t}} \text{ s.t. } x = h_\lambda(i,y,p_{\hat{t}\wedge\lambda})$$

where h_λ is the canonical Σ_1-Skolem function for $J_{\rho_{\hat{t}\wedge\lambda}}^{A_{\hat{t}\wedge\lambda}}$. Clearly this makes x $\Sigma_1^{\langle b_t, C_t\rangle}$ from y, and, noting that $p_{\hat{t}\wedge\lambda}$, $A_{\hat{t}\wedge\lambda}$ etc. are uniformly p.r. in $\hat{t}\wedge\lambda$, together with the fact that $g : \omega\alpha \longleftrightarrow J_\alpha$ is $\Delta_1^{J_\alpha}$, we may define a $h' : \alpha \twoheadrightarrow b_t$ which is $\Sigma_1^{\langle b_t, C_t\rangle}$.

QED

Definition Set $h_t = h'$ of the last lemma.

$$\ell_t = \{\langle \xi, \zeta, i\rangle \mid (i = 0 \wedge h_t(\xi) \in h_t(\zeta)) \vee (i = 1 \wedge h(\xi) \in \hat{a}_t)\}$$

Thus $\ell_t \subset \alpha_t$ and h_t is uniformly recoverable from ℓ_t. Note that ℓ_t is also uniformly $\Sigma_1^{\langle b_t, C_t\rangle}$.

Definition Let $\delta < \alpha \in U$. $\mathbf{Q}_\alpha^\delta =_{df} \mathbf{Q}_\alpha^\delta \cap L_\alpha^A$.

Hence $s \in T_\alpha \rightarrow \mathbf{Q}_\alpha^\delta = (\mathbf{Q}^\delta)_{b_s} \in b_s$.

Definition Let $\langle S_\alpha \mid \alpha < \beta^+\rangle$ be our $\diamondsuit$-sequence. Suppose $\alpha \in U$, $t \in T_\delta$. Suppose further that $s, \Delta, g \in b_t$ and that $h_t(\dot{s}) = s$, $h_t(\dot{\Delta}) = \Delta$ and $h_t(\dot{g}) = g$ with

a) $s \leq t$.

b) g bijects a $\delta < \alpha$ into $\{r \mid s \leq r \leq t\}$.

c) $\Delta \subset \mathbf{Q}_\alpha^\delta$ is dense in $\mathbf{Q}_\alpha^\delta$.

Suppose yet further that there exist such s,Δ,g,t such that

$$S_\alpha = \{\langle \nu,i\rangle \mid (i=0 \wedge \nu \in \ell_t) \vee (i=1 \wedge \nu = \dot{s}) \vee (i=2 \wedge \nu = \dot{\Delta}) \vee$$
$$\vee (i=3 \wedge \nu = \dot{g}) \quad =_{df} V_{t,s,\Delta,g} .$$

Then set $\Gamma_\alpha = \langle t,s,\Delta,g\rangle$.
In all other cases Γ_α is undefined.

We can now state our promised diagonalisation requirements whose necessity should hopefully be clear by the end of the proof:

h) Let $\bar{s} \prec_* s$, where s is initial in $<$. Then if $\Gamma_{\alpha_{\bar{s}}}$ is defined and equals $\langle t,\bar{s},\Delta,g\rangle$ where $\delta = \mathrm{dom}(g)$. Suppose that $\langle b_{g(i)} \mid i < \delta\rangle \in b_s$. Then $\langle b_{s,g(i)} \mid i < \delta\rangle$ meets $\pi_{\bar{s}s}(\Delta)$.

i) Let $\bar{s} \prec_* s$, where s is a successor in $<$. Let r immediately precede s in $\dashv$. Let $\Gamma_{\alpha_{\bar{s}}}, g, \delta$ be as above. Suppose that $\langle b_{r,g(i)} \mid i < \delta\rangle \in b_s^{\bar{s}}$. Then $\langle b_{s,g(i)} \mid i < \delta\rangle$ meets $\pi_{\bar{s}s}(\Delta)$.

By induction on ν_s we can construct b_s ,

$\langle b_{st} \mid \exists \bar{s} \prec s \wedge t \leq \bar{s} \vee \bar{s} \leq t\rangle$,

simultaneously verifying at each step that a)-i) hold for the points constructed. These verifications are mostly trivial and are left to the reader.

I We may define $\langle b_{st} \mid \exists \bar{s} \prec s \wedge t < \bar{s}\rangle$ outright since $\nu_{\pi_{\bar{s}s}(t)} < \nu_s$ and we may use the induction hypothesis to define $b_{st} = b_{\pi_{\bar{s}s}(t)}$ thus satisfying e) directly.

II Case 1 s a limit in $\prec$.

Set $b_s = \bigcup_{\bar{s}\prec s} b_{\bar{s}}$; $b_{st} = \bigcup_{\substack{\bar{s}\prec s \\ t\in W(\bar{s})}} b_{\bar{s}t}$.

Case 2 s is initial in $\prec$.

Set $b_s = \emptyset$.

Case 3 $\bar{s} \prec_* s$.

Case 3.1 s is initial in $<$.

Suppose that $\Gamma_{\alpha_{\bar{s}}}$ is defined and equals $\langle t,\bar{s},\Delta,g\rangle$, $\delta = \mathrm{dom}(g)$ and that $\tilde{b} = \langle b_{g(i)} \mid i < \delta\rangle \in b_s$. Set:

$<b_i' \mid i < \delta> = L^A_{\alpha_s}$-least extension of $\tilde{b}$

meeting $\pi_{\bar{s}s}(\Delta)$. Set $b_{s,g(i)} = b_i'$ and in all other cases set $b_{s,t} = b_{\bar{s},t}$. Finally set $b_s = b_{s,\bar{s}}$.

Case 3.2 s is a successor in $<$.

Let r immediately precede s in $\dashv$. If $\Gamma_{\alpha_{\bar{s}}}, g, \delta$ are as above and $\tilde{b} = \langle b_{r,g(i)} \mid i < \delta\rangle \in b_s$, define $\langle b_i' \mid i < \delta\rangle$, $b_{s,g(i)}$ as above. Otherwise set $b_{st} = b_{rt}$ for $\bar{s} \le t$. For $s' \prec s$, $s' < t$, set $b_{st} = b_{\bar{s}t}$. Finally set $b_s = b_{s\bar{s}}$.

Case 3.3 s is a limit in $<$ and $\sup \pi_{\bar{s}s}"\nu_s = \nu_s$.

For $\bar{s} \le t$ set: $b_{st} = \bigcup_{r \dashv s} b_{rt}$. For $s' \prec s$, $s' < t$ set $b_{st} = b_{\bar{s}t}$. Finally set $b_s = b_{s\bar{s}}$.

Case 3.4 All other cases fail.

Set $\lambda = \sup \pi_{\bar{s}s}"\nu_{\bar{s}} < \nu_s$. $s_0 = s{\restriction}\lambda$. Let $\bar{s} \prec_* s' \prec s_0$.

Then $\{u \mid u \prec s'\} = \{u \mid u \prec s\}$ and $\{r \mid r \dashv s'\} = \{r \mid r \dashv s\}$.

In particular $W(s') = W(s)$.

Set: $b_{st} = b_{s't}$; $b_s = b_{s\bar{s}} = b_{s'}$.

This completes the constructions and we have the easy corollary:

j) $\langle b_{s,t} \mid s \in T\setminus T_{\beta+} \wedge t \in W(s) \rangle$ is uniformly $\langle H_{\beta+}, A \cap \beta+ \rangle$-definable.

Now we shall use a)-j) to prove:

Lemma 7.6 $\langle b_s \mid s \in T_{\beta+} \rangle$ *satisfies Theorem 7.1.*

Proof

a) is immediate.

c) Notice that $\{\bar{s} \mid \bar{s} \prec s\} \in L_\eta^{\hat{a}_s}$ is $L_\eta^{\hat{a}_s}$-definable in ν_s . The same is true for $\langle b_t \mid t \in T\setminus T_{\beta+} \rangle$ by j). Hence it is true for $b_s = \bigcup_{\bar{s} \prec s} b_{\bar{s}}$.

b) Let s,t,δ,g be as in Theorem 7.1 b). We may suppose w.l.o.g. that t is chosen large enough so that $g \in b_t$. Let $\Delta \in b_s$ be dense in $\mathbf{Q}^\delta$.

Claim $\langle b_{g(i)} \mid i < \delta \rangle$ meets Δ .

Proof Obviously we may assume that s = the $\prec$-least $s \prec t$ s.t. $\Delta \in b_s$. So then s is initial or a successor in $\prec$. By axiom II b) of the definition of a quasi-morass, as t is a $\prec$-limit $\exists\, t_0 \prec t$ s.t. $s,\Delta,g \in \mathrm{ran}(\pi_{t_0 t})$, and hence $s,\Delta,g \in \mathrm{ran}(\pi_{t't})$ $\forall\, t'$ s.t. $t_0 \prec t' \prec t$. Let

$$V = \bigcup_{t_0 \preccurlyeq t' \preccurlyeq t} V_{t',s',\Delta',g'}$$

where

$$\langle s',\Delta',g' \rangle = \pi_{t_0 t'}(\langle s_0,\Delta_0,g_0 \rangle) = \pi_{t_0 t}^{-1}(\langle s,\Delta,g \rangle) .$$

Then $V \subset \beta^+$. By the definition of a $\diamond$-sequence $\{\alpha \mid S_\alpha = V \cap \alpha\}$ is stationary beneath β^+. But by axiom I d) $\{\alpha_{t'} \mid t_0 \preccurlyeq t' \preccurlyeq t\}$ is cub beneath β^+. Hence $\exists\, \alpha_{\bar{t}}$ s.t. $S_{\alpha_{\bar{t}}} = V \cap \alpha_{\bar{t}} = V_{\bar{t},\bar{s},\bar{\Delta},\bar{g}}$ by the uniformity of our definitions, where:

(*) $\pi_{\bar{t}t}(\langle \bar{s},\bar{\Delta},\bar{g}\rangle) = \langle s,\Delta,g\rangle$.

Hence $\Gamma_{\alpha_{\bar{t}}} = \langle \bar{t},\bar{s},\bar{\Delta},\bar{g}\rangle$ is defined.

Let $\bar{t} \prec_* t' \prec t$, $\bar{s} \prec_* s' \prec s$. Then $s' \dashv t'$. Since $\pi_{s's} : b_{s'} \xrightarrow[\Sigma_1]{} b_s$ it follows by j) that $B \in b_{s'}$ where

$$B = \langle b_{r,u} \mid r,u \in L^A_{\alpha_{s'}} \wedge r \in T \wedge u \in W(r)\rangle .$$

Since $\langle S_\alpha \mid \alpha < \beta^+\rangle$ is uniformly $\langle H_{\beta^+}, A \cap \beta^+\rangle$-definable, it follows that $\langle S_\alpha \mid \alpha < \alpha_{s'}\rangle \in b_{s'}$. But then $S_{\alpha_{\bar{t}}} \in b_{s'}$ and hence $\Gamma_{\alpha_{\bar{t}}} \in b_{s'}$.

If s is initial in $<$, then so is s' and we can conclude that the hypothesis of h) holds: i.e. $\langle b_{\bar{g}(i)} \mid i < \delta\rangle \in b_{s'}$. Likewise if s is a successor in $<$ then so is s' and if r immediately precedes s' in $\dashv$, then we know that $r \in b_{s'}$ and conclude that the hypothesis of i) holds, i.e. $\langle b_{r,\bar{g}(i)} \mid i < \delta\rangle \in b_{s'}$. In either case, by our construction $\langle b_{s',\bar{g}(i)} \mid i < \delta\rangle$ meets $\pi_{\bar{s}s'}(\Delta)$. But $b_{\pi_{\bar{t}t'}(\bar{g}(i))} = b_{t',\bar{g}(i)}$ extends $b_{s',\bar{g}(i)}$, since $s' \dashv t'$ by g). Hence $\langle b_{\pi_{\bar{t}t'}(\bar{g}(i))} \mid i < \delta\rangle$ meets $\pi_{\bar{t}t'}(\bar{\Delta})$. Hence by (*) above, $\langle b_{g(i)} \mid i < \delta\rangle$ meets Δ.

QED

8 . How to get rid of "$\neg 0^{\#}$"

In this chapter we shall describe the methods used to rid ourselves of the assumption that $\langle M,A\rangle \models \neg 0^{\#}$. In fact what we shall do in §1 is use the assumption that

$\langle M,A\rangle \models$ "Every set of ordinals has a sharp" ;

we shall then prove Theorem 1 of the Introduction using this alternative assumption. In case $\langle M,A\rangle$ is not closed under sharps we describe a strategy for circumventing this in §2: if β is least such that $\beta \in \mathrm{card}\backslash\omega$ and $\exists\, a \subset \beta^{+}$ s.t. $\neg a^{\#}$ in M. Then we use the methods of Chapters 1-4 to extend M to an M' s.t.

I) $M' \models ZF + \exists\, b \subset \beta^{+}$ s.t. $V = L[b]$.

II) $H_{\beta}^{M'} = H_{\beta}^{M}$.

Then (with a little juggling about at the "boundary point" β) we use the methods of this chapter to code down from β^{+} to ω . Of course, we ensure that the large cardinal properties of Theorem 0.1 continue to be preserved. So assume for the moment:

I) $\langle M,A\rangle \models ZF + GCH +$ "Every set of ordinals has a sharp".

II) Facts 1-4 of 2.1 hold for $\langle M,A\rangle$.

[As before 1-4 can be achieved by the preliminary forcing with the class $\mathbf{Q}$. This still keeps M closed under sharps, see the Appendix.]

Looking at the statement of Theorem 1 of the Introduction there seems little intrinsic reason that the assumption of

$\neg 0^{\#}$ is necessary for the proof, and in Chapters 1-4 it appears to be used just to help things fall into place rather than for any deeper reason. We shall now look at the points it is used and see how, in this chapter, we shall overcome the obstacle of its absence.

I. Firstly we assumed $\neg 0^{\#}$ during the definition of the $\mathbf{S}_\alpha$ at Clause 2 in 2.2. This means for $s \in \mathbf{S}_\alpha$ that μ^1_s is "not too far away" (in L-terms) from $|s|$ and that we can code s fairly "low down". Ultimately this is used in the crucial 3.22(b) to show that μ' is small enough to get $b \in \mathfrak{A}^1_{p_\gamma}$, which will in turn show that $p/\gamma \in \mathbf{P}^{p_\gamma}_\tau$.

II. We used the fact that singular cardinals of M are singular in L in the definition on p. 41 and so on p. 42 where we defined the canonical sequences of singular cardinals $\langle \gamma^\beta_i : i < \lambda_\beta \rangle$ which were in turn used in coding.

III. We insisted on p. 28 that μ^1_s be such that

$$\mathfrak{A}^1_s \models \text{"}\forall x \subseteq \alpha(\bar{\bar{x}} \geq \omega_2 \longrightarrow \exists y \in L(x \subseteq y \wedge \bar{\bar{x}} = \bar{\bar{y}}))\text{" if } \alpha \geq \omega_2 .$$

This was just so that we could prove Lemma 3.40.7 to obtain the τ-distributivity of $\mathbf{P}^s_\tau$ where α was singular, $s \in \mathbf{S}_\alpha$.

II and III turn out to be superfluous for that proof. Instead of using $\langle \gamma_i \mid i < \lambda_\beta \rangle$ in II we may utilise the set of singular cardinals D_p guaranteed by Lemma 2.4. For each $s \in \mathbf{S}_\beta$ and $p \in \mathbf{P}^s_\tau$ we may define a closed (if $cf(\alpha) > \omega$, unbounded) subset of D_s, D^p_s, on which the burden of (C) (iv) and (v) of the previous definition of $\mathbf{P}^\tau_s$ will fall, and which we shall use during the lemmas on extensions of conditions.

In III the trick of defining the set $\Pi^{s,p}_f$ was used where $\omega < cf(\alpha) < \alpha$. Having done away with C_α in this context a new proof is needed which will be a construction along the lines of 2.14.3 (for extending conditions) and the "central lemma" 3.13, but performed simultaneously.

Whereas these two alterations could have been done in the earlier proof, I cannot be so easily altered. We change the definition of $\mathbf{S}_\alpha$, by simply ignoring clause 2) of the first definition in §2.2. Roughly speaking, for $p \in \mathbf{S}_\alpha$ we ensure that $\mathfrak{A}^1_p$ contains $(A \cap \alpha \cup p)^{\#}$. Then, when we come to our version of 3.21 (that $b \in L_{\mu'+\omega}[A \cap \gamma, p_\gamma]$) and in 3.22, whereas there we had to show that μ' was not too large in order that b may be seen to be a member of $\mathfrak{A}^1_{p_\gamma}$, the mere fact that $b \in L[A \cap \gamma, p_\gamma]$ (and $b \in H^M_{\gamma^+}$) will ensure that p/γ is in $\mathfrak{A}^1_{p_\gamma}$ (which will contain $H^{L[A \cap \gamma, p_\gamma]}_{\gamma'}$, where $\gamma' = \gamma^{+L[A \cap \gamma, p_\gamma]}$).

We in turn modify the definition of $\mathbb{R}^s$ so that not only does it code s, but also $(s/\xi)^{\#}$ $(\alpha \leq \xi < |s|)$ in some suitable form. This is so, in order to be able in our decoding process to have the sharps available to formulate the mechanism and definitions described above in some model or other.

<u>The first case:</u>

8.1 M CLOSED UNDER SHARPS

As intimated we shall assume in this section that

I. $\langle M, A\rangle \models$ ZF + GCH + "Every set of ordinals has a sharp".

II. $\langle M, A\rangle \models$ "$V = L[A]$", where $H_\kappa = L_\kappa[A]$ for infinite cardinals κ.

III. There is a $\diamondsuit$-sequence on κ^+, uniformly $\langle H_{\kappa^+}, A \cap \kappa^+\rangle$-definable for infinite cardinals κ.

IV. Facts 4.2.13 and 14 hold.

As remarked earlier a prior forcing can achieve this if need be.

As before we take card $= \{0\} \cup$ the infinite cardinals. We shall now give our definition of the modified $\mathbf{S}_\alpha$'s .

Definition Let $\alpha \in$ card . $\mathbf{S}_\alpha$ is the set of $p: [\alpha, |p|) \longrightarrow 2$ such that $\alpha \leq |p| < \alpha^+$ and

$$\alpha \leq \xi \leq |p| \longrightarrow \bar{\bar{\xi}}^{L[A \cap \alpha, p \restriction \xi]} = \alpha , \quad \text{if } \alpha \geq \omega .$$

In the following definition $\hat{s}$ will be analogous to the s^* defined on p.29, but we have also thrown in indiscernibles for "initial segments of s ".

Definition Let $\alpha \in \text{card} \backslash \omega$, $s \in \mathbf{S}_\alpha$. Define by induction the following:

$$\hat{s} = (A \cap \alpha) \cup \{\mu^0_{s \restriction \xi} \mid s(\xi) = 1\} \cup \bigcup_{\alpha \leq \xi < |s|} e_{s \restriction \xi} .$$

e_s = the first ω canonical indiscernibles for $L[\hat{s}]$.

$\hat{e}_s$ = $\hat{s} \cup e_s$.

μ^0_s = $\sup(\alpha \cup \{\mu^\omega_{s \restriction \xi} \mid \alpha \leq \xi < |s|\})$.

μ^{i+1}_s = the least $\mu > \mu^i_s$ s.t. $\mu > \sup e_s$ and $L_\mu[\hat{e}_s] \models$ "ZF$^-$" . For $i \leq \omega$

$\mathfrak{A}^i_s$ = $\langle L_{\mu^i_s}[\hat{e}_s], \hat{e}_s \cap \mu^i_s \rangle$.

Thus $\mathfrak{A}^0_s = \langle L_{\mu^0_s}[\hat{s}], \hat{s} \rangle$, $\mathfrak{A}^i_s = \langle L_{\mu^i_s}[\hat{e}_s], \hat{e}_s \rangle$ $(i > 0)$.

As before put $\mathfrak{A}_s = \mathfrak{A}^\omega_s$, $\mu_s = \mu^\omega_s$, and for $\alpha \leq \xi \leq |s|$, put $\hat{\xi}, e_\xi, \mathfrak{A}_\xi$ etc. for $\hat{s \restriction \xi}, e_{s \restriction \xi}, \mathfrak{A}_{s \restriction \xi}$ etc.

Definition

$\tilde{\mu}_s = \sup e_s$; $\tilde{\mathfrak{A}}_s = \langle L_{\tilde{\mu}_s}[\hat{e}_s], \hat{e}_s \rangle$.

μ'_s = the least $\mu' > \mu^0$ s.t.:

I. $L_{\mu'}[\hat{s}] \models$ "$ZF^- + \forall\nu\ \bar{\bar{\nu}} \le \alpha$" .

II. $cf(\mu') = \alpha$ if α is a successor cardinal.

$\mathfrak{A}'_s = \langle L_{\mu'_s}[\hat{s}], \hat{s} \rangle$.

Consequently $\mu^o_s < \mu'_s < \alpha^{+L[\hat{s}]} < \tilde{\mu}_s < \mu^1_s$. At successor cardinals we shall, virtually everywhere, utilise $\mathfrak{A}'_s$ where we formerly used $\mathfrak{A}^1_s$.

We shall turn now to some alterations of the fine structural lemma statements. Lemma 2.3 (dealing with the C_s-sequences) is unchanged except that (iv) is replaced by iv':

(iv)' $C_p \in \mathfrak{A}'_p$ is uniformly $\Sigma_1(\mathfrak{A}'_p)$ in α .

The proof is the same; those of the last two are essentially unchanged. Define $\mathbf{S}^*_\alpha$ as before but:

Definition $\mathbf{S}^o_\alpha = \{s \in \mathbf{S}^*_\alpha \mid L[\hat{s}] \models$ " α is singular"$\}$.

$\mathbf{S}^1_\alpha = \mathbf{S}^*_\alpha \setminus \mathbf{S}^o_\alpha$.

Then (v) of 2.4 is replaced by:

(v)' If $p \in \mathbf{S}^o_\alpha(\mathbf{S}^1_\alpha)$, then D_p, $\langle p_\nu \mid \beta \in D_p \rangle$, $\langle \pi^p_\beta \mid \beta \in D_p \rangle \in L[\hat{p}]$ $(\mathfrak{A}_p)$ and are uniformly $\Sigma_1(L[\hat{p}])$ $(\Sigma_1(\mathfrak{A}_p))$ in α, p . Also $p_\beta \in \mathbf{S}^o_\beta(\mathbf{S}^1_\beta)$ for $\beta \in D_p$. Moreover in the latter case ($p \in \mathbf{S}^1_\alpha$) if $\beta \in D_p$ then π^p_β extends uniquely to a $\tilde{\pi}^p_\beta : \mathfrak{A}_{p_\beta} \xrightarrow[\Sigma_1]{} \tilde{\mathfrak{A}}_p$.

6.56 works as before but note for $p \in \mathbf{S}^1_\beta$ we are defining $\mathring{p}$ as $\langle \tilde{\mu}_p, \tilde{e}_p \rangle$. Setting $\tilde{k} = \tilde{k}_{\beta\mathring{p}}$, $\eta = \sup \tilde{k}''\ \mu^o_{p_\beta}$, we want $\langle\langle \mu^o_{p_\beta}, \hat{p}_\beta \rangle , \pi^p_\beta \restriction \mu^o_{p_\beta} , \langle \mu^o_p, \hat{p} \restriction \eta \rangle\rangle \in \mathbb{F}$. But this follows as in 6.55 case 2. $\tilde{\pi}^p_\beta$ is the usual extension of π^p_β defined by $\tilde{\pi}^p_\beta(i'_m) = i_m$ where $\langle i'_m \rangle$, $\langle i_m \rangle$ enumerate e_{p_β} and e_p.

Definition e^*_s = the complete Σ_1-theory of $\langle \tilde{\mathfrak{A}}_s, \xi_{(\xi<\alpha)} \rangle$, coded as a subset of α.

By (v)' we have: if $s \in \mathbf{S}^1_\alpha$, $\beta \in D_s$ then $e^*_{s_\beta} = \beta \cap e^*_s$.
In 2.5 replace ii by:

(ii)' $b_p \in L[\hat{p}]$, and is uniformly $\Sigma_1(L[\hat{p}])$ in p,α.

Definition For b_p as in 2.5 set $b^\nu_p = \{\xi \mid \{\nu,\xi\} \in b_p\}$ for $\nu < \alpha$. Then replace (iii) by:

(iii)' If $\delta < \alpha$ and f injects δ into $[\xi, |p|]$. Then: $\langle b^\nu_{p/f(i)} \mid i < \delta \wedge \nu < \alpha \rangle$ is Cohen generic over $\mathfrak{A}'_\xi$.

This is just a well-known property of Cohen generic sets.

Define Z_ν, V_ν, $S(b)$ as in 2.4.1. The following new definition of $\mathbf{R}^s$ reveals the rôle that the sets e^*_s will play in coding "sharps":

Definition Let $s \in \mathbf{S}_{\alpha^+}$. Let $\tilde{s} = \{\xi \mid s(\xi) = 1\}$.
$\mathbf{R}^s$ = set of pairs $\langle \dot{r}, r \rangle$ s.t.

(I) $r \in \mathbf{S}_\alpha$.

(II) $\dot{r} = \dot{r}^0 \times \{0\} \cup \dot{r}^1 \times \{1\}$ where
$$\dot{r}^0 \subset (A \cap \alpha^+) \times [\alpha, \alpha^+)$$
$$\dot{r}^1 \subset (P_{<\alpha^+}(\alpha^+) \cup \{b^0_{s/\xi} \mid \xi \in \tilde{s}\} \cup$$
$$\cup \{b^{i+1}_{s/\xi} \mid \alpha^+ \leq \xi < |s| \wedge i \in e^*_\xi\}) \times [\alpha, \alpha^+).$$

(III) $\overset{=}{\dot{r}} \leq \alpha$.

(IV) $\langle \nu, \eta \rangle \in \dot{r}^0 \longrightarrow (V_\nu \setminus \eta) \cap \tilde{r} = \emptyset$.

(V) $\langle b, \eta \rangle \in \dot{r}^1 \longrightarrow (S(b) \setminus \eta) \cap \tilde{r} = \emptyset$.

$\mathbf{R}^s \in \mathfrak{A}'_s$ and is α-distributive in $\mathfrak{A}'_s$. In reproving 2.6 over $\mathfrak{A}'_s$ we decode e^*_ξ from the b^{i+1}_ξ to get e_ξ, so as to be able to define $\mathfrak{A}^1_\xi$ in $L_{\mu'_s}[D_0 \cup D_1]$.

$\mathbf{S}_\alpha^+$ = set of $D \subset [\alpha,\alpha^+)$ s.t.

(I) $\delta \in [\alpha,\alpha^+) \longrightarrow \chi_D \restriction \delta \in \mathbf{S}_\alpha$.

(II) $\langle L_{\alpha^+}[\hat{D}],\hat{D}\rangle \models ZF^-$, where $\hat{D} = \bigcup_{\alpha\le\delta<\alpha^+} \chi_D \restriction \delta$.

Then $\mathfrak{A}_D = \langle L_{\alpha^+}[\hat{D}],\hat{D}\rangle$. The other definitions and proofs (2.7-8) are verbatim.

Definition Let $\alpha \in \text{card}\backslash\omega$, $s \in \mathbf{S}_\alpha$.

$\mathbf{A} = \mathbf{A}_s$ = the set of $b = \langle L_\eta[\hat{e}_s],\hat{e}_s \cap \eta\rangle$ s.t. $\mu_s^o < \eta \le \mu_s$ and:

(I) Every element of b is b-definable with parameters from α .

(II) Either $b \models ZF^-$ or there are arbitrarily large $\nu < \eta$ s.t. $\langle L_\nu[\hat{e}_s],\hat{e}_s \cap \nu\rangle \models ZF^-$.

Definition $\mathbf{A}^o = \mathbf{A}_s^o \cap L[\hat{s}]$.

So $\mathfrak{A}_s' \in \mathbf{A}_s^o$ and $\mathbf{A}\backslash\mathbf{A}^o = \{\tilde{\mathfrak{A}}_s,\mathfrak{A}_s^1,\ldots,\mathfrak{A}_s^\omega\}$.

Note that $\mathfrak{A}_s^o \notin \mathbf{A}_s^o$. Also notice that $\forall\, \xi < \alpha^{+L[\hat{s}]}$ $\exists\eta(\xi < \eta < \alpha^{+L[\hat{s}]})$ s.t. $\langle L_\eta[\hat{s}],\hat{s}\rangle \in \mathbf{A}_s^o$.

Hence $\forall\, x \in H_{\alpha^+}^{L[\hat{s}]}\ \exists b \in \mathbf{A}_s^o\ (x \in b)$.

Definition Let α be a limit cardinal, $s \in \mathbf{S}_\alpha$. Let $b \in \mathbf{A}_s$, $\tau \le \alpha$, then define $X_\tau^b, b_\tau, \pi_\tau^b$ by:

X_τ^b = smallest $X \prec b$ s.t. $\tau \subset X$.

$\pi_\tau^b = b_\tau \longleftrightarrow X_\tau^b$, where b_τ is transitive.

Fact Let $b \in \mathbf{A}$. Then

(I) $b = \bigcup_{\tau<\alpha} X_\tau^b = b_\alpha$ (since each $x \in b$ is definable w.p.f. α).

(II) Let $\bar{\alpha} = \alpha \cap X_\alpha^b$. Set $\bar{s} = \pi_\alpha^{b-1}(s)$.

a) $\bar{s} \in \mathbf{S}_{\bar{\alpha}}$ and $b_{\bar{\alpha}} \in \mathbf{A}_{\bar{s}}$.

b) $b_{\bar\alpha} \in \mathbf{A}^o_{\bar s}$ if $b \in \mathbf{A}^o_s$.

c) If $b = \mathfrak{A}'_s, \tilde{\mathfrak{A}}_s, \mathfrak{A}^i_s$ then $b_{\bar\alpha} = \mathfrak{A}'_{\bar s}, \tilde{\mathfrak{A}}_{\bar s}, \mathfrak{A}^i_{\bar s}$ respectively for $1 \le i \le \omega$ (cf. §2.5).

Definition Let α be a limit cardinal, $s \in \mathbf{S}_\alpha$, $\gamma \in \text{card} \cap \alpha$.

a) For $1 \le n \le \omega$ set

$$\rho^n_{s,\gamma} = \{\rho', 2\}, \text{ where } \rho' = L[A]\text{-code of } \mathfrak{A}^n_{s,\gamma^+}$$

[so $\rho^n_{s,\gamma} \in Z_2 \cap (\gamma^+, \gamma^{++})$].

b) For $b \in \mathbf{A}^o_s$, $\nu < \gamma$ set

$$\rho^b_\gamma(\nu) = \{\{\rho', \nu\}, 3\}, \text{ where } \rho' = L[A]\text{-code of } b_{\gamma^+}$$

[so $\rho^b_\gamma(\nu) \in Z_3 \cap (\gamma^+, \gamma^{++})$].

We shall use the $\rho^n_{s,\gamma}$ as before: using a $\rho^n_{\xi,\gamma}$-sequence on which to code $s(\xi)$ (in certain circumstances). Similarly we shall use a sequence of $\rho^b_\gamma(\nu)$'s (for one b) to code e^*_ξ .

We shall now make this explicit; as before we shall be coding an $s \in \mathbf{S}_\alpha$ by a partial map $p: \alpha \longrightarrow 2$ (α a limit). This coding will ensure that $\hat{s} \in L[p]$, and is canonically definable from p in $L[p]$.

We first dispose of those $\xi \in [\alpha, |s|)$ where α is regular in $\mathfrak{A}_\xi$ (and in fact the above ρ's are needed):

Definition $B^n_s = \{\bar\alpha < \alpha \mid \bar\alpha = \alpha \cap X^{\mathfrak{A}^n_s}_\alpha\}$ $(1 \le n \le \omega)$.

So if α is regular in $\mathfrak{A}_s$, then B^n_s is cub in α. In any case $B^n_s \in \mathfrak{A}^{n+1}_s$ and $B^n_s \supset B^{n+1}_n$.

Definition Let α be a limit cardinal, $s \in \mathbf{S}_\alpha$. Let p be a partial map $p: \alpha \longrightarrow 2$.

I) Let $\alpha \le \xi < |s|$ s.t. α is regular in $\mathfrak{A}_\xi$.

p s-codes ξ iff $\exists n < \omega, \eta < \alpha$ s.t. $\forall m \geq n$ we have:

$$\gamma \in B^m_\xi \setminus \eta \longrightarrow p(\rho^m_{\xi\gamma}) = s(\xi) .$$

II) Let $\alpha \leq \xi \leq |s|$ s.t. α is regular in $L[\hat{\xi}]$. Let $b \in \mathbf{A}^o_\xi$

p s-codes ξ at b iff $\forall \nu < \alpha$ $\exists C \in L[\hat{\xi}]$ s.t. C is cub in α and

$$\gamma \in C \longrightarrow p(\rho^b_\gamma(\nu)) = \chi_{e^*_\xi}(\nu) .$$

III) Let ξ be as in II).

p s-codes e_ξ iff p s-codes e_ξ at some $b \in \mathbf{A}^o_s$.

Definition $\Omega = \Omega_s \simeq$ the least $\xi \in [\alpha, |s|]$ s.t. $\mathfrak{A}_\xi \models$ "α is singular", for α a limit cardinal and $s \in \mathbf{S}_\alpha$.

So, $|s| = \Omega$ iff $s \in \mathbf{S}^*_\alpha$ and $|s| \nmid \Omega$ implies that α is regular in $\mathfrak{A}_\xi$ $\forall \xi < |s|$.

Definition Let α, s, p be as above, with $|s| \nmid \Omega$.

p codes s iff p s-codes ξ, e_ξ $\forall \xi \in [\alpha, |s|)$.

Remark p codes at most one s of given 'length'. For, suppose p coded s,s' with $s \neq s'$, but both of the same length. Let ξ be least such that $s(\xi) \neq s'(\xi)$. Then $B^n_{s \restriction \xi} = B^n_{s' \restriction \xi}$, $\rho^m_{s \restriction \xi} = \rho^m_{s' \restriction \xi}$ $\forall m,n < \omega$. But then for a sufficiently large γ (by I) above) $p(\rho^m_{s\restriction\xi,\gamma}) = p(\rho^m_{s'\restriction\xi,\gamma}) = s(\xi) = s'(\xi)$. Contradiction. Likewise it is easy to see that if p codes s , so does p' where $p' \supset p$ or $p' = p \restriction [\eta, \alpha)$ for some $\eta < \alpha$.

We would now like to work out a definition of coding for s with $|s| > \Omega_s$. We give a few preliminary statements with a view to defining a sequence to take the place of the $\langle \gamma^\beta_i : i < \lambda_\beta \rangle$ sequence on p.42.

Definition Let α, s, p be as above with $s \in \mathbf{S}^*_\alpha$ (i.e. $|s| = \Omega$) let D_s, $\langle s_\beta \mid \beta \in D_s \rangle$, $\langle \pi^s_\beta \mid \beta \in D_s \rangle$ be as in 2.4 (with modification (v)')

p <u>supports</u> s iff

I) p codes s.

II) For sufficiently large $\gamma \in D_s$, $p \restriction \gamma$ codes an r_γ s.t. either $|r_\gamma| \geq |s_\gamma|$ or $r_\gamma \in \mathbf{S}^*_\gamma$.

III) If $cf(\alpha) > \omega$, there are arbitrarily large $\gamma \in D_s$ s.t. $r_\gamma \supset s_\gamma$.

Remark Again, if p supports s so does any extension or final segment of p.

Lemma 8.1 *Let* α, s, p *be as in the above definition, with* p *supporting* s. *Let* r_γ *be as there defined for* $\gamma \in D_s \setminus \eta$ *for some* $\eta < \alpha$.

Set: $D = \{\gamma \in D_s \setminus \eta \mid r_\gamma \supset s_\gamma\}$. *Then* D *is closed in* α *(and hence cub in* α *if* $cf(\alpha) > \omega$ *)*.

Proof Let γ be a limit point of D. We show that $r_\gamma \restriction \xi = s_\gamma \restriction \xi$ by induction on $\xi \in [\alpha, |s_\gamma|]$. $\xi = \gamma$ or ξ a limit are trivial. So let $\xi = \nu+1$ with $r_\gamma \restriction \nu = s_\gamma \restriction \nu$. But then $\nu < |s_\gamma|$ so that $r_\gamma \restriction \nu \notin \mathbf{S}^*_\gamma$. Thus $p \restriction \gamma$ r_γ-codes ν. Suppose $\eta < \gamma$ and $n < \omega$ s.t. $\forall m \geq n$, $\forall \delta \geq \eta$ we have

$$\delta \in B^m_\gamma \longrightarrow p(\rho^m_{\nu\delta}) = r_\gamma(\nu).$$

Now γ's a limit point of D_s and $D_{s_\gamma} = \gamma \cap D_s$ so $\sup D_{s_\gamma} = \gamma$, thus by 2.4 (iv) $\mathfrak{N}_{s_\gamma} = \bigcup_{\beta \in D_{s_\gamma}} \pi^{s_\gamma}_\beta {}'' \mathfrak{N}^o_{s_\beta}$.

Thus we may pick $\bar{\gamma} \in D$ s.t. $\eta < \bar{\gamma} < \gamma$ and $\nu \in ran(\pi^{s_\gamma}_{\bar{\gamma}})$.

Set $\pi = \pi^{s_\gamma}_{\bar{\gamma}}$, $\bar{\nu} = \pi^{-1}(\nu)$. Then $\pi(r_{\bar{\gamma}} \restriction \bar{\nu}) = \pi(s_{\bar{\gamma}} \restriction \bar{\nu}) = s_\gamma \restriction \nu = r_\gamma \restriction \nu$.

Hence $\pi(\mathfrak{A}^m_{s_{\bar\gamma}\wedge\bar\nu}) = \mathfrak{A}^m_{s_\gamma\wedge\nu}$ for $m < \omega$. Hence $B^m_{s_{\bar\gamma}\wedge\bar\nu} = B^m_{s_\gamma\wedge\nu} \cap \bar\gamma$ and $\rho^m_{\nu\tau} = \rho^m_{\bar\nu\tau}$ for $\tau < \bar\gamma$. But $p\wedge\bar\gamma$ codes $s_{\bar\gamma}$. Hence there is $m \geq n$, $\delta \in B^m_{\bar\nu}\setminus\eta$ s.t. $p(\rho^m_{\bar\nu\delta}) = s_{\bar\gamma}(\bar\nu)$. Hence $r_\gamma(\nu) = p(\rho^m_{\nu\delta}) = p(\rho^m_{\bar\nu\delta}) = s_{\bar\gamma}(\bar\nu) = s_\gamma(\nu)$.

Q.E.D.

Definition Let α, s, p be as in Lemma 8.1. Let η = least $\eta < \alpha$ s.t. r_γ is defined for $\gamma \in D_s\setminus\eta$. Set $D = \{\gamma \in D_s\setminus\eta \mid r_\gamma \supset s_\gamma\}$. We define a set $D^p_s \subset D_s$ as follows:

Case 1 sup $D = \alpha$. Then $D^p_s = D$.

Case 2 sup $D < \alpha$. Then $\mathrm{cf}(\alpha) = \omega$ and we take D^p_s as the $L[\hat e_s]$-least ω-sequence of cardinals with supremum α. [Note in this case $\mathfrak{A}_s = L_{\mu_s}[\hat e_s] \models$ " α singular" .]

The following are easily checked:

Fact If $\gamma < \alpha$ is a limit point of D^p_s then $p\wedge\gamma$ supports s_γ and $D^{p\wedge\gamma}_{s_\gamma} = \gamma \cap D^p_s$.

Proof Follow the definitions noting $D_{s_\gamma} = \gamma \cap D_s$.

Fact If $s \in \mathbf{S}^0_\alpha$, then $D^p_s \in L[\hat s]$.

Proof If Case 1 above holds then $D_s \in L[\hat s]$. But $D^p_s \subset D_s$, $|D_s| < \alpha$, and $H_\alpha \subset L[\hat s]$. Case 2 is trivial.

Fact If $s \in \mathbf{S}^1_\alpha$ then $D^p_s \in \mathfrak{A}_s$.

Proof Similar.

Fact If $p' \supset p$, then $D_s^p \subset D_s^{p'}$ and differs from $D_s^{p'}$ only by an initial segment.

Proof That $D_s^p \subset D_s^{p'}$ is trivial. Suppose that there existed an (arbitrarily large) $\gamma \in D_s^{p'} \setminus D_s^p$. Let p'/γ code $r'_\gamma \supset s_\gamma$ according to the definition of supports (p.265). Let p/γ code an $r_\gamma \not\supset s_\gamma$. But p'/γ also codes $r'_\gamma/|r_\gamma|$ and so this latter must equal r_γ (<u>and</u> $\neq r'_\gamma$). So $r'_\gamma \supset s_\gamma$ and $r'_\gamma \supset r_\gamma$ with $r_\gamma \not\supset s_\gamma$. Thus $s_\gamma \supset r_\gamma$. But then $|r_\gamma| < |s_\gamma|$, and $r_\gamma \in \mathbf{S}^*_\gamma$ contradicting $s_\gamma \in \mathbf{S}^*_\gamma$.

The next definition extends the notation and definitions for our convenience.

Definition Let $|s| \geq \Omega$. p <u>supports</u> s iff p supports s/Ω

$$D_s = D_{s/\Omega},\ D_s^p = D_{s/\Omega}^p,\ s_\gamma = (s/\Omega)_\gamma \quad \text{for} \quad \gamma \in D_s .$$

Then we set by way of a parallel to the definition of ρ_{si} on p.42.

Definition Let $|s| \geq \Omega$. For $\gamma \in \text{card} \cap \alpha$ set:

$$\tilde{\rho}_{s\gamma} = \left\{\left\{\rho^\omega_{s\gamma}, |D_s|\right\}, 4\right\}.$$

For $b \in \mathbf{A}^o_s$, $\nu < \gamma$ set:

$$\tilde{\rho}^b_\gamma(\nu) = \left\{\left\{\rho^b_\gamma(\nu), |D_s|\right\}, 5\right\}.$$

We are now in a position to extend the notion of coding to arbitrary s.

Definition Let α, s, p be as above.

(I) Let $\Omega \leq \xi < |s|$

p <u>s-codes</u> ξ iff p supports s and

$p(\tilde{\rho}_{\xi\gamma}) = s(\xi)$ for sufficiently large $\gamma \in D^p_s$.

(II) Let $\Omega \le \xi \le |s|$ s.t. α is singular in $L[\hat{\xi}]$.
Let $b \in \mathbf{A}^o_\xi$. p <u>s-codes</u> e_ξ <u>at *b*</u> iff p supports s and $\forall \nu < \alpha$ $p(\tilde{\rho}^b_\gamma(\nu)) = \chi_{e^*_\xi}(\nu)$, for sufficiently large $\gamma \in D^p_s$.

(III) p <u>s-codes</u> e_ξ iff p s-codes e_ξ at some $b \in \mathbf{A}^o_\xi$.

Definition Let α, s, p be as above. p <u>codes</u> s iff p s-codes ξ, e_ξ $\forall \xi \in [\alpha, |s|)$.

The remarks made after the last definition carry over *mutatis mutandis* here.

We now give the definition of the sets $\mathbb{P}^s_\tau$ $(\tau \in \text{card} \cap \alpha)$ designed to code s by a subset of $[\tau, \tau^+)$. As before $p \in \mathbb{P}^s_\tau \longrightarrow p\colon \text{card} \cap [\tau, \alpha) \longrightarrow V^2$, and $p(\gamma) = \langle \dot{p}_\gamma, p_\gamma \rangle$, $\overset{\frown}{p} = \bigcup_{\gamma \in \text{dom}(p)} p_\gamma$.

Definition By induction on $\alpha \in \text{card} \setminus \omega$ we define $\mathbb{P}^s_\tau$ $(\tau \in \text{card} \cap \alpha)$ as the set of $p\colon [\tau, \alpha) \longrightarrow V^2$ s.t.

(A) $p \restriction \gamma \in \mathbb{P}^{p_\gamma}_\tau$ for $\gamma \in (\tau, \alpha)$.

(B) $p(\beta) \in \mathbf{R}^s$ if $\alpha = \beta^+$.

(C) If $\lim(\alpha)$ then

I. $p \in \mathfrak{A}_s$.

II. $\overset{\frown}{p}$ codes $s \restriction |p|$ [where $|p|$ = least $\xi \le |s|$ s.t. $p \in \mathfrak{A}_\xi$].

III. $p \notin L[|\hat{p}|] \longrightarrow \overset{\frown}{p}$ s-codes $e_{|p|}$.

IV. If α's regular in $L[|\hat{p}|]$ and $p \in L[|\hat{p}|]$, then $\exists$ cub $C \in L[|\hat{p}|]$ s.t.

(*) $\gamma \in C \longrightarrow |p/\gamma| = |p_\gamma| \wedge \dot{p}_\gamma = \emptyset$.

V. If α's regular in $\mathfrak{A}_{|p|}$, $\exists$ cub $C \in \mathfrak{A}_{|p|}$ satisfying (*).

VI. $cf(\alpha) > \omega$ $s/|p| \in \mathbf{S}^o_\alpha \wedge p \in L[|\hat{p}|] \longrightarrow \exists$ cub $D \subset D_s$ s.t.

$\gamma \in D \longrightarrow p_\gamma \supset s_\gamma \wedge |p/\gamma| \geq |s_\gamma|$.

VII. $cf(\alpha) > \omega \wedge s/\Omega \in \mathbf{S}^1_\alpha \wedge p \notin L[\hat{\Omega}] \longrightarrow \exists$ cub $D \subset D_s$ s.t.

$\gamma \in D \longrightarrow p_\gamma \supset s_\gamma \wedge p/\gamma \notin L[\hat{s}_\gamma]$.

Several remarks are in order:

Note

1) The remarks of §2.6 are still true:
$\xi < |p| \longrightarrow s(\xi)$ still determined by p and s/ξ etc.

2) By (C)III we have for $\xi \in [\alpha, |s|]$
$p \notin L[\hat{\xi}]$ iff p codes e_ξ .

3) The conclusion of VII will always hold for $|p| \geq \Omega$ (cf. II) of the definition of 'supports').

4) The reader will note that iii and v of §2.6 have been subsumed by the definition of 'codes', i.e. into (C)II above, and that we use D^p_s instead of $\langle \gamma^\alpha_i \mid i < \lambda_\alpha \rangle$.

Lemma 2.9 goes through verbatim; in 2.10 just change $\mathfrak{A}^1_s$ to $\mathfrak{A}'_s$ as remarked earlier. The collapsing lemma will be given a more general formulation.

Lemma 8.2 (The Collapsing Lemma) *Let* α *be a limit cardinal, regular in* $\mathfrak{A}_s$ $(L[\hat{s}])$ *, where* $s \in \mathbf{S}_\alpha$ *. Let* $p \in \mathbb{P}^s_\tau$ $(\tau < \alpha)$ *s.t.* $|p| = |s|$ $(|p| = |s| \wedge p \in L[\hat{s}])$ *. Let* $b \in \mathbf{A}_s$ $(b \in \mathbf{A}^o_s)$ *s.t.* $\exists$ *a cub* $C \subseteq \mathrm{card} \cap [\tau, \alpha)$ *s.t.* $C \in b \wedge \gamma \in C \longrightarrow \dot{p}_\gamma = \emptyset \wedge |p/\gamma| = |p_\gamma|$.

Let $\bar{\alpha} < \alpha$ *s.t.* $\bar{\alpha} = \alpha \cap X^b_{\bar{\alpha}} \wedge p \in X^b_{\bar{\alpha}}$ *. Then:*

a) $\dot{p}_{\bar{\alpha}} = \emptyset \wedge p_{\bar{\alpha}} = \pi_{\bar{\alpha}}^{b-1}(s)$.

b) $b_{\bar{\alpha}} \in \mathbf{A}_{p_{\bar{\alpha}}}$ $(b_{\bar{\alpha}} \in \mathbf{A}^{o}_{p_{\bar{\alpha}}})$

c) $|p/\bar{\alpha}| = |p_{\bar{\alpha}}| \wedge (p/\bar{\alpha} \in L[\hat{p}_{\alpha}] \longleftrightarrow p \in L[\hat{s}])$.

Proof Set $\langle \pi, \bar{b}, \bar{s} \rangle = \langle \pi_{\bar{\alpha}}^{b}, b_{\bar{\alpha}}, \pi^{-1}(s) \rangle$. We know by Fact II p.262 that

a) $\bar{s} \in \mathbf{S}_{\bar{\alpha}}$.

b) $\bar{b} \in \mathbf{A}_{\bar{s}} \wedge \bar{b} \in \mathbf{A}^{o}_{\bar{s}}$ if $b \in \mathbf{A}^{o}_{s}$,

As usual we may assume that $C \in X_{\bar{\alpha}}^{b}$, since some such C is definable (by (C) IV or V above). But then $\pi(C \cap \bar{\alpha}) = C$ and $C \cap \bar{\alpha}$ is cub in $\bar{\alpha}$. Hence $\bar{\alpha} \in C$ and

c) $\dot{p}_{\bar{\alpha}} = \emptyset \wedge |p/\bar{\alpha}| = |p_{\bar{\alpha}}|$.

Clearly $\pi(p/\bar{\alpha}) = p$. Since $\overset{\frown}{p}$ codes s :

d) $\overset{\frown}{p/\bar{\alpha}}$ codes $\bar{s}$.

e) $p/\bar{\alpha} \in L[\hat{\bar{s}}] \longleftrightarrow p \in L[\hat{s}]$.

But $p/\bar{\alpha} \in \mathbf{P}_{\tau}^{p_{\bar{\alpha}}}$, and as $|p/\bar{\alpha}| = |p_{\bar{\alpha}}|$, $\overset{\frown}{p/\bar{\alpha}}$ codes $p_{\bar{\alpha}}$ by (C) II. So $|p_{\bar{\alpha}}| = |p/\bar{\alpha}| = \pi^{-1}(|p|) = \pi^{-1}(|s|) = |\bar{s}|$. Hence

f) $p_{\bar{\alpha}} = \bar{s}$

since $\overset{\frown}{p/\bar{\alpha}}$ cannot code two different $\bar{s} \in \mathbf{S}_{\alpha}$ of the same length.

2.13 goes through as before. Q.E.D.

Now we have to turn our attention to §2.7 which include the lemmas on the extendibility of conditions. The changes required are extensive and although the parallel arguments are very similar in parts we shall rewrite the whole of that section.

Recall first the definition of "thin" (p.50).

Definition $X \subset On$ is thin iff $X \cap \kappa$ is not stationary in κ for any regular κ.

We shall now give a name to a notion commonly occurring.

Definition Let $s \in \mathbf{S}_\alpha$, $\tau \in \text{card} \cap \alpha$, $p \in \mathbf{P}^s_\tau$.

p is smooth iff whenever γ is a limit point of $\text{card} \cap [\tau, \alpha)$ we have $|p_\gamma| = |p \restriction \gamma|$.

In fact, all we really need for the Coding Theorem to go through is the following Extension Theorem:

Let $s \in \mathbf{S}_{\alpha^+}$, $\tau \in \text{card} \cap \alpha^+$, $p \in \mathbf{P}^s_\tau$. *Let* $X \subseteq \text{card} \cap [\tau, \alpha^+)$ *be thin and* $\xi_\gamma \in [\gamma, \gamma^+)$ *for* $\gamma \in X$. *There is a* $q \le p$ *in* $\mathbf{P}^s_\tau$ *s.t.* q *is smooth and* $|q_\gamma| \ge \xi_\gamma$ *for* $\gamma \in X$.

However, in order to carry through its inductive proof, we shall have to prove the full extension lemma which we shall restate in this later setting.

(The Extension Lemma) *Let* α *be a limit cardinal,* $s \in \mathbf{S}_\alpha$, $\tau \in \text{card} \cap \alpha$, $p \in \mathbf{P}^s_\tau$. *Let* $X \subset \text{card} \cap [\tau, \alpha)$ *be thin in* $\mathfrak{A}_s$. *Let* $\langle \xi_\gamma \mid \gamma \in X \rangle \in \mathfrak{A}_s$ *s.t.* $\xi_\gamma \in [\gamma, \gamma^+)$ *for* $\gamma \in X$. *Then there is a* $q \le p$ *in* $\mathbf{P}^s_\tau$ *s.t.*

a) $|q| = |s|$.

b) q *supports* s, *if* $|s| = \Omega$.

c) q *is smooth.*

d) $|q_\gamma| \ge \xi_\gamma$ *for* $\gamma \in X$.

By induction on α we prove that the Extension Theorem holds and the Extension Lemma holds if α is a limit cardinal. The cases $\alpha = 0$ and α a successor are trivial. Let α be a limit cardinal. Notice that the lemma trivially implies the theorem at α, so it suffices to prove the lemma in this case. So suppose from now on that α is a limit cardinal and that the theorem and lemma hold below α. As a further simplification we can of course just prove the lemma for the case $\tau = 0$, since if $p \in \mathbf{P}^s_\tau$, $r \in \mathbf{P}^{p_\tau}$ and $q \le p \cup r$ satisfying the conclusion of the lemma, so does $q/\!\upharpoonright[\tau,\alpha) \le p$. As before the proof will stretch over several sublemmas, some of which are somewhat lengthy.

Lemma 8.3 *If* $p \in L[\hat{s}]$, $|p| = |s|$ *and* $s \in \mathbf{S}^1_\alpha$, *then* p *supports* s.

Proof Let D_s, $\langle s_\beta \mid \beta \in D_s\rangle$, $\langle \pi^s_\beta \mid \beta \in D_s\rangle$ be as in (the modified) Lemma 2.4. Recall further that π^s_β extended uniquely to a $\tilde{\pi}^s_\beta : \tilde{\mathfrak{A}}_{s_\beta} \xrightarrow[\Sigma_1]{} \tilde{\mathfrak{A}}_s$. We may assume (via the ind. hypoth.) that $\sup D_s = \alpha$, since otherwise there is nothing to prove. In which case (by the obvious extension of 2.4(IV)) $\tilde{\mathfrak{A}}_s = \bigcup_{\beta \in D_s} \tilde{\pi}^s_\beta{}''\tilde{\mathfrak{A}}_{s_\beta}$. Let $\beta_0 \in D_s$ s.t. $p \in \mathrm{ran}(\tilde{\pi}^s_{\beta_0})$. It will suffice to show:

<u>Claim</u> $\beta \in D_s \setminus \beta_0 \longrightarrow s_\beta = p_\beta \wedge |s_\beta| = |p/\!\upharpoonright\beta|$.

Proof By (C) IV there is a cub $C \in L[\hat{s}]$ s.t.

$$\gamma \in C \longrightarrow \dot{p}_\gamma = \emptyset \wedge |p_\gamma| = |p/\!\upharpoonright\gamma|.$$

But then $p, C \in b$ for some $b \in \mathbf{A}^0_s$. We may assume further that $b \in \mathrm{ran}(\tilde{\pi}^s_{\beta_0})$ since some such $b \in \mathbf{A}^0_s$ will be. But now put $\bar{b} = \tilde{\pi}^{s-1}_{\beta_0}(b)$ and we have $\tilde{\pi}^s_{\beta_0}/\!\upharpoonright\bar{b} = \pi^b_{\beta_0}$ and so $\bar{b} = b_{\beta_0}$; noting that $\tilde{\pi}^s_{\beta_0}/\!\upharpoonright\beta_0 = \mathrm{id}/\!\upharpoonright\beta_0$ etc. we may use the Collapsing Lemma to conclude that $|p_\beta| = |p/\!\upharpoonright\beta|$ and $\tilde{\pi}^{s-1}_\beta(s) = s_\beta = \pi^b_\beta(s) = p_\beta$ for $\beta \in D_s \setminus \beta_0$ as required.

That p supports s follows from the fact that $\forall\beta \in D_s\setminus\beta_0$, $p/\!\upharpoonright\beta \in \mathbb{P}^{p_\beta} = \mathbf{P}^{s_\beta}$ and $p/\!\upharpoonright\beta$ codes (i.e. supports)

$p_\beta \wedge |p \wedge \beta| = p_\beta = s_\beta \in \mathbf{S}^1_\beta$.

Q.E.D.

Lemma 8.4 *Let* α *be regular in* $L[\hat{s}]$. *Let* $p \in L[\hat{s}]$, $|p| = |s|$. *Then there is a* $q \leq p$ *in* $\mathbb{P}^s$ *s.t.* $q \notin L[\hat{s}]$.

Proof It will suffice to construct a $q \leq p$ in $\mathbb{P}^s$ s.t. q codes e_s (which cannot be in $L[\hat{s}]$) . In the course of the proof we shall prove in addition the following:

Lemma 8.4.1 *Let* p, s, α *be as above. Let* $C \in L[s]$ *be cub in* α *s.t.*

$$\gamma \in C \longrightarrow \dot{p}_\gamma = \emptyset \wedge |p_\gamma| = |p \wedge \gamma| .$$

There is a $q \leq p$ *in* $\mathbb{P}^s$ *s.t.* $q \notin L[\hat{s}]$ *and*

I. $\dot{p}_\gamma = \dot{q}_\gamma \ \forall \gamma < \alpha$.

II. q *differs from* p *only on* $\{\gamma^+ \mid \gamma \in C\}$.

Now for the proof itself: let $b \in \mathbf{A}^o_s$ s.t. $p, C \in b$. We shall code e_s at b . Set

$$B = \{\gamma \mid \gamma = X^b_\gamma \cap \alpha \wedge p, C \in X^b_\gamma\} .$$

Then $B \in L[\hat{s}]$ is cub in α . If $\gamma \in B$ set $\pi_\gamma = \pi^b_\gamma$. Then $\gamma \in C$ since $\pi_\gamma(C \cap \gamma) = C$ and $C \cap \gamma$ is cub in and thus $\gamma \in C$. Also by utilising the Collapsing Lemma we have:

$$p_\gamma = \pi^{-1}_\gamma(s) , \ |p \wedge \gamma| = |p_\gamma| , \ p \wedge \gamma = \pi^{-1}_\gamma(p) , \ \dot{p}_\gamma = \emptyset .$$

Note further that $b_\gamma \in \mathbf{A}^o_{p_\gamma} \subset L[\hat{p}_\gamma]$ and that $B \cap \gamma \in L[\hat{p}_\gamma]$ since $B \cap \gamma$ can be constructed from $b_\gamma , p \wedge \gamma$ as B was constructed from b, p .

<u>Case 1</u> α is regular in $\mathfrak{N}_s$.

Set $B^* = \{\gamma \mid \gamma = \alpha \cap X_\gamma^{\mathfrak{A}_s^1} \wedge p, b \in X_\gamma^{\mathfrak{A}_s^1}\}$.

Set $\tilde{\pi}_\gamma = \pi_\gamma^{\mathfrak{A}_s^1}$ for $\gamma \in B^*$. Then, for $\gamma \in B^*$ still, we have:

$$\mathfrak{A}^1_{s,\gamma} = \mathfrak{A}^1_{p_\gamma} , \ \pi_\gamma(B \cap \gamma) = B \quad \text{(and hence } \gamma \in B\text{)},$$

$$\tilde{\pi}_\gamma(b_\gamma) = b , \ \tilde{\pi}_\gamma \upharpoonright b_\gamma = \pi_\gamma . \ \text{But then } \tilde{\pi}_\gamma(e_{p_\gamma}) = e_s .$$

Hence $e^*_{p_\gamma} = \tilde{\pi}_\gamma^{-1}(e^*_s) = \gamma \cap e^*_s$. $B^* \cap \gamma \in \mathfrak{A}_{p_\gamma}$, since $B^* \cap \gamma$ is definable from $\mathfrak{A}^1_{p_\gamma}$, $p \upharpoonright \gamma$, b_γ as B was constructed from $\mathfrak{A}^1_s$, p , b . For $\gamma \in B$ set $\gamma^* = \sup \gamma \cap B^*$. Define $r_\gamma \in \mathbf{S}^+_\gamma$ for $\gamma \in B$ by:

(1) $|r_\gamma| = \text{lub}\ ((|p_{\gamma^+}| + \gamma^+) \cup \{\rho^b_\gamma(\nu) \mid \nu < \gamma^*\})$.
This is just to give ourselve enough room.

(2) $r_\gamma \upharpoonright |p_{\gamma^+}| = p_{\gamma^+}$.

(3) $r_\gamma \upharpoonright ([\,|p_{\gamma^+}|, |p_{\gamma^+}| + \gamma^+) \cap \bigcup_{i>5} Z_i)$ canonically codes a well-ordering of type $|r_\gamma|$. This ensures that $r_\gamma \in \mathbf{S}_{\gamma^+}$.

(4) $r_\gamma(\rho^b_\gamma(\nu)) = \chi_{e^*_s}(\nu)$ for $\nu < \gamma^*$.
This clause will lead to the coding of e_{p_γ} (in certain cases) and eventually to e_s .

(5) $r_\gamma(\xi) = 0$ otherwise.

Clearly, $\langle \dot{p}_{\gamma^+}, r_\gamma \rangle \le p(\gamma^+)$ in $\mathbf{R}^{p_{\gamma^{++}}}$. Define

q: card $\cap\, \alpha \longrightarrow V^2$ by:

$$q(\gamma^+) = \langle \dot{p}_{\gamma^+}, r_\gamma \rangle \quad \text{if } \gamma \in B$$
$$q(\beta) = p(\beta) \quad \text{otherwise.}$$

We shall show that q has the desired properties. As usual we must first show that q is a condition, i.e. that $q \upharpoonright \gamma \in \mathbf{P}^{q_\gamma}$, by induction on γ . For $\gamma \notin B$, or γ not a limit point of B this is easily observed. So suppose γ

is a limit point of B, and further that $\gamma^* < \gamma$. Then $q \upharpoonright \gamma \in L[\hat{p}_\gamma]$, since $q \upharpoonright \gamma$ may be defined (using the fact that $\rho_\delta^{b_\gamma}(\nu) = \rho_\delta^{b}(\nu)$ for $\nu < \delta < \gamma$) from b_γ, $B \cap \gamma$, $p \upharpoonright \gamma$, $q \upharpoonright \gamma^*$, $e_s^* \cap \gamma^*$, and $q \upharpoonright \gamma^*, e_s^* \cap \gamma^* \in H_\gamma \subset L_\gamma[A] \cap L[\hat{p}_\gamma]$. Note that as $p \upharpoonright \gamma$ codes p_γ so does $q \upharpoonright \gamma$. If γ is regular in $L[\hat{p}_\gamma]$ we must verify (C) IV in the definition of $\mathbf{P}^s$. But for this just take $C = B \cap \gamma$. If γ is singular we must verify (C) VI. But this holds for $q \upharpoonright \gamma$ since it does for $p \gamma$. Note that (C) VII does not apply here since $q \upharpoonright \gamma \in L[\hat{p}_\gamma]$. Thus $q \upharpoonright \gamma \in \mathbf{P}^{p_\gamma}$, where $p_\gamma = q_\gamma$.

Now suppose $\gamma = \gamma^*$. Then γ is a limit point of B^*. $q \upharpoonright \gamma \in \mathfrak{A}_{p_\gamma}$, since $q \upharpoonright \gamma$ is definable from $B \cap \gamma$, $B^* \cap \gamma$, b_γ, $e^*_{p_\gamma} = \gamma \cap e^*_s$ as q was defined from B, B^*, p, b, e^*_s. $q \upharpoonright \gamma$ codes $p_\gamma = q_\gamma$, since $p \upharpoonright \gamma$ does. But $q \upharpoonright \gamma$ p_γ-codes e_{p_γ} at b_γ by virtue of clause (4). If γ is regular in $\mathfrak{A}_{p_\gamma}$, then $q \upharpoonright \gamma$ satisfies (C) V, taking $C = B \cap \gamma$. Now if γ is singular in $\mathfrak{A}_{p_\gamma}$ we must verify (C) VII. [(C) VI is inapplicable since $q \upharpoonright \gamma$ p_γ-coding $e_{p_\gamma} \longrightarrow q \upharpoonright \gamma \notin L[p_\gamma]$.] Assume $\mathrm{cf}(\gamma) > \omega$ since otherwise there is nothing to prove. By Lemma 8.3 $p \upharpoonright \gamma$ supports p_γ. Set $D = D^{p \upharpoonright \gamma}_{p_\gamma}$. Then by definition of D we may define D' as

$$\delta \in D' \longleftrightarrow q_\delta = p_\delta \supset (p_\gamma)_\delta \wedge \delta \in D .$$

If D^* is the intersection of D with the limit points of B^*, then D^* is cub γ. If $\delta \in D^*$ then $q \upharpoonright \delta$ p_δ-codes e_{p_δ}. Hence $\delta \in D^* \longrightarrow q \upharpoonright \delta \notin L[\hat{p}_\delta] \supset L[\widehat{(p_\gamma)}_\delta]$ which proves (C) VII, and completes the verification of $q \upharpoonright \gamma \in \mathbf{P}^{q_\gamma}$ for $\gamma < \alpha$. But a repetition of the same arguments give that $q \in \mathfrak{A}_s$, q s-codes e_s, and q satisfies (C) V with $C = B$. Hence $q \leq p$ in $\mathbf{P}^s$, $q \notin L[\hat{s}]$.

Q.E.D. (Case 1)

Case 2 α is singular in $\mathfrak{A}_s$ and $\mathrm{cf}\,\alpha > \omega$.

Let $D_s, \langle s_\gamma \rangle, \langle \pi^s_\gamma \rangle$ be as in Lemma 2.4'. π^s_γ, as remarked earlier, extends uniquely to a $\tilde{\pi}_\gamma : \tilde{\mathfrak{A}}_{s_\gamma} \xrightarrow[\Sigma_1]{} \tilde{\mathfrak{A}}_s$ with $\tilde{\mathfrak{A}}_s = \bigcup_{\gamma \in D_s} \mathrm{ran}(\tilde{\pi}_\gamma)$.

Let $B^* = \{\gamma \in D_s \mid p, b \in \mathrm{ran}(\tilde{\pi}_\gamma)\}$. Then B^* is cub in α. For $\gamma \in B^*$ we have: $\tilde{\pi}_\gamma(\langle p_\gamma, b_\gamma, B \cap \gamma, p/\gamma \rangle) = \langle p, b, B, p \rangle$ and $\tilde{\pi}_\gamma / b_\gamma = \pi_\gamma$. Then $B^* \cap \gamma \in \mathfrak{A}_{p_\gamma}$ and $e^*_\gamma = e^*_s \cap \gamma$. The verification of $q/\gamma \in \mathbb{P}^{q_\gamma}$ is as before, except that for $\gamma = \gamma^*$, the verification of (C) VII is immediate, taking D = limit points of $B^* \cap \gamma$. Again we just have (C) VII to consider. As before $q \in \mathfrak{A}_s$, q codes s, q s-codes e_s and q satisfies (C) VII, taking D = limit points of B^*.

Q.E.D. (Case 2)

Case 3 α is singular in $\mathfrak{A}_s$ and $\mathrm{cf}(\alpha) = \omega$.

Choose $B^* \in \mathfrak{A}_s$ s.t. $B^* \subset B$, $|B^*| = \omega$, $\sup B^* = \alpha$. Repeat the construction with this B^*; the verifications are trivial.

Q.E.D.

Definition C is good for p if

a) $C \in \mathfrak{A}_{|p|}$.

b) $C \subseteq \mathrm{card} \cap \alpha$ is cub in α.

c) $\gamma \in C \longrightarrow \dot{p}_\gamma = \emptyset \wedge |p/\gamma| = |p_\gamma|$.

Definition Let $C \in \mathfrak{A}_{|p|}$ be good for p.

$q \leq_C p$ iff a) $q \leq p$ in $\mathbb{P}^s$.

b) $\dot{q}_\gamma = \dot{p}_\gamma \ \forall \gamma$.

c) q differs from p only on $C \cup \{\gamma^+ \mid \gamma \in C\}$.

Note: $q' \leq_D q \leq_C p \wedge D \subseteq C \longrightarrow q' \leq_C p$.

The next lemma is really Lemma 2.14.5 modified.

Lemma 8.5 *Let* $|s| \nmid \Omega$. *There is a* $q \leq p$ *s.t.* $|q| = |s|$.

Proof We assume that $|p| < |s|$ otherwise there is nothing to prove. In fact we prove the following by induction on $|s|$:

Claim Let $|s| \nmid \Omega$, $|p| < |s|$ and $C \in \mathfrak{A}_{|p|}$ be good for p .

There is a $q \leq_C p$ s.t. $|q| = |s|$ and C is good for q .

Case 1 $|s| = \alpha$. There is nothing to prove!

Case 2 $|s| = \xi + 1$.

By the induction hypothesis we may assume that we have extended p so that $|p| = \xi$. By Lemma 8.4 we may also assume that $p \notin L[\hat{\xi}]$. We may consider all of the following to be performed within $L[\hat{s}]$.

Let m = the least $n \geq 1$ s.t. $p, C \in \mathfrak{A}^m_\xi$. Recall the definition $B^n = B^n_\xi = \{\gamma < \alpha \mid \gamma = \alpha \cap X^{\mathfrak{A}^n_\xi}_\gamma\}$.
Set η = the least $\eta \in B^m$ s.t. $p, C \in X^{\mathfrak{A}^m_\xi}_\eta$. Clearly if $m \geq n$, $\gamma \in B^m$, then $\gamma \in B^n$ and $X^{\mathfrak{A}^n_\xi}_\gamma \subset X^{\mathfrak{A}^m_\xi}_\gamma$. Hence $p, C \in X^{\mathfrak{A}^m_\xi}_\gamma$ for $\eta \leq \gamma \in B^m$, $m \geq n$.

We define a sequence of conditions q_i $(m-1 \leq i < \omega)$ s.t. $q_i \in \mathfrak{A}^{i+1}_\xi$, $q_{m-1} = p$, and $q_{i+1} \leq_C q_i$, $|q_i| = \xi$, as follows by induction on i .

q_{m-1} is defined as p . Now let $i = j+1 \geq m$. For each $\gamma \in B^i \setminus \eta$ choose canonically an $r_\gamma \in \mathbf{S}_{\gamma^+}$ as the $L_{\gamma^{++}}[A]$-least s.t.

a) $\langle \dot{p}_{\gamma^+}, r_\gamma \rangle \leq q_j(\gamma^+)$ in $\mathbb{R}^{p_{\gamma^{++}}}$.

b) $r_\gamma(\rho^i_{\xi\gamma}) = s(\xi)$.

Such an r_γ certainly exists (cf. the corresponding statement of 2.14.5).

c) $Z_6 \cap (\tilde{r}_\gamma \setminus \tilde{q}_{j\gamma^+})$ has exactly one element.

To ensure this last condition, suppose r' satisfies a) and b). Replace r' by r_γ by defining

$|r_\gamma|$ = the least p.r. closed $\rho > |r'|$.

$r_\gamma \restriction |q_{j\gamma^+}| = q_{j\gamma^+}$.

$r_\gamma(\nu) = r'(\nu)$ for $\nu \in |r'| \cap \bigcup_{i \geq 5} Z_5$.

$r_\gamma(\{\eta, 7+i\}) = r'(\{\eta, 6+i\})$ $(i \geq 0, \{\eta, 6+i\} < |r'|)$.

$r_\gamma(\{|r'|, 6\}) = 1$.

$r_\gamma(\nu) = 0$ in all other cases.

It is evident that this r_γ has all the required properties.

Define $q\colon \mathrm{card} \cap \alpha \longrightarrow V^2$ by

$q(\gamma^+) = \langle \dot{p}_{\gamma^+}, r_\gamma \rangle$ if $\gamma \in B^i$;

$q(\gamma) = q_j(\gamma)$ if not.

Set

$$q_i = \begin{cases} q & \text{if } q \in \mathbf{P}^s ; \\ \text{undefined} & \text{if not.} \end{cases}$$

<u>Claim 1</u> a) q_i is defined.

b) $q_i \leq_C q_{i-1}$ in $\mathbf{P}^{s \restriction \xi}$.

c) C is good for q_i .

We prove firstly by induction on $\gamma \in \mathrm{card} \cap \alpha$:

<u>Claim 1.1</u> a) $q_\gamma \in \mathbf{S}_\gamma$.

b) $q/\gamma \in \mathbf{P}^{q_\gamma}$ if $\gamma > 0$.

c) $|q_\gamma| = |q/\gamma|$ if $\gamma \in C$.

a) is trivial by our construction. b) and c) are simple as long as γ is not a limit point of $B^i\setminus\eta$. So suppose otherwise. Set $\pi_\gamma = \pi_\gamma^{\mathfrak{A}^i_\xi}$. Then:

$$(*) \quad \mathfrak{A}^i_{p_\gamma} = \mathfrak{A}^i_{\xi,\gamma} \; ; \; \pi_\gamma(\langle p/\gamma, p_\gamma, e_{p_\gamma}, C\cap\gamma, q_{i-1}/\gamma\rangle) =$$
$$= \langle p, s/\xi, e_\xi, C, q_{i-1}\rangle \; ; \; B^i_{p_\gamma} = \gamma \cap B^i_\xi \; ;$$
$$\rho^i_{\xi\nu} = \rho^i_{p_\gamma \nu} \quad \text{for} \quad \nu < \gamma .$$

Since p codes s/ξ and $p \notin L[\hat{\xi}]$ we obtain by (*) that p/γ codes p_γ , e_{p_γ} and hence so does q/γ , with $q/\gamma \in \mathfrak{A}_{p_\gamma}$ (in fact $\in \mathfrak{A}^{i+1}_{p_\gamma}$), since q/γ is defined from $\mathfrak{A}^i_{p_\gamma}, p/\gamma, q_{i-1}/\gamma$ as q was defined from $\mathfrak{A}^i_\xi, p, q_{i-1}$. Hence $|q/\gamma| = |p_\gamma| = |q_\gamma|$. Since $q/\gamma \notin L[\hat{p}_\gamma]$ we must show that if γ is regular in $\mathfrak{A}_{p_\gamma}$, q/γ satisfies (C) V. But this it does easily: $C \cap \gamma$ is good for q/γ . If γ is singular in $\mathfrak{A}_{p_\gamma}$, q/γ satisfies (C) VII since p/γ does. Thus $q/\gamma \in \mathbb{P}^{p_\gamma}$ where $p_\gamma = q_\gamma$.

Q.E.D. (Claim 1.1)

Observe that $q \in \mathfrak{A}_\xi$ and q codes s/ξ , e_ξ since p does and that q satisfies (C) V, noting that $p \notin L[\hat{\xi}]$, since C is good for q by Claim 1.1. Hence $q \in \mathbf{P}^{s/\xi}$, but then $q_i = q$ is defined. Trivially $q_i \leq_C q_{i-1}$.

Q.E.D. (Claim 1)

Set $B \cdot = \bigcap_{m \le i < \omega} B^i_\xi \setminus \eta$. Define $q\colon \mathrm{card} \cap \alpha \longrightarrow V^2$ as follows:

For $\gamma \in B$ set $q(\gamma) = \langle \dot{p}_\gamma, q_\gamma \rangle$, where $|q_\gamma| = |p_\gamma| + 1$, $q_\gamma(|p_\gamma|) = s(\xi)$, $q_\gamma \restriction |p_\gamma| = p_\gamma$. Since $\gamma \in \bigcap_i B^{i\gamma}$, b) of the definition of q_i will ensure that $q \restriction \gamma$ codes q_γ and that q codes s .

Otherwise set $q(\gamma) = \langle \dot{p}_\gamma, \bigcup_{m \le i < \omega} q_{i\gamma} \rangle$.

Claim 2 a) $q \in \mathbb{P}^s$ and $|q| = |s|$.

b) $q \le_C p$.

c) C is good for q .

We first prove:

Claim 2.1 Let $\gamma \in \mathrm{card} \cap \alpha$:

a) $q_\gamma \in \mathbf{S}_\gamma$.

b) $q \restriction \gamma \in \mathbb{P}^{q_\gamma}$ if $\gamma > 0$.

c) $|q_\gamma| = |q \restriction \gamma|$ if $\gamma \in C$.

Firstly, a): for $\gamma \notin B \cup \{\tau^+ \mid \tau \in B\}$ this follows easily from 1.1; for $\gamma \in B$ it is immediate from: $p_\gamma \in \mathbf{S}_\gamma$, $q_\gamma \supset p_\gamma$ and $|q_\gamma| = |p_\gamma| + 1$. Now let $\gamma \in \beta^+$, $\beta \in B$. We are required to show that $\bar{\bar{\xi}}^{L[q_\gamma \restriction \xi]} = \alpha$ for $\xi \in [\alpha, |q_\gamma|]$. If $\xi < |q_\gamma|$ then $\xi < |q_{i\gamma}|$ for some i , with $q_{i\gamma} \in \mathbf{S}_\gamma$ and $q_\gamma = \bigcup_i q_{i\gamma}$. If ξ equals $|q_\gamma|$ it follows from the easily observed fact that $Z_6 \cap (\tilde{q}_\gamma \setminus \tilde{p}_\gamma)$ is cofinal in $|q_\gamma|$ and has order type ω . Now for b) and c) proven by induction on γ . The case $\gamma = 0$ is trivial. If $\gamma = \beta^+$ we must show that $q(\beta) \in \mathbf{R}^{q_\gamma}$. If $\beta \notin B$ this follows by: $\dot{q}_\beta = \dot{p}_\beta$, $q_\beta = \bigcup_i q_{i\beta}$, with $\langle \dot{p}_\beta, q_{i\beta} \rangle \in \mathbf{R}^{p_\gamma}$ for $m \le i < k$ for some $k < \omega$, and $q_\gamma \supset p_\gamma$. If $\beta \in B$ it is trivial by the fact that $\dot{q}_\beta = \emptyset$. Now let γ be a limit cardinal. If $\gamma \notin B$, then there is an $i < \omega$ s.t. $q_\gamma = q_{i\gamma}$ and q/γ differs from q_i/γ only on an initial segment. The conclusion is immediate. Now suppose $\gamma \in B$.

Note that $\pi_\gamma^{\mathfrak{A}_\xi^i} = \pi_\gamma^{\mathfrak{A}_\xi^i} / \mathfrak{A}_{\xi\gamma}^i$ for $\gamma \in B$, $m \leq i \leq j$.

set: $\pi = \pi_\gamma = \bigcup_{m \leq i < \omega} \pi_\gamma^{\mathfrak{A}_\xi^i}$.

As before this is well defined (see p.60). Since $\mathfrak{A}_{\xi\gamma}^i = \mathfrak{A}_{p_\gamma}^i$, we obtain:

(*) $\pi: \mathfrak{A}_{p_\gamma} \xrightarrow[\Sigma_1]{} \mathfrak{A}_\xi$ cofinally;

$\pi(\langle p/\gamma, p_\gamma, e_{p_\gamma}, \mathfrak{A}_{p_\gamma}^i, C\cap\gamma, q_i/\gamma\rangle) =$

$= \langle p, s/\xi, e_\xi, \mathfrak{A}_\xi^i, C, q_i\rangle$ $(m \leq i < \omega)$;

hence $B_{p_\gamma}^i = B_\xi^i \cap \gamma$; $\rho_{\xi\nu}^i = \rho_{p_\gamma\nu}^i$ for $\nu < \gamma$.

p/γ codes p_γ, e_{p_γ} by (*); hence so does q/γ . But q/γ q_γ-codes $|p_\gamma|$ by the construction and by (*) since

$\gamma \in B \longrightarrow \forall\tau \in B^i \setminus \eta \cap \gamma\ (\widehat{q/\gamma}(\rho_{\xi\tau}^i) = s(\xi))$.

Hence q/γ codes q_γ .

$q/\gamma \in L[\hat{q}_\gamma] = L[\hat{e}_{p_\gamma}]$, since q/γ is definable from $p/\gamma, C\cap\gamma, \mathfrak{A}_{p_\gamma}$ in $L[\hat{q}_\gamma]$ as q was defined from $p, C, \mathfrak{A}_\xi$ in $L[\hat{s}]$ by (*).

Thus $|q/\gamma| = |q_\gamma|$, for suppose otherwise: if $|q/\gamma| < |q_\gamma|$ then $q/\gamma \in \mathfrak{A}_{p_\gamma}^n$ for some $n > m$, but q/γ extends q^n/γ and so $\rho_{p_\gamma,\tau}^m \in \mathrm{dom}(q/\gamma)$ – a contradiction (cf. Claim 3d) of 2.14.5). If γ is regular in $L[\hat{q}_\gamma]$, (C) IV is satisfied, since $C \cap \gamma$ is good for q/γ . Now let γ be singular in $L[\hat{q}_\gamma]$. We must verify (C) VI . Assume $\mathrm{cf}(\gamma) > \omega$ otherwise there is nothing to do. Set $r = q_\gamma$ and let D_r, $\langle \pi_\delta^r \mid \delta \in D_r\rangle$, $\langle r_\delta \mid \delta \in D_r\rangle$ be as in

Lemma 2.4'. D_r is unbounded in γ and so $\mathfrak{A}^o_r = \bigcup_{\delta \in D_r} \pi^r_\delta{}''\mathfrak{A}^o_{r_\delta}$, so now set

$$D = \{\delta \in B \cap D_r \mid \mathfrak{A}^1_{p_\gamma} \in \mathrm{ran}(\pi^r_\delta)\} . \quad \text{Let } \bar{\gamma} \in D .$$

Set $\bar{\pi} = \pi^r_{\bar{\gamma}}$, $\bar{\mathfrak{A}} = \bar{\pi}^{-1}(\mathfrak{A}^1_{p_\gamma})$. Then

$$\bar{\pi}\restriction\bar{\mathfrak{A}}: \bar{\mathfrak{A}} \xrightarrow[\Sigma_\omega]{} \mathfrak{A}^1_{p_\gamma} , \quad \bar{\pi}\restriction\bar{\gamma} = \mathrm{id}\restriction\bar{\gamma} , \ \pi(\bar{\gamma}) = \gamma .$$

Hence $\bar{\pi}\restriction\bar{\mathfrak{A}} = \pi^{\mathfrak{A}^1_{p_\gamma}}_{\gamma}$ and $\bar{\mathfrak{A}} = \mathfrak{A}^1_{p_\gamma,\bar{\gamma}} = \mathfrak{A}^1_{p_{\bar{\gamma}}}$, since $\delta \in B \subseteq C$ [by the Collapsing Lemma] and $\mathfrak{A}^1_{p_\gamma,\bar{\gamma}} = \mathfrak{A}^1_{\xi,\bar{\gamma}}$. But then $\bar{\pi}(q_{\bar{\gamma}}) = q_\gamma = r$. Hence $q_{\bar{\gamma}} = r_{\bar{\gamma}}$ and $|q\restriction\gamma| = |r|$.

Q.E.D. (Claim 2.1)

Now to finish off Claim 2: q codes s by the construction and remembering our remark at the outset we have that $q \in L[\hat{s}]$, hence $|q| = |s|$. C is good for q by Claim 2.1. (C) IV & VI may be verified just as above.

Q.E.D. (Claim 2 & Case 2)

<u>Case 3</u> $\mathrm{Lim}(|s|)$.

Let C^* = limit points of C_s . Suppose that C^* is cofinal in C_s . Let $\langle \xi_i \mid i \le \theta \rangle$ be the monotone enumeration of

$$\{\xi \mid \xi = |p| \vee (\xi > |p| \wedge \mu^o_\xi \in C^*)\}$$

for some θ . [Recall that by 2.3(iii) every element of C^* is of the form μ^o_ξ .]

If C^* is not cofinal in C_s , then $\mathrm{cf}(\alpha) = \omega$, so take $\theta = \omega$ and define $\langle \xi_i \mid i \le \omega \rangle$ by:

$\xi_o = |p|$.

ξ_{i+1} = the least $\xi > \xi_i$ s.t. $C^* \setminus \mu^o_\xi = \emptyset \wedge C_s \cap (\mu^o_{\xi_i}, \mu^o_\xi) \neq \emptyset$.

$$\xi_\omega = \sup \xi_i = |s| .$$

So in both cases $\xi_\theta = |s|$ and the sequence is normal.

We now use this sequence to define $q_i \in \mathbb{P}^s$ with $|q_i| = \xi_i$. The union of this sequence of q_i's will be our desired q. Simultaneously we define cub sets $B_i \in \mathfrak{A}_{\xi_i}$ for $i < \theta$.

First set:

$m = m_i =$ the least $m \geq 1$ s.t.

$$\langle C_s \cap \mu^o_{\xi_i}, C, p, \langle q_h \mid h < i \rangle, \langle B_h \mid h < i \rangle\rangle \in \mathfrak{A}^m_{\xi_i} .$$

Since both i and $|C_s \cap \mu^o_{\xi_i}|$ are less than α, this m_i exists

$$B_i = \{\gamma < \alpha \mid \gamma = \alpha \cap X_\gamma^{\mathfrak{A}^{m_i}_{\xi_i}} \wedge$$

$$\langle C_s \cap \mu^o_{\xi_i}, C, p, \langle q_h \mid h < i \rangle, \langle B_h \mid h < i \rangle\rangle \in X_\gamma^{\mathfrak{A}^{m_i}_{\xi_i}}\} .$$

Define q_i $(i \leq \theta)$:

$q_0 = p$

$q_{i+1} = \mathfrak{A}_{\xi_{i+1}}$-least q s.t.

a) $q \leq_{B_i} q_i$.

b) $|q| = \xi_{i+1}$.

c) C is good for q .

d) $\beta \in B_i \longrightarrow Z_6 \cap (\tilde{q}_{\beta^+} \setminus \tilde{q}_{i\beta^+})$ contains exactly one element.

q_{i+1} is undefined if no such q exists.

For $\lambda \leq \theta$ $\lim(\lambda)$ set:

$$q = \bigcup_{i<\lambda} q_i \quad (\text{i.e. } q(\gamma) = \langle \dot{p}_\gamma, \bigcup_{i<\lambda} q_{i\gamma} \rangle).$$

Set $q_\lambda = q$ if $q \in \mathbb{P}^{\xi_\lambda}$. Otherwise not defined.

Claim 1 Let $i \le \theta$.

a) q_i is defined and $q_i \le_C p$.

b) $|q_i| = \xi_i$.

c) C is good for q_i.

Proof By induction on i. $i = 0$ is trivial, so suppose $i = j+1$. In this case it will be enough to show the existence of a q_i, i.e. one satisfying a) - d) in the definition above. By the Induction Hypothesis we can easily find a q' satisfying a) - c) there. But we may replace each $q'_{\beta+}$ for $\beta \in B_i$ by a $q_{\beta+}$ using the technique of p.278, to obtain d).

Now consider $i = \lambda < \theta$, $\lim(\lambda)$. Clearly $B_i \supset B_j$ for $i \le j < \lambda$. Set $B = B^\lambda = \bigcap_{i<\lambda} B_i$. Let $\gamma \in B$, evidently for $i \le j < \lambda$

$$\pi_\gamma^{\mathfrak{A}_{\xi_j}^{m_j} \restriction \mathfrak{A}_{\xi_i,\gamma}^{m_i}} = \pi_\gamma^{\mathfrak{A}_{\xi_i}^{m_i}} .$$

Set $\pi_\gamma = \pi_\gamma^\lambda = \bigcup_{i<\lambda} \mathfrak{A}_{\xi_i,\gamma}^{m_i}$

$$\bar{\mathfrak{A}} = \bigcup_{i<\lambda} \mathfrak{A}_{\xi_i,\gamma}^{m_i} .$$

Then $\pi_\gamma : \bar{\mathfrak{A}} \xrightarrow[\Sigma_1]{} \mathfrak{A}_{\xi_\lambda}^{o}$ cofinally, since $\mu_{\xi_i}^{m_i} < \mu_{\xi_i+1}^{\omega} < \mu_{\xi_{i+1}}^{m_{i+1}}$.

We then have:

$$\pi_\gamma(\langle q_i \restriction \gamma, q_{i\gamma}, \mathfrak{A}_{q_{i\gamma}}, C \cap \gamma, B_i \cap \gamma \rangle) = \langle q_i, s \restriction \xi_i, \mathfrak{A}_{\xi_i}, C, B_i \rangle \quad (i < \lambda) .$$

Hence $q_\gamma = \pi_\gamma^{-1} {}'' s \restriction \xi_\lambda$ since $q_\gamma = \bigcup_{i<\lambda} q_{i\gamma}$.

Our definitions of m_i, B_i ensure that $C_{\xi_\lambda} \cap \mu_{\xi_i}^{o} \in \operatorname{ran}(\pi_\gamma)$, for $i < \lambda$. Hence we may set $\bar{C} = \pi_\gamma^{-1} {}'' C_{\xi_\lambda}$, and obtain

$$\pi_\gamma : \langle \bar{\mathfrak{A}}, \bar{C} \rangle \xrightarrow[\Sigma_1]{} \langle \mathfrak{A}^o_{\xi_\lambda}, C_{\xi_\lambda} \rangle \quad \text{cofinally.}$$

By 2.3 we get:

$$q_\gamma \in \mathbf{S}_\gamma \ , \quad \bar{\mathfrak{A}} = \mathfrak{A}^o_{q_\gamma} \ , \quad \bar{C} = C_{q_\gamma} \ .$$

Note also that $q/\gamma = \bigcup_{i<\lambda} \widehat{q}_i/\gamma$ codes q_γ since $\widehat{q}_i/\gamma$ codes $q_{i\gamma}$.

We are now in a position to prove the analogous statement to Claim 1.1 of Case 2, i.e.

Claim 1.1 Let $\gamma \in \text{card} \cap \alpha$:

a) $q_\gamma \in \mathbf{S}_\gamma$.

b) $q/\gamma = \mathbb{P}^{q_\gamma}$ if $\gamma > 0$.

c) $|q_\gamma| = |q/\gamma|$ if $\gamma \in C$.

Proof First a); for $\gamma \in B$ it is proven already by the above remarks. Suppose $\gamma = \beta^+$ where $\beta \in B$. Then we may observe that $q_\gamma/\xi \in \mathbf{S}_\gamma$ for $\gamma \le \xi < |q_\gamma|$ and that $Z_6 \cap (\tilde{q}_\gamma \setminus \tilde{p}_\gamma)$ is cofinal in $|q_\gamma|$ and has order type λ where $\lambda \le |C_{q_\beta}| \le \beta < \gamma$. In all other cases we have that $q_\gamma = q_{i\gamma}$ for an $i < \lambda$.

By induction on γ we prove b) and c). $\gamma = 0$ and γ a successor are trivial as usual. So suppose γ is a limit cardinal. If $\gamma \notin B$, then for some $i < \lambda$ we have: $q_\gamma = q_{i\gamma}$ and q_γ differs from $q_{i\gamma}$ only by an initial segment. Hence b) and c) hold since they hold at i . So suppose $\gamma \in B$. Again by the above remarks q/γ codes q_γ . $\langle q_i/\gamma \mid i < \lambda \rangle$ is definable from $p/\gamma, C\cap\gamma, C_{q_\gamma}, q_\gamma$ by the same inductive definition that defines $\langle q_i \mid i < \lambda \rangle$ from $p, C, C_{\xi_\lambda}, s/\xi_\lambda$. Hence $\langle q_i/\gamma \mid i < \lambda \rangle \in L[\hat{q}_\gamma]$. Hence $q/\gamma = \bigcup_{i<\lambda} q_i/\gamma \in L[\hat{q}_\gamma]$. But then $|q/\gamma| \ge \sup |q_i/\gamma| = \sup |q_{i\gamma}| = |q_\gamma|$. If γ is regular in $L[\hat{q}_\gamma]$, (C) IV holds since $C \cap \gamma$ is good for q/γ. Now let γ be singular in $L[\hat{q}_\gamma]$. Assume $cf(\gamma) > \omega$, since otherwise there is nothing to prove. Set $r = q_\gamma$ and let $D_r, \langle r_\gamma \rangle, \langle \pi^r_\delta \rangle$ be as given by Lemma 2.4'. Set $r^i = q_{i\gamma}$ $(i \leq \lambda)$. Let D = the set of $\delta \in D_r$ s.t. for some limit $\rho \leq \lambda$, $\sup ran(\pi^r_\delta) = \mu^o_{r\rho}$ and $C \cap \gamma, p/\gamma, r^i \in ran(\pi^r_\delta)$ for $i < \rho$.

<u>Claim</u> D is cub in γ (and in fact is a final segment).

Proof Obviously D is closed in γ. Let C^*, as before, be the set of limit points of C_s. If C^* is not cofinal in μ^o_s, then $\theta = \lambda = \omega$, and the claim follows by $cf(\gamma) > \omega$. Otherwise $\langle \mu^o_{\xi_i} \mid i \leq \theta \rangle$ enumerates $\{\mu^o_{|p|}\} \cup (C^* \setminus \mu^o_{|p|})$. Hence $\pi^{-1}_\gamma{}'' \langle \mu^o_{\xi_i} \mid i \leq \theta \rangle$ enumerates $\{\mu^o_{|p/\gamma|}\} \cup (C^*_r \setminus \mu^o_{|p/\gamma|})$, where C^*_r = set of limit points of C_r. As $cf(\gamma) > \omega$, let $\delta \in D_r$ s.t. $C \cap \gamma, p/\gamma \in ran(\pi^r_\delta)$. Then $\mu^o_{ro} = \mu^o_{|p/\gamma|} \in ran(\pi^r_\delta)$ and hence $\sup ran(\pi^r_\delta) = \mu^o_{r\rho}$ for a limit $\rho \leq \lambda$. But then $\mu^o_{ri} \in C_r \subset ran(\pi^r_\delta)$ for a $0 < i < \rho$, since $\pi^r_\delta{}'' C_{r_\delta}$ is an initial segment of C_r (cf. Remark 3 of 2.4). Hence $\hat{r}^i = \hat{r} \cap \mu^o_{ri} \in ran(\pi^r_\delta)$ and so $r^i \in ran(\pi^r_\delta)$ for $i < \rho$. Hence $\delta \in D$.

Q.E.D. (Claim)

Now let $\bar{\gamma} \in D$, set $\bar{r} = r_{\bar{\gamma}}$, $\bar{\pi} = \pi^r_{\bar{\gamma}}$, $\bar{r}^i = \bar{\pi}^{-1}(r^i)$. Then $\bar{r} = \bigcup_{i<\rho} \bar{r}^i$. But

$$\bar{\pi}/\mathfrak{A}_{\bar{r}^i}: \mathfrak{A}_{\bar{r}^i} \xrightarrow[\Sigma_\omega]{} \mathfrak{A}_{r^i},$$

$\bar{\pi}/\bar{\gamma} = id/\bar{\gamma}$ and $\bar{\pi}(\bar{\gamma}) = \gamma$. Hence $\bar{\pi}/\mathfrak{A}_{\bar{r}^i} = \pi^{\mathfrak{A}_{r^i}}_{\bar{\gamma}}$ for $i < \rho$.

But $q_i/\gamma \in ran(\pi^{\mathfrak{A}_{r^i}}_{\bar{\gamma}})$ since q_i/γ is $\mathfrak{A}_{r^i}$-definable from

$C \cap \gamma, p \wedge \gamma$. Hence, by the collapsing lemma, $\bar{r}^i = q_{i\bar{\gamma}}$ and $|q_i \wedge \bar{\gamma}| = |\bar{r}^i|$. Hence $\bar{r} = \bigcup_{i<\rho} q_{i\bar{\gamma}} = q_{\rho\bar{\gamma}}$ and $|q_\rho \wedge \bar{\gamma}| = |\bar{r}|$, since $q_\rho \wedge \bar{\gamma} = \bigcup_{i<\rho} q_i \wedge \bar{\gamma}$. Since $\lambda \geq \rho$ we may conclude that $q_{\lambda\bar{\gamma}} \supset \bar{r}$ and $|q_\lambda \wedge \bar{\gamma}| \geq \bar{r}$.

Hence (C) VI is satisfied.

Q.E.D. (Claim 1.1)

We can now complete the proof of Claim 1 for $i = \lambda$ and thus complete Case 3. $\overset{\frown}{q}$ codes $s \wedge \xi_\lambda$, since $\overset{\frown}{q_i}$ codes $s \wedge \xi_i$ and $q = \bigcup_{i<\lambda} q_i$. $q \in L[\hat{\xi}_\lambda]$, since the inductive definition of q from p, C, C_{ξ_λ} can be carried out in $L[\hat{\xi}_\lambda]$. Hence $|q| = \xi_\lambda$. Again we need only verify (C) IV or (C) VI (as $q \in L[\hat{\xi}_\lambda]$) and thus is done just as above. Hence $q = q_\lambda \in \mathbb{P}^{\xi_\lambda}$. But then $q \leq_C p$, since $q_i \leq_C p$ for $i < \lambda$ and $q = \bigcup_{i<\lambda} q_i$. C is good for q by Claim 1.1. In case $\lambda = \theta = \alpha$ we finish off as in 2.14.5 Case 2.2.

Q.E.D. (Lemma 8.5)

Having concluded Lemma 8.5 we may now polish off the case of α being regular in $\mathfrak{A}_s$ - to wit:

Lemma 8.6 *Let α be regular in $\mathfrak{A}_s$. Then the conclusion of the Extension Lemma holds.*

Proof By the last lemma we may safely assume that $|p| = |s|$. But in addition Lemma 8.4 gives us that we may also take $p \notin L[\hat{s}]$.

So now suppose $X \subseteq \mathrm{card} \cap \alpha$ and X is thin in $\mathfrak{A}_s$ and further that $\langle \xi_\nu \mid \nu \in X \rangle \in \mathfrak{A}_s$ with $\nu \leq \xi_\nu < \nu^+$ for $\nu \in X$.

But now we may follow through the proof of 2.14.1 verbatim. When we wish to verify that $q \in \mathfrak{A}_s$, the only non-trivial case of proving that $q \wedge \gamma \in \mathbb{P}^{q_\gamma}$ for $\gamma \in \mathrm{card} \cap \alpha$ is when

γ is a limit point of D. Now we have that q/γ codes $p_\gamma = q_\gamma$ and e_{p_γ} since p/γ does. $q/\gamma \in \mathfrak{A}_{p_\gamma}$ for the same reasons. If γ is regular in $\mathfrak{A}_{p_\gamma}$ then (C) V holds, using $C = D \cap \gamma$. If not then (C) VII is satisfied by q/γ because p/γ does. Hence $q/\gamma \in \mathfrak{A}_{p_\gamma}$.

Q.E.D.

We shall now prove a lemma analogous to that of 2.14.2, but first let us take stock of what we have achieved so far, summarised in the following Assumption as a starting point for further work.

Assumption 1 $\Omega \leq |s| \wedge |p| \geq \Omega \wedge$

$$(p \notin L[\hat{\Omega}] \text{ if } \alpha \text{ is regular in } L[\hat{\Omega}]).$$

The last lemma justifies the first clause, the second may be attained by a prior application of Lemma 8.5 to s/Ω, and the third by a prior application of Lemma 8.4 also to s/Ω.

Recall that for $s \geq \Omega$, we defined $D_s = D_{s/\Omega}$, $s_\gamma = (s/\Omega)_\gamma$ for $\gamma \in D_s$.

Lemma 8.7 *Let* $|p| = |s|$. *Then* $\exists q \leq p$ *s.t.* $\forall \gamma \in D_s$.

a) $|q/\gamma| \geq |s_\gamma|$. *(Thus* q *supports* s.*)*

b) If α *is regular in* $L[\hat{s}]$ *then*

$$q_\gamma \supset s_\gamma \longrightarrow q/\gamma \notin L[\hat{s}_\gamma].$$

Proof Set $X = D_s$. Since $D_s \in \mathfrak{A}_\Omega$ and D_s consists only of singular cardinals, by the definition of "thin", X is thin in $\mathfrak{A}_\Omega$ and is also thin in $L[\hat{\Omega}]$ if α is singular in $L[\hat{\Omega}]$. A fortiori X is thin in $\mathfrak{A}_s$, and in $L[\hat{s}]$ if $p \in L[\hat{s}]$ (by Assumption 1). For $\nu \in X$ set $\xi_\nu = |s_\nu| + 1$.

Then $\langle \xi_\nu \mid \nu \in X \rangle \in \mathfrak{A}_s$ (and $\in L[\hat{s}]$ if $p \in L[\hat{s}]$).

Case 1 $\quad cf(\alpha) = \omega$.

We proceed in a similar fashion to Case 1 of 2.14.2: Choose a monotone sequence $\langle \gamma_i \mid i < \omega \rangle \in \mathfrak{A}_s$ with $\sup \gamma_i = \alpha$. If $p \in L[\hat{s}]$ then $L[\hat{s}] \models$ "$cf(\alpha) = \omega$" and we may take $\langle \gamma_i \mid i < \omega \rangle \in L[\hat{s}]$. Since the Extension Theorem holds below α , define:

q_o = the $\mathfrak{A}'_{p_{\gamma_o^+}}$-least $q \leq p \restriction \gamma_o^+$ in $\mathbb{P}^{p_{\gamma_o^+}}$ s.t.

$\qquad q$ is smooth and $|q_\nu| \geq \xi_\nu$ for $\nu \in X \cap \gamma_o^+$.

q_{i+1} = the $\mathfrak{A}'_{p_{\gamma_{i+1}^+}}$-least $q \leq p \restriction [\gamma_i^+, \gamma_{i+1}^+)$ in $\mathbb{P}_{\gamma_i^+}^{p_{\gamma_{i+1}^+}}$ s.t.

$\qquad q$ is smooth and $|q_\nu| \geq \xi_\nu$ for $\nu \in X \cap [\gamma_i^+, \gamma_{i+1}^+)$.

$q = \bigcup_{i<\omega} q_i$.

It is trivial to observe that $q \restriction \gamma \in \mathbb{P}^{q_\gamma}$ for $\gamma \in \text{card} \cap \alpha$. Since p codes $\hat{s}$ (and possibly $\hat{e}_s$) then so does q ; $q \in \mathfrak{A}_s$ (and $L[\hat{s}]$ if $p \in L[\hat{s}]$) by the construction and satisfies (C) VI or VII since p does. Thus $q \in \mathbb{P}^s$ and $q \leq p$ in $\mathbb{P}^s$; it clearly satisfies a) and b) of the lemma's conclusion.

Q.E.D. (Case 1)

Case 2 $\quad cf(\alpha) > \omega$.

Let $D \subset D_s$ be cub in α s.t.

a) $p_\gamma \supset s_\gamma \wedge |p \restriction \gamma| \geq |s_\gamma|$.

b) If $s \restriction \Omega \in \mathbf{S}_\alpha^1$, then $p \restriction \gamma \notin L[\hat{s}_\gamma]$.

Since $|D| \leq |D_s| < \alpha$ and $H_\alpha \subset L[\hat{s}] \subset \mathfrak{A}_s$ we have:

c) $D \in \mathfrak{A}_s$.

d) $D \in L[\hat{s}]$ if $p \in L[\hat{s}]$ (since they are true for D_s).

Similarly putting $p'_\gamma = p_\gamma \restriction |p \restriction \gamma|$ for $\gamma \in D$:

e) $D \cap \gamma \in \mathfrak{A}_{p'_\gamma}$.

f) $D \cap \gamma \in L[\hat{p}'_\gamma]$ if $p \restriction \gamma \in L[\hat{p}'_\gamma]$.

Finally, since $X \cap \gamma = D_{s_\gamma}$ for $\gamma \in D$:

g) $X \cap \gamma$ is thin in $\mathfrak{A}_{p'_\gamma}$ and $\langle \xi_\nu \mid \nu \in X \cap \gamma \rangle \in \mathfrak{A}_{p'_\gamma}$.

h) $X \cap \gamma$ is thin in $L[\hat{p}'_\gamma]$ and $\langle \xi_\nu \mid \nu \in X \cap \gamma \rangle \in L[\hat{p}'_\gamma]$

if $p \restriction \gamma \in L[\hat{p}'_\gamma]$.

Let $\langle \gamma_i \mid i < \lambda \rangle$ be the monotone enumeration of $\{0\} \cup D$. Then set:

$$q_i = \mathfrak{A}_{p'_{\gamma_{i+1}}}\text{-least } q \le p \restriction (\gamma_i, \gamma_{i+1}) \text{ in } \mathbb{P}^{p'_{\gamma_{i+1}^+}}_{\gamma_i^+} \text{ s.t.}$$
$$q \text{ is smooth and } |q_\nu| \ge \xi_\nu \text{ for } \nu \in X \cap (\gamma_i, \gamma_{i+1}) .$$

Again since the Extension Theorem holds below α , q_i exists. Set $q = \bigcup_{i<\lambda} (\{\langle p(\gamma_i), \gamma_i \rangle\} \cup q_i)$. We shall show that q is as required. It is trivial to observe that $q \restriction \gamma \in \mathbb{P}^{q_\gamma}$ for $\gamma \in \text{card} \cap \alpha$, unless γ is a limit point of D . In that case $q \restriction \gamma \in \mathbb{P}^{p'_\gamma}$ since $q \restriction \gamma$ codes p'_γ (and $e_{p'_\gamma}$ if $p \restriction \gamma \notin L[\hat{p}'_\gamma]$) , because $p \restriction \gamma$ does, and since e) - h) imply that $q \restriction \gamma$ is constructed in $\mathfrak{A}_{p'_\gamma}$ (and in $L[p'_\gamma]$ if $p \restriction \gamma \in L[p'_\gamma]$) and that $|q \restriction \gamma| = |p \restriction \gamma| = |p'_\gamma|$; finally $q \restriction \gamma$ satisfies (C) VI or VII since $p \restriction \gamma$ does. Similarly $q \in \mathbb{P}^S$. Again a) or b) are clearly satisfied for $\gamma \in D$. But by the construction they also hold for $\gamma \in X \setminus D$ since then $|q \restriction \gamma| > |s_\gamma|$.

Q.E.D.

By applying this last lemma to $s \restriction |p|$ we can add the following to our Assumption 1:

Assumption 2 If $\gamma \in D_s$ then:

a) $|p/\gamma| \geq |s_\gamma|$.

b) If α is regular in $L[\widehat{s/\Omega}]$ then $p/\gamma \notin L[\hat{s}_\gamma]$.

But this implies that p supports s .

For the next three lemmata set:

$D = D_s^p$; D^* = limit points of D .

We are now going to show that there is a $q \leq p$ s.t. $|q| = |s|$. We shall slightly modify the proof of 2.14.3 and make some definitions which are (virtually) the same as those preceding that lemma. However, here we divide into two cases.

Lemma 8.8 *Let* D^* *be non-cofinal in* α . *Then there is a* $q \leq p$ *s.t.* $|q| = |s| \wedge q \notin L[\hat{s}]$.

Proof Let $\langle \beta_i \mid i < \omega \rangle$ be the monotone enumeration of a final segment of D . Assume (w.l.o.g.) that $\beta_0 > |D_s|$.

Definition Let U_α be the set of $\{\{\eta, |D_s|\}, i\}$ s.t. $\eta \in \bigcup_{i<\omega} [\beta_i^+, \beta_i^{++})$ and $i = 4,5$.

Definition $\tilde{\mathbb{P}} = \tilde{\mathbb{P}}^s$ = the set of partial maps $\tilde{q}: U_\alpha \longrightarrow 2$ s.t.

a) $\tilde{q} \supset \overset{\frown}{p}/U$.

b) $\tilde{q} \in \mathfrak{A}_s$.

c) [Setting $|\tilde{q}|$ = the least $\xi \leq |s|$ s.t. $\tilde{q} \in \mathfrak{A}_\xi$]
$\tilde{q}$ s-codes ξ, e_ξ for $\Omega < \xi < |\tilde{q}|$.

d) $\tilde{q}$ s-codes e_Ω if $s/\Omega \in \mathbb{S}_\alpha^o$, $\tilde{q} \notin L[\widehat{s\ \Omega}]$.

e) $\tilde{q}$ s-codes $e_{|\tilde{q}|}$ if $|\tilde{q}| > \Omega$ and $\tilde{q} \notin L[|\tilde{q}|]$.

f) $\overline{\overline{\tilde{q}/[\beta_i^+, \beta_i^{++})}} \leq \beta_i^+$ for $i < \omega$.

Just as remarked on p.54 we may use 1.5's methods to show that there exists a $\tilde{q} \leq p \upharpoonright U_\alpha$ in $\tilde{\mathbb{P}}$ s.t. $|\tilde{q}| = |s|$ and $\tilde{q} \notin L[\hat{s}]$.

Now pick $q'_i \in \mathbf{S}_{\beta_i^+}$ $(i < \omega)$ s.t.

$$\langle \dot{p}_{\beta_i^+}, q'_i \rangle \leq p(\beta_i^+) \text{ in } \mathbf{R}^{p_{\beta_i^{++}}} ;$$

$$\tilde{q} \upharpoonright [\beta_i^+, \beta_i^{++}) \subseteq q'_i .$$

Set

$$q(\beta) = \begin{cases} \langle \dot{p}_{\beta_i^+}, q'_i \rangle & \text{if } \beta \in \beta_i^+ \\ p(\beta) & \text{if not.} \end{cases}$$

Clearly $q \upharpoonright \gamma \in \mathbf{P}^{q_\gamma}$ for $\gamma \in \text{card} \cap \alpha$. But $q \in \mathfrak{A}_s$ by the construction. q codes $s \upharpoonright |p|$ and also e_p if $p \notin L[|\hat{p}|]$, since p does. But $\tilde{q} \subseteq \overset{\frown}{q}$, and so q codes s, e_s since we have c), d), e) together with $|\tilde{q}| = |s|$, $\tilde{q} \notin L[\hat{s}]$. Thus $|q| = |s|$, $q \notin L[\hat{s}]$. If (C) VII applies, it holds trivially since it does so for p.

Q.E.D.

The other case will parallel 2.14.3.

Lemma 8.9 *Let* D^* *be cofinal in* α. *Let* $\langle \beta_i \mid i < \rho \rangle$ *be the monotone enumeration of* $D^* \cap (|D_s|, \alpha)$. *Let*

$$\langle \dot{r}_i, r_i \rangle \leq p(\beta_i) \text{ in } \mathbf{R}^{p_{\beta_i^+}} \text{ s.t. } \langle \langle \dot{r}_i, r_i \rangle \mid i < \rho \rangle \in \mathfrak{A}_s .$$

There is a $q \leq p$ *s.t.*

a) $|q| = |s|$ *and* $q \notin L[\hat{s}]$.

b) $q(\beta_i) \leq \langle \dot{r}_i, r_i \rangle$ *in* $\mathbf{R}^{p_{\beta_i^+}}$ $(i < \rho)$.

c) $|q \upharpoonright \beta_i| = |q_{\beta_i}|$ $(i < \rho)$.

Proof (cf. 2.14.3).

Define $U_\alpha, \mathbb{P}, \mathbb{P}^s$ as in the last lemma and further, pick $\tilde{q} \le \overset{\frown}{p}/U_\alpha$ as there. [This performs a).]

Definition (cf. p.53)

$\forall i < \rho \quad \tilde{U}_{\beta_i} =_{df} \{\{\{\eta, |D_{s_{\gamma_i}}|\}, h\} \mid \eta \in \bigcup_{\gamma \in D \cap \beta_i} [\gamma^+, \gamma^{++}) \quad h = 4,5\}$

Definition (cf. p.54) Define $\tilde{\mathbb{P}}_i^r$ like $\tilde{\mathbb{P}}^s$ with $\tilde{U}_{\beta_i}, p/\tilde{U}_{\beta_i}, r$ in place of $U, \overset{\frown}{p}/U, s$ where $r \in \mathbf{S}_{\beta_i}$, $r \supset p_{\beta_i}$. Again note that $\tilde{U}_{\beta_i} \cap \tilde{U}_{\beta_j} = \emptyset$ and $\tilde{U}_{\beta_i} \cap U = \emptyset$ for $i < j < \rho$. We now proceed as in the proof of 2.14.3 (except of course we have the notational change of r_i for r_i' there). We shall define $r_{ij} \in \mathbf{S}_{\beta_i}$ as before, together with $\tilde{q}_{ij} \in \tilde{\mathbb{P}}_i^{r_{ij}}$ for $j < \omega$, $i < \rho$ s.t. $\tilde{q}_{ij} \le \tilde{q}_{ih} \le \overset{\frown}{p}/\tilde{U}_{\beta_i}$ in $\tilde{\mathbb{P}}_i^{r_{ij}}$ for $h \le j \le \omega$.

The clauses of the definition are the same except that we may take

$$r_{i,j+1} = \text{the } \mathfrak{M}_{p_{\beta_i^+}}\text{-least } r \in \mathbf{S}_{\beta_i} \text{ s.t. } |r| = |r_{ij}| + \beta_i$$

and $r/Z_6 \cap [|r_{ij}|, \beta_i)$ performs our coding now (rather than Z_4). We may take $r(\xi) = 0$, $\forall \xi \in \bigcup_{h>6} Z_h \cap [|r_{ij}|, |r|)$.

The rest of the proof may be finished off in the same manner: Claim 1 is now trivial to prove. Claim 2 holds for much the same reasons: $\tilde{q}_i \in L[\hat{r}_i^*]$, since r_i^* canonically codes, in particular, $\langle \tilde{q}_{ij}/\beta_i \mid j < \omega \rangle$, and $\tilde{q}_i$ has the requisite coding properties, since $\tilde{q}_{i,j+1}$ has them for r_{ij}. As there we may canonically define a

$q\colon \text{card} \cap [\tau, \alpha) \longrightarrow V^2$ (using $H = \{\gamma^+ \mid \gamma \in D\}$,

and the other obvious modifications; Claim 3 is verbatim.

<u>Claim 4</u> $q \in \mathbb{P}_\tau^s$, $|q| = |s|$, $q \notin L[\hat{s}]$, and

$|q \upharpoonright \beta_i| = |q_{\beta_i}|$ for $i < \rho$.

Proof We first show by induction on $\delta \in \text{card} \cap [\tau,\alpha)$ that $q \upharpoonright \delta \in \mathbb{P}_\tau^{q_\delta}$ and $|q \upharpoonright \delta| = |q_\delta|$ for $\delta = \beta_i$. As before Cases 1-3 are easy.

Case 4 $\delta = \beta_i$.

$q \upharpoonright \beta_i$ codes $r_i^* = q_{\beta_i}$, since $\hat{p} \upharpoonright \beta_i, \tilde{q}_i \subset \hat{q} \upharpoonright \beta_i$ and $\tilde{q}_i$ r_{ij}-codes ξ, e_ξ for $\Omega < \xi < |q_{ij}|$ and

$|\tilde{q}_i| = |r_i^*| = \sup_j |\tilde{q}_{ij}|$.

$\langle p \upharpoonright \beta_i, \tilde{q} \upharpoonright \beta_i, \langle \tilde{q}_{hj} \upharpoonright \beta_i \mid h < \rho, j < \omega \rangle \rangle \in L[\hat{r}_i^*]$ by the coding properties of r_i^* . But then the construction of $q \upharpoonright \beta_i$ can be carried out in $L[\hat{r}_i^*]$. Hence $q \upharpoonright \beta_i \in L[r_i^*]$. It follows easily that $q \upharpoonright \beta_i \in \mathbb{P}^{r_i^*}$ and $|q \upharpoonright \beta_i| = |r_i^*| = |q_{\beta_i}|$.

Q.E.D. (Case 4)

That $q \notin L[\hat{s}]$ follows from that property for $\tilde{q}$ and the fact that $\hat{q} \supset \tilde{q}$. Similarly $|q| = |s|$; notice that the construction of q can be carried out in $\mathfrak{A}_s$. Thus $q \in \mathbb{P}^s$ and $|q| = |s|$.

Q.E.D.

Lemma 8.10 *There is a* $q \le p$ *satisfying the conclusion of the Extension Lemma.*

Proof By the last lemma we may now assume that $|p| = |s|$ and $p \notin L[\hat{s}]$. Let $X \subseteq \text{card} \cap [\tau,\alpha)$ be thin in $\mathfrak{A}_s$. Let $\langle \xi_\gamma \mid \gamma \in X \rangle \in \mathfrak{A}_s$ with $\xi_\gamma \in [\gamma,\gamma^+)$ for $\gamma \in X$.

Case 1 $\text{cf}(\alpha) = \omega$.

Just use the same construction as Case 1 of Lemma 8.7.

Case 2 $cf(\alpha) > \omega$.

Then D^* is cofinal in α . Let $\langle \beta_i \mid i < \rho \rangle$ be as in Lemma 8.9. We may assume:

a) $X \cap \beta_i$ is thin in $\mathfrak{A}_{p_{\beta_i}}$.

b) $\langle \xi_\nu \mid \nu \in X \cap \beta_i \rangle \in \mathfrak{A}_{p_{\beta_i}}$.

c) $|p \restriction \beta_i| = |p_{\beta_i}|$

again by Lemma 8.9. Then using the induction hypothesis that EL holds below α , we repeat the construction of Lemma 8.7, Case 2, using $\langle \beta_i \mid i < \rho \rangle$ in place of $\langle \gamma_i \mid i < \lambda \rangle$ there.

Q.E.D.

and Q.E.D. (Extension Theorem)

* * * *

This completes our discussion of the necessary alterations to Chapter 2. Turning now to Chapter 3, recall that we remarked on p.260 that $\mathfrak{A}^1_s$ should be replaced by $\mathfrak{A}'_s$ more or less throughout (and consequently μ^1_s by μ'_s). §3.1 is as before with this change. Likewise §3.2, however Lemma 3.7 requires some additions. Noting the changes already remarked upon, Clause 3 of the definition of $\mathbf{S}'_{\gamma_{i+1}}$ is now superfluous and we have in (2)

$$b_r \in H^{L[\hat{r}]}_{(\alpha^+)^{L[\hat{r}]}} \subseteq L_{\mu'_s}[a_{i+1}, d_i] \subseteq L_{\mu'_s}[a_o, d_o]$$

As before we define $r = f \restriction \xi$, and also $\hat{r}$ this time, by induction on ξ . Assuming this is done we define $f(\xi)$ as before using b^o_r but note that we also need $\hat{r}^{\#}$ to obtain e_r . But we ensured our capability to obtain $\hat{r}^{\#}$ by using b^{j+1}_r $(j < \alpha^+)$ to code e^*_r :

$$j \in e^*_r \longleftrightarrow S(b^{j+1}_r) \cap d_i \text{ is bounded in } \gamma_{i+1} .$$

So $\hat{r}^{\#} \in L[\hat{r}, d_i]$. It should be remarked that this yields the true $\hat{r}^{\#}$ since the formula "$y = x^{\#}$" is absolute in inner models.

(3) must also be changed. Hopefully it is clear what we must do: we define $\langle \mathfrak{A}_r^i \mid i < \omega \rangle$ as it originally was. If $\mathfrak{A}_r \models$ "γ is regular" we may define

$$\langle \rho^i_{r,\eta} \mid \eta \in (\mathrm{card} \cap \gamma_\lambda)_{\mathfrak{A}_r}, i < \omega \rangle , \ B^n_r$$

again as before. We set

$$f(\xi) = 1 \quad \text{iff } \exists n < \omega, \delta < \alpha \text{ s.t. } \forall m \geq n$$
$$\eta \in B^m_r \setminus \delta \longrightarrow \rho^i_{r,\eta} \in d$$
$$0 \quad \text{otherwise.}$$

If $\mathfrak{A}_r \nvDash$ "γ is singular", then, defining D_r and $\tilde{\rho}^b_\eta$, $\langle \tilde{\rho}_{r,\eta} \mid \eta \in (\mathrm{card} \cap \gamma_\lambda)_{\mathfrak{A}_r} \rangle$ again as in the original definitions, set

$$f(\xi) = 1 \quad \text{iff } \exists \delta \in D_r \text{ s.t. } \forall \eta > \delta, \eta \in D_r$$
$$\tilde{\rho}_{r,\eta} \in d$$
$$0 \quad \text{otherwise.}$$

Again set $d_\lambda = \{\xi \mid f(\xi) = 1\}$, and this may all be carried out in $L_{\mu_s'}[a_o, d_o]$.

The rest of §3.2 and §3.3 go through as before. This brings us to the all important Lemma 3.13. Lemmas 3.14-19 have just the notational changes already mentioned. In 3.20 again Clause 3 of the definition of $\mathbf{S}'_{\gamma_i}$ is superfluous. The crucial 3.22 should now read:

Lemma 3.22' *a)* $p_\gamma \in \mathbf{S}_\gamma$.

b) $b \in L[\hat{p}_\gamma]$.

a) is now easier to prove [just the first few lines of the

former proof suffice] and b) is completely trivial. 3.23 is as before. In 3.24 the proof that part (C) I of the definition on p.268 holds is identical. Note, by b) above, that $p/\gamma \in L[p_\gamma]$ so III holds vacuously. II and IV are identically shown, and we are only left with (C) VI: the arguments of "Case 1" and "Case 2" prove this ["Case 3" was shown to be impossible]. Lemmas 3.13, 3.37 and 3.38 are the same *mutatis mutandis*. The definition following 3.38 has no analogue in our present context and so we need a new proof for 3.40. But we can prove this just like 3.13 once we have:

Lemma 8.11 *Let* $\omega < cf(\alpha) < \alpha$, $p \in \mathbb{P}^s_\tau$, $f \in \mathbb{F}(p)$, $dom(f) \subset \alpha$. *Then* $\sum^{s,p}_f$ *is dense in* $\mathbb{P}^s_\tau$.

Proof This proof will combine the proof of Lemma 3.13 with that of Lemma 8.9. A few preliminaries: Suppose that $p_o \in \mathbb{P}^s_\tau$.

Note that if $p_o \mid p$ we are already finished, so we may assume $p_o \le p$ and we must find a $p' \le p_o$ s.t. $p' \in \sum^{s,p}_f$. We may also assume w.l.o.g. that

(i) $(p_o)^\alpha$ supports p_α. Let $D = D^{(p_o)^\alpha}_{p_{o\alpha}}$.

(ii) $|(p_o)^\alpha| = |p_{o\alpha}| \ge \Omega$.

(iii) $|(p_o)^\gamma| = |p_{o\gamma}| \ge \Omega$ for $\gamma \in D$.

Further we shall assume that w.l.o.g. $|D| < \tau$. [If not let $\eta^+ > |D|$, $\eta^+ < \alpha$. Defining as following Fact 3.9.1 $f^* \in \mathbb{F}((p_o)_{\eta^+})$ by replacing each $\Delta \in f(\gamma)$ with $\Delta^* = \{q \in \mathbb{P}^s_{\eta^+} \mid \Delta^q$ is dense in $\mathbb{P}^{q_{\eta^+}}_\tau\}$, it suffices by Fact 3.9.2 to find $p' \le (p_o)_{\eta^+}$ s.t. $p' \in \sum^{s,(p_o)_{\eta^+}}_{f^*}$.]

Let $\langle \beta_i \mid i \le \rho \rangle$ enumerate the limit points of $D \cap (\tau, \alpha]$. Thus $\beta_\rho = \alpha$.

By induction on $i \le \rho$ we construct:

(a) p_i s.t. $p_j \le p_i \le p_o$ for $i \le j \le \rho$.

$p_{i+1} \in \sum^{s,p}_{f/\beta_i^+}$, $(p_{i+1})_{\beta_i^+} = (p_o)_{\beta_i^+}$ for $i < \rho$.

Thus p_ρ will be our desired p' . We also construct:

(b) $\langle Y_i^\gamma \mid \gamma \in [\tau, \beta_i^+)\rangle$ s.t. $Y_i^\gamma \prec \mathfrak{A}_s'$.

(c) $g_i \in \mathbb{F}(p_i)$ s.t. $\mathrm{dom}(g_i) \subset \beta_i^+$ and $p_{i+1} \in \sum^{s,p_i}_{g_i}$ for $i < \rho$.

Simultaneously we shall be performing a construction along the lines of Lemma 8.9. Define $\tilde{U}_{p_i}$, $\tilde{\mathbb{P}}_i^r$ $(r \in \mathbf{S}_{\beta_i}$, $r \supset p_{o\beta_i}$, $i \le \rho)$ as before (p.293). We construct:

(d) $\langle r_{hi} \mid i \le h \le \rho\rangle$ $(i \le \rho)$ s.t. $r_{hi} \in \mathbf{S}_{\beta_h}$ and

$\langle \dot{p}_{o\beta_h}, r_{hj}\rangle \le \langle \dot{p}_{o\beta_h}, r_{hi}\rangle$ for $i \le j \le h \le \rho$.

(e) $\langle \tilde{q}_{hi} \mid i \le h \le \rho\rangle$ for $i \le \rho$ s.t. $\tilde{q}_{hi} \in \mathbb{P}_h^{r_{hi}}$,

$\tilde{q}_{hj} \le \tilde{q}_{hi} \le \hat{p}_{o}/\tilde{U}_{\beta_h}$ in $\tilde{\mathbb{P}}_h^{r_{hj}}$ and

$\mathrm{dom}(\tilde{q}_{hj} \setminus \tilde{q}_{hi}) \subset [\beta_i^+, \beta_h)$ for $i \le j \le h \le \rho$.

In Lemma 8.9 we used sequences such as in (d) and (e) to "smooth out" our required function into a condition. They will perform this rôle here in going from p_i to p_{i+1} . (b) and (c) will ensure that at limit stages we can define (with the help of $r_{\lambda\lambda}$) a suitable p_λ . In fact they will be required for showing that $p_\lambda/\beta_\lambda = \bigcup_{i<\lambda} p_i/\beta_\lambda$ is a condition. The exact definitions are as follows. Suppose that p_j is defined for $j \le i$.

(1) Y_i^γ = the smallest $Y \prec \mathfrak{A}_s'$ s.t. $\gamma \subset Y$ and

$\gamma \cup \{p_j, \langle Y_j^\nu \mid \nu \in \mathrm{card} \cap \alpha^+\rangle, \langle r_{hj} \mid j \le h \le \rho$,

$\langle \tilde{q}_{hj} \mid j \le h \le \rho\rangle\} \subseteq X$, for $j < i$.

Thus $Y_\lambda^\gamma = \bigcup_{i<\lambda} Y_i^\gamma$ for limit λ .

(2) $g_i(\gamma) = Y_i^\gamma \cap \mathfrak{A}'_{p_{i\gamma^+}}$ if $\gamma^+ \in Y_i^\gamma$, $\gamma < \beta_i^+$

undefined if not.

(3) (a) $r_{ho} = p_{o\beta_h}$.

(b) $r_{h,i+1}$ = the $\mathfrak{A}'_{p_{o\beta_h^+}}$-least $r \in \mathbb{S}_{\beta_h}$ s.t.

I $\langle \dot{p}_{o\beta_h}, r\rangle \leq \langle \dot{p}_{o\beta_h}, r_{hi}\rangle$ in $\mathbb{R}^{p_{o\beta_h^+}}$.

II $|r| = |r_{hi}| + \beta_h$.

III $[|r_{hi}|, |r|) \cap \bigcup_{n>5} Z_n$ canonically codes:

$\langle \tilde{q}_{kj}/\!\!\wedge\beta_h \mid j \leq i, k \leq \rho\rangle$, $\langle p_j/\!\!\wedge\beta_h \mid j \leq i\rangle$.

[As before card $\langle \tilde{q}_{kj}/\!\!\wedge\beta_h \mid j \leq i, k \leq \rho\rangle$ etc. is $\leq \beta_h$.]

(c) $r_{h\lambda} = \bigcup_{i<\lambda} r_{hi}$ for $\lim(\lambda)$.

(4) (a) $\tilde{q}_{ho} = \overset{\frown}{p}_o/\!\!\wedge\tilde{U}_{\beta_h}$.

(b) $\tilde{q}_{h,i+1}$ = the $\mathfrak{A}'_{r_{h,i+1}}$-least $\tilde{q} \leq \tilde{q}_{h,i}$ in $\tilde{\mathbb{P}}_h^{r_{h,i+1}}$

s.t. $|\tilde{q}| = |r_{h,i+1}|$ and $\beta_i^+ \cap \operatorname{dom}(\tilde{q}\setminus\tilde{q}_{hi}) = \emptyset$.

(c) $\tilde{q}_{h\lambda} = \bigcup_{i<\lambda} \tilde{q}_{hi}$.

(5) To define p_{i+1} we first define a $p_i^* \leq (p_i)^{\beta_i^+}$ in $\mathbb{P}^{p_{o\beta_i^+}}$ and set:

$(p_{i+1})^{\beta_i^+}$ = the $\mathfrak{A}'_{p_{o\beta_i^+}}$-least $p' \leq p_i^*$ in $\mathbf{P}^{p_{o\beta_i^+}}$ s.t.

$$p' \in \sum_{f/\!\wedge\beta_i^+}^{p_{\beta_i^+}, (p)^{\beta_i^+}} \cap \sum_{g_i}^{p_{i\beta_i^+}, (p_i)^{\beta_i^+}} .$$

Note that there is no conflict here: $p_{o\beta_i^+} = p_{i\beta_i^+} \supseteq p_{\beta_i^+}$ and $\sum_{f/\!\wedge\beta_i^+}^{p_{\beta_i^+}, (p)^{\beta_i^+}}$ is dense in $\mathbf{P}_\tau^{p_{\beta_i^+}}$ (by Induction Hypothesis) and hence also in $\mathbb{P}_\tau^{p_{o\beta_i^+}}$ by 2.10. We put $p_{i+1} = (p_o)_{\beta_i^+} \cup p'$.

Hence $p_{i+1} \in \sum_{g_i}^{s,p_i} \cap \sum_{f/\beta_i^+}^{s,p}$. $\overset{*}{p}_i$ is defined as follows:

$\overset{*}{p}_i = (p_i)^{\beta_i^+}$ for $i = 0$ or a limit.

Let $i = j+1$. For $\gamma \in D \cap [\beta_j^+, \beta_i)$, we set:

$\overset{*}{p}_i(\gamma^+) = \langle \dot{p}_{o\gamma^+}, r \rangle$ where r is the $\mathfrak{A}_{p_{o\gamma^+}}$-least r s.t.

$\langle \dot{p}_{o\gamma^+}, r \rangle \le p_o(\gamma^+)$ in $\mathbb{R}^{p_{o\gamma^{++}}}$ and $\bigcup_{i \le h \le \rho} \tilde{q}_{hi} \restriction [\gamma^+, \gamma^{++}) \subset r$.

$\overset{*}{p}_i(\beta_i) = \langle \dot{p}_{o\beta_i}, r_{ii} \rangle$.

$\overset{*}{p}_i(\nu) = p_i(\nu)$ in all other cases.

<u>Claim 1</u> $\overset{*}{p}_i \restriction \gamma \in \mathbb{P}^{\overset{*}{p}_{i\gamma}}$ for $\gamma \in \text{card} \cap [\tau, \beta_i]$ and $\overset{*}{p}_i \in \mathbb{P}^{p_{o\beta_i^+}}$.

Proof We argue as in Lemma 8.9, Claim 4: Again, the only problematical case is $\gamma = \beta_i$ itself. But r_{ii} has coded up $\langle \tilde{q}_{h\ell} \restriction \beta_i \mid h \le \rho, \ell \le j \rangle$ and $\langle p_\ell \restriction \beta_i \mid \ell \le j \rangle$. Thus $\overset{*}{p}_i \restriction \beta_i$ can be constructed in $L[\widehat{r_{ii}}]$. But $\overset{*}{p}_{i\beta_i} = r_{ii}$. Thus $\overset{*}{p}_i \restriction \beta_i \in \mathbb{P}^{\overset{*}{p}_{i\beta_i}}$. But by (3)(b) $r_{ii} \in \mathbf{S}_{\beta_i}$. So $\overset{*}{p}_i(\beta_i) \in \mathbb{R}^{p_{o\beta_i^+}}$ and thus $\overset{*}{p}_i \in \mathbb{P}^{p_{o\beta_i^+}}$.

Q.E.D. (Claim 1)

<u>Claim 2</u> $p_\lambda \in \mathbb{P}_\tau^s$.

Proof Since $p_\lambda = (p_\lambda)^{\beta_\lambda^+} \cup (p_o)_{\beta_\lambda^+}$ it suffices to show

(*) $p_\lambda \restriction \beta_\lambda^+ \in \mathbb{P}_\tau^{p_{o\beta_\lambda^+}}$.

<u>Claim 2.1</u> $p \restriction \gamma \in \mathbb{P}_\tau^{p_\gamma}$, $\tau < \gamma < \beta_\lambda$ where $p = (p_\lambda)^{\beta_\lambda^+}$.

Proof 2.1 We prove this by the methods of Lemma 3.13 with

β_λ replacing α, 3.14 is the same as before, but 3.15 has no analogue here.

Define: $\sigma = \sigma_\lambda^\gamma$, $b = b_\lambda^\gamma$ $(\gamma \in [\tau, \beta_\lambda])$ as before. Likewise $A', \mu', p_i', \alpha', \alpha^*, D_\delta'$. Set $\beta' = \sigma^{-1}(\beta_\lambda)$. This β' plays the rôle of α' in the Lemmas 3.16-22. Also put $r' = \sigma^{-1} '' r_{\lambda\lambda} = \bigcup_{i<\lambda} \sigma^{-1}(r_{\lambda i})$. Then we obtain the following sequences of lemmas, putting β' for α' in their statements and proofs:

Lemma $D' \cap (\delta^+)_b$ is $\mathbb{P}_\gamma^{(\delta^+)^b}$ - generic over $(\mathfrak{A}'_{(\delta^+)^b})_b$.

Lemma 3.17 has no analogue of course. We get 3.18 as:

Lemma Let $\delta < \beta'$, $\delta \in \mathrm{card}_b$. Then

$L_{\beta'}[A' \cap \beta', D' \cap \beta'] \models$ "δ is a cardinal".

Lemma $p_{i\gamma} = p'_{i\gamma} = p^\gamma_{i\lambda\gamma}$ for $i < \lambda$ and $\gamma \in \mathrm{card} \cap [\tau, \beta_\lambda]$.

Lemma $A' \cap \beta'$, $r' \in L[\hat{p}_\gamma]$.

Now $r_{\lambda\lambda}$ codes $\langle p_i \restriction \beta_\lambda \mid i < \lambda \rangle$. So r' codes $\langle p_i' \restriction \beta' \mid i < \lambda \rangle$ and we have (by the above):

Lemma $\langle p_i' \restriction \beta' \mid i < \lambda \rangle \in L[\hat{p}_\gamma]$.

But this in turn gives us:

Lemma $p_\gamma \in \mathbb{S}_\gamma$, $p \restriction \gamma \in L[\hat{p}_\gamma]$ $(\tau \le \gamma < \beta_\lambda)$.

And finally we get:

Lemma $p(\gamma) \in \mathbb{R}^{p_{\gamma^+}}$ for $\gamma \in [\tau, \beta_\lambda)$.

Lemma $p/\gamma \in \mathbb{P}_\tau^{p_\gamma}$ for $\gamma \in [\tau,\beta_\lambda)$.

Q.E.D. (Claim 2.1)

<u>Claim 2.2</u> $p/\beta_\lambda \in \mathbb{P}^{r_{\lambda\lambda}}$.

Proof p/β_λ codes $r_{\lambda\lambda}$ since $\widehat{p/\beta_\lambda} \cup \tilde{q}_{\lambda\lambda} \subset p$. Further, $p/\beta_\lambda \in L[r_{\lambda\lambda}]$ by $r_{\lambda\lambda}$'s coding properties, and the conclusion follows.

Q.E.D. (Claim 2.2)

Finally $\langle \dot{p}_{o\lambda}, r_{\lambda\lambda}\rangle \in \mathbb{R}^{p_{o\beta_\lambda^+}}$, hence $p/\beta_\lambda^+ \in \mathbb{P}^{p_{o\beta_\lambda^+}}$ which is (*).

Q.E.D. (Claim 2 and the Lemma)

This completes Theorems 3.1 and 3.2. But 3.3 goes through with the usual notational changes, as do §§4.1-4.3, thus the aim of this section is completed, namely proving Theorem 0.1 when $\langle M,A\rangle$ is closed under sharps.

8.2 THE INTERMEDIATE CASE

It should be obvious that Chapters 1-4 prove the following:

Theorem 8.12 *Let* $\langle M,A\rangle \models ZF + GCH$. *Suppose* $a \subseteq \alpha^+$ *s.t.* "$\neg a^{\#}$" *holds in* M, *and further for* $\tau \geq \alpha^+$, $\tau \in \text{card}^M$ *that* $H_\tau^M = L_\tau[A]$, *then there is a class of conditions* $\mathbb{P}$ *which is* $\langle M,A\rangle$*-definable in* α, a *s.t. if* N *is* $\mathbb{P}$*-generic over* M *then*

I $N \models ZF + \exists b \subseteq \alpha^+$ *s.t.* $V = L[b]$.

II $H_{\alpha^+}^M = H_{\alpha^+}^N$.

III *Cardinals and cofinalities are preserved, as are "the large cardinal properties".*

As before, a preliminary Easton forcing may be needed to ensure that Facts 1-4 of §2.1 hold above α. This may be combined with our forcing just as outlined in §2.1.

Now suppose that $\langle M,A\rangle$ satisfies the following (different) assumptions:

(*) a) $A \subseteq \alpha$, where $\alpha \in \text{card}\setminus\omega$.

b) $H_\tau = L_\tau[A]$ for $\tau \in \text{card}$.

c) Facts 1-4 of §2.1 hold for $\kappa^+ < \alpha$.

d) $(A \cap \xi)^{\#}$ exists for $\xi < \alpha$.

Assuming (*) holds we may define a simplified form of $\mathbb{P}$ in the following manner: let $\mathbb{P}^{[\alpha]}$ be the set of $p: \text{card} \cap \alpha \longrightarrow V^2$ s.t.

I $p{\restriction}\tau \in \mathbb{P}^{p_\tau}$ for $\tau < \alpha$.

II $p \in \mathbb{P}^{\alpha}$ if α is a successor.

III If α is inaccessible then there is a cub

$C \subseteq \text{card} \cap \alpha$ s.t. $\gamma \in C \longrightarrow \dot{p}_\gamma = \emptyset \wedge |p_\gamma| = |p{\restriction}\gamma| = \gamma$.

It is clear that the lemmas on the extendibility of conditions, in a slightly modified form, go through for $\mathbb{P}^{[\alpha]}$, *a fortiori*. Likewise, repeating the argument of §3.7 yields:

Theorem 8.13 *Let* $\langle M,A\rangle$ *satisfy* (*). *Then* $\mathbb{P}_\tau^{[\alpha]}$ *is* τ*-distributive over* M *for* $\tau \in \text{card} \cap \alpha$.

And in turn §1 gives:

Corollary 8.14 *Let* $\langle M,A\rangle$ *satisfy* (*). *Let* N *be* $\mathbb{P}^{[\alpha]}$*-generic over* M. *Then:*

a) $N \models ZF + \exists a \subset \omega V = L[a]$.

b) Cardinals and cofinalities are preserved as are the "large cardinal properties".

Note that "large cardinal properties" are automatically preserved for cardinals $\kappa > \alpha$, since $\mathbb{P}^{[\alpha]}$ is a small set of conditions with respect to such cardinals.

We now turn to the task of proving Theorem 0.1 when M is not closed under sharps. Let β be the least $\beta \in \mathrm{card}\setminus\omega$ s.t. $\exists a \subseteq \beta$ with $\neg a^{\#}$ in M.

Consider the following three cases:

<u>Case 1</u> $\beta = \omega = 0^{+}$. Then we may simply apply Theorem 8.12 to obtain 0.1.

<u>Case 2</u> $\beta = \gamma^{+} > \omega$.

By Theorem 8.12, first extend M to an M' satisfying $V = L[b]$ where $b \subseteq \beta$. Our efforts are now directed towards getting M' into a form where (*) is applicable. We first of all replace b by a more suitable Cohen generic subset of $[\gamma,\beta)$. Define $\mathcal{b}$ to be $L_\rho(H_\beta^{M'})$ where ρ = the least ρ s.t. $L_\rho(H_\beta^{M'}) \models ZF^-$. The round brackets are to indicate constructibility <u>over</u> $H_\beta^{M'}$ rather than relative constructibility. That is, here we are taking

$$L_0(H_\beta^{M'}) = H_\beta^{M'} .$$

$$L_{\alpha+1}(H_\beta^{M'}) = \mathrm{Def}\ L_\alpha(H_\beta^{M'}) .\quad L_\lambda(H_\beta^{M'}) = \bigcup_{r<\lambda} L_r(H_\beta^{M'})$$

for $\lim(\lambda)$.

Definition Let $\mathbb{C}_\tau$ be the set of Cohen conditions of size $< \tau^{+}$ for adding a subset of $[\tau,\tau^{+})$. Then $\mathbb{C}_\gamma \in \mathcal{b}$. Note also that $\bar{\bar{\mathcal{b}}} < \beta^{+}$.

__Claim__ There is $A_\gamma \subset [\gamma,\beta)$ s.t.

I A_γ is $\mathbb{C}_\gamma$-generic over b.

II M' satisfies $V = L[A_\gamma]$, and b is $L[A_\gamma]$-definable in A_γ.

Proof Notice that since $\bar{\bar{b}} < \beta^+$ there are only β many dense sets in b; further $\mathbf{C}_\gamma$ is closed under chains of length $< \beta$. Hence we may inductively choose (in M') a $\mathbf{C}_\gamma$-generic set over b. [If $\langle \Delta_\nu \mid \nu < \beta\rangle$ enumerates those sets, pick $p_1 \in \Delta_o$, $p_{i+1} \le p_i$ with $p_{i+1} \in \Delta_i$, and $p_\lambda = \bigcup_{r<\lambda} p_\nu$ for $\lim(\lambda)$; then p_i is defined for $i < \beta$.] Let a^o be such a generic set. Notice also that $H_\beta^{M'}$ must equal $L_\beta[a^o]$. Now let $\langle \Delta_\nu \mid \nu < \beta\rangle$ be the $L[a^o]$-least enumeration of the $\mathbf{C}_\gamma$-dense sets in $b[a^o]$. In a familiar fashion construct by induction a tree:

$$\langle p_s \mid s : |s| \longrightarrow 2,\ |s| < \beta\rangle$$

of conditions in $\mathbf{C}_\gamma$ such that:

I $s \supset s' \longrightarrow p_s \le p_{s'}$ in $\mathbb{C}_\gamma$.

II $p_{s\frown o} \mid p_{s\frown 1}$ in $\mathbf{C}_\gamma$.

III $p_{s\frown i} \in \Delta_{|s|}$ $(i = 0,1)$.

Let $a^1 \subset \beta$ be s.t. $\chi_{a^1} = \bigcup_{\xi<\beta} p_{\chi_b \restriction \xi}$. Then a^1 is also $\mathbf{C}_\gamma$-generic over $b[a^o]$ and $b \in L[a^o, a^1]$.

Finally for $i = 0,1$ set:

$$\tilde{a}^i = \{\omega\tau + 2n + i \mid \omega\tau + n \in a^i\}.$$

$$A_\gamma = \tilde{a}^o \cup \tilde{a}^1 .$$

Then A_γ is easily seen to have the properties required by the claim.

Q.E.D. (Claim)

Now define $\mathbb{Q}^\tau$ as the set of ordinary Easton conditions for adding an $A \subset \tau$ s.t. $A_\nu = A \cap [\nu, \nu^+)$ is $\mathbb{C}_\nu$-generic for $\nu \in \text{card} \cap \tau$. Let A' be $\mathbb{Q}^\gamma$-generic over M'. Set $M'' = M'[A']$ and let $A = A' \cup A_\gamma$. Then $M'' \models$ "$V = L[A]$". Notice that $H_\beta^{M''} = L_\beta[A]$ will remain closed under #. [In fact the indiscernibles for $\mathbb{C}_\nu$ (coded as a set of ordinals) will remain indiscernibles for A_ν, $\nu < \gamma$.] But A is $\mathbb{Q}^\beta$-generic over b. But then it is easily checked that A satisfies (*). Case 2 is then finished off by applying Corollary 8.14.

Q.E.D. (Case 2)

Case 3 β is a limit cardinal.

Then, *a fortiori*, there is a subset of β^+ without a sharp. We then use Theorem 8.12 again to extend M to $M' = M[b]$ satisfying $V = L[b]$ with $b \subseteq \beta^+$. Now we may utilise the methods of §1.4 to extend M' to $M'' = M'[d]$ again satisfying $V = L[d]$, with $d \subseteq \beta$. Note that $H_\beta^{M''} = L_\beta[d]$ remains closed under #: (with the notation of §1.4) if $\mathbb{P}$ is the notion of forcing there $\mathbb{P} = \mathbb{P}_\tau \times \mathbb{P}^\tau$ for $\tau < \beta$. It suffices to show that $(d \cap \gamma)^\#$ exists for arbitrary $\gamma < \beta$; but since $\mathbb{P}_\gamma$ is γ-distributive, we may just utilise the indiscernibles for $\mathbb{P}^\gamma$ to give us indiscernibles for $d \cap \gamma$, as remarked above.

Yet again extend M'' to $M''' = M''[A']$ where $A' \subset \beta$ is $\mathbb{Q}^\beta$-generic over M''.

Defining $A'_\tau = A \cap [\tau, \tau^+)$, replace A' by an A made up of $A_\tau = \{\tau + \nu \mid \nu \in d \cap \tau \cup A'_\tau\}$.

$\langle M''', A'\rangle$ satisfies Facts 4.2.14 and 15 (alluded to as Fact 4 of §2.1) since A' is $\mathbb{Q}^\beta$-generic (see Appendix). It is easily seen, since A is simply defined from A', that they will continue to hold for $\langle M''', A\rangle$. Fact 2 of §2.1

(the existence of $\diamondsuit$-sequences) is true since each A_τ is $\mathbf{C}_\tau$-generic over M'' . Hence $\langle M''', A\rangle$ satisfies (*) and we may apply Corollary 8.14 to extend, finally, M''' to an N satisfying $V = L[a]$ with $a \subseteq \omega$. Cardinals and cofinalities are preserved, and the large cardinal properties above and below β . It remains to show that the properties of β , if any, are preserved in the transition from M to N . Splitting this up into $M \longrightarrow M' \longrightarrow M'' \longrightarrow M''' \longrightarrow N$: for $M \longrightarrow M'$ this is catered for by Theorem 8.12 itself. The reader may easily see that Mahloness, α-Erdös ($\alpha < \omega_1$) and subtlety are preserved using the set $\mathbb{P}^{[\beta]}$ of Corollary 8.14, thus disposing of the $M''' \longrightarrow N$ extension. The results of the Appendix show the preservation of these properties from $M'' \longrightarrow M'''$, leaving us with the extension $M' \longrightarrow M''$ of the type given by 1.5. To show that Mahloness is preserved we may prove (using the notation of 1.5) an easy version of Lemma 4.3, where we replace $\mathfrak{N}^1_{\kappa^+}$ by $L[b]$, κ by β , and $\mathbf{P}^{\kappa^+}$ by $\mathbb{P}$. We redefine, taking advantage of the fact here that $\mathbb{P} = \mathbf{P}_{\tau^+} \times \mathbf{P}^{\tau^+}$ for $\tau \in \mathrm{card} \cap \beta$,

$$\Delta^q = \{p \in \mathbb{P}^{\tau^+} \mid (p,q) \in \Delta_\tau\} .$$

That $\Delta = \{p \in \mathbb{P} \mid \forall\tau (p)_{\tau^+} \in \Delta^*_\tau\}$ is dense in $\mathbb{P}$ is now given by Lemma 1.8. The argument goes through as before. Likewise with similar adaptations, 4.6 shows that subtlety is preserved. Notice that β could not have been weakly compact, since $\neg B^\#$ whilst $(B \cap \gamma)^\#$ exists in M . For the α-Erdös property we can use (a simplified version of) the method of 4.11 for this $\mathbb{P}$.

Q.E.D. (Theorem 0.1)

9 . Some further applications

Here we shall consider some extensions and applications of the previous methods and results. The first theorem will take a certain iterable inner model M , and will, with the methods of §4.4 and using M's iteration points (rather than the indiscernibles provided by $0^{\#}$) provide a coding real for that model. The second theorem can be regarded as a refinement of the first when, in addition, M contains a measurable cardinal.

Theorem 9.1 *Let* M *be an inner model (i.e. is transitive and contains all ordinals) of* ZFC + GCH . *Let* U *be an ultrafilter on* $P(\kappa) \cap M$ *(not necessarily in* M *) such that, setting* $\kappa^{+} = \kappa^{+M}$ *, we have:*

(a) $\langle H^{M}_{\kappa+}, U\rangle$ *is amenable (i.e.* $\forall x \in H^{M}_{\kappa+}$, $x \cap U \in H^{M}_{\kappa+}$ *).*

(b) U *is normal on* κ *in* $\langle H^{M}_{\kappa+}, U\rangle$.

(c) $\langle M, U, \kappa\rangle$ *is iterable (i.e. all the successive ultrapowers starting with* M *and* U *are well founded) yielding iterates* $\langle M, U_i, \kappa_i\rangle$ *and iteration maps* $\pi_{ij}: M \xrightarrow[\Sigma_1]{} M$.

(d) $\bar{\bar{\kappa}}^{+} = \aleph_0$.

Then $\exists\, a \subseteq \omega$ *such that*

(i) $M \subseteq L[a]$.

(ii) *The cardinals and cofinalities of* M *are preserved in* L[a] *as are the aforementioned large cardinal properties of Theorem 0.1.*

(iii) *Each* π_{ij} *has a unique extension* $\tilde{\pi}_{ij} : L[a] \xrightarrow[\Sigma_1]{} L[a]$.

Indeed (i)-(iii) provide:

(iv) $a^{\#}$ *exists.*

(iii) shows that clearly the κ_i belong to the indiscernibles for $L[a]$. Let V_i be the ultrafilter on $P(\kappa_i) \cap L[a]$ given by $a^{\#}$:

$$x \in V_i \longleftrightarrow \kappa_i \in \tilde{\pi}_{ij}(x) \qquad (i < j < \infty)$$

As is known here

$$L[a] = \Sigma_o(L[a])\text{-closure of } \mathrm{ran}(\tilde{\pi}_{ij}) \cup \{\kappa_\ell \mid i \leq \ell < j\}$$

This yields:

(v) $\langle L[a], V_o\rangle$ *is iterable, yielding iterates* $\langle L[a], V_i\rangle$ *and iteration maps* $\tilde{\pi}_{ij}$ $(i \leq j < \infty)$.

But the indiscernibles given by $a^{\#}$ above κ_o are just exactly the iteration points of $\langle L[a], V_o\rangle$. So:

(vi) *Let* I^a *be the canonical indiscernibles for* $L[a]$. *Then* $I^a \backslash \kappa_o = \{\kappa_i \mid i < \infty\}$.

A typical application of this would be the following. Let U be a normal measure on κ in L^U , and suppose further that κ^{+L^U} is countable. Let L^{U_i} be the iterates of L^U, with U_i normal on κ_i in L^{U_i} . Set M of the theorem equal to the core model (see [DoJ]) $K = \bigcap_i L^{U_i}$. Then $M, U, \langle U_i \mid i < \infty\rangle$ satisfy the hypothesis of the theorem. Let a be as given there, then $a^{\#}$ "codes" U in the sense that κ is a canonical indiscernible for $L[a]$, and again defining V on $L[a]$, using $a^{\#}$ and its resultant embeddings as above, we have that $V \cap K = U$. Note also that as cofinalities are preserved between K and $L[a]$, the covering lemma holds with respect to K in $L[a]$.

Another application is to *critical mice*. In [J5] a critical mouse was defined as a core mouse (see [DoJ]) $N = J_\alpha^U$ such that $H_\kappa^N \in N$, where U is normal on κ in N. There it is shown that N is critical iff $\tilde{N} = \bigcup_{i<\infty} H_{\kappa_i}^{N_i}$ is a ZF model, where $N_i = J_{\alpha_i}^{U_i}$ is the i-th iterate of N and U_i is normal on κ_i in N_i. It is also proved there that conversely, if M is an inner model of ZF such that $K \nsubseteq M$, i.e. there is some mouse not in M, and if N is the $<_*$-least core mouse with $N \notin M$, then N is critical and $\tilde{N} = (K)^M$. [Here the $<_*$ well-ordering is that defined in [J5]: $N <_* P \longleftrightarrow \exists\theta$ regular, $\theta > \bar{\bar{N}}, \bar{\bar{P}}$ with $N_\theta \in P_\theta$.]

Now suppose N is a countable critical mouse. Let $M = \tilde{N}$. Then $M, U, \langle U_i \mid i < \infty\rangle$ satisfy the hypotheses of Theorem 9.2. Hence there is $a \subseteq \omega$ such that $(K)^{L[a]} = \tilde{N}$, and $a^\#$ "codes" U (and hence N), in the above sense.

(Whilst on the subject of the core model K, the third author has proved, under the assumption $M \models \neg L^U$ (there is no inner model with a measurable cardinal) a coding theorem over K, in place of L, using a parallel development to Chapters 1-4,6,7. This provides a model $N \supseteq M$, with $N \models (\exists a \subseteq \omega, V = K[a]) + ZF + GCH$; cardinals, cofinalities and the large cardinal properties of Theorem 0.1 are preserved into N as are additionally (e.g.) statements of the form "all subsets of a cardinal κ have a sharp". [It is to be remarked that this involved a heavier use of $\neg L^U$, than that of $\neg 0^\#$ in the L case.])

The second theorem proven in this section is the following:

Theorem 9.2 *Let* M *be a transitive model of* $ZF + GCH +$ "κ *is measurable". Let* U *be normal on* κ *in* M. *Let*

$\langle M_i, U_i, \kappa_i \rangle$ *be the iterations of* $\langle M, U, \kappa \rangle$ *with iteration maps* π_{ij} . *Set* $H = \bigcap_{i < \infty_M} M_i$.

There is a class of conditions $\mathbb{P}$ *such that if* N *is* $\mathbb{P}$-*generic over* M , *then* $N = M[a]$, *where* $a \subseteq \omega$ *and the following hold in* N :

(i) $H[a] = L[a]$.

(ii) *Cardinals, cofinalities and the aforementioned large cardinal properties of* H *are preserved in* $L[a]$.

(iii) $\pi_{ij} \restriction H$ *has a unique extension to a*

$$\tilde{\pi}_{ij} : L[a] \xrightarrow[\Sigma_1]{} L[a] .$$

(iv) $M_i[a] = L[a^{\#}]$ $(i < \infty_M)$.

Thus $a^{\#}$ not only "codes" U but also the entire model M . As mentioned to prove Theorem 9.1 we shall imitate quite closely the construction of §4.4. As a preliminary we shall use the iteration maps π_{ij} and iteration points κ_j to provide actually what we would normally force to obtain, namely an initial A which is $\mathbb{Q}$-generic over M . Let $M, U, \kappa, U_i, \kappa_i$ be as in Theorem 9.1.

Lemma 9.3 *There is* $A \subseteq On$ *s.t.*

I A *is* $\mathbb{Q}$-*generic over* M .

II π_{ij} *extends uniquely to a*

$$\tilde{\pi}_{ij} : N \xrightarrow[\Sigma_1]{} N \text{ , where } N = \langle M[A], A \rangle .$$

Proof Choose a $b \in M$ s.t. $b \subseteq \kappa_0$ and $H_{\kappa_0} = L_{\kappa_0}[b]$ in M . Set $B = \bigcup_{i<\infty} \pi_{0i}(b)$. Then $H_{\kappa_i} = L_{\kappa_i}[B]$ and $\pi_{ij} : \langle M, B \rangle \xrightarrow[\Sigma_1]{} \langle M, B \rangle$. [That b exists follows from the GCH ; the rest follows from the fact that we have $B \cap \kappa_i = \pi_{oi}(b)$.] Let $\mathbf{Q}_\alpha^\beta$ be the conditions for adding an $A \cap [\alpha,\beta)$, for $\alpha,\beta \in \mathrm{Card} \cup \{\infty\}$, $0 \le \alpha \le \beta \le \infty$ (see p.328). Set:

$$\mathbb{Q}_\alpha = \mathbf{Q}_\alpha^\infty \quad ; \quad \mathbf{Q}^\beta = \mathbf{Q}_o^\beta .$$

Let $g: \omega \longleftrightarrow \kappa_o^+$ and set:

$$X_i^h = \text{the smallest } X \prec \langle H_{\kappa_{i+1}}, B \cap \kappa_{i+1} \rangle \text{ s.t.}$$

$$\kappa_i \cup \{\kappa_i\} \cup \pi_{o,i+1}{}''g''h \subseteq X \quad \text{for } h < \omega \text{ , } i < \infty .$$

Then $\bigcup_h X_i^h = H_{\kappa_{i+1}}$.

Set:

$$\Delta_i^h = \bigcap \{\Delta \mid \Delta \in X_i^h \wedge \Delta \text{ is strongly dense in } \mathbf{Q}_{\kappa_i}^{\kappa_{i+1}}\} .$$

Then Δ_i^h is strongly dense in $\mathbf{Q}_{\kappa_i}^{\kappa_{i+1}}$ since the latter is (κ_i,∞)-distributive and $\overline{\overline{X_i^h}} = \kappa_i$ in M .

Define $p_{h,i} \in \mathbb{Q}_{\kappa_i}^{\kappa_{i+1}}$ by:

$$p_{o,i} = \emptyset \text{ ; } p_{h+1,i} = \text{the } L_{\kappa_{i+1}}[B]\text{-least } p \le p_{h,i} \text{ s.t. } p \in \Delta_i^h .$$

Then $\pi_{ij}(p_{h,i}) = p_{h,j}$ $(h < \omega,\ i \le j < \infty)$.

Set:

$$G_h' = \text{the class of } p_{h,i_1} \cup \ldots \cup p_{h,i_n} \quad (i_1 < \ldots < i_n < \infty) .$$

<u>Claim 1</u> Let Δ be an $\langle M,B\rangle$-definable dense class in $\mathbf{Q}_{\kappa_o}$. Then $\exists h \exists q \in G_h'$ s.t. $q \in \Delta$.

Proof Note first that $\mathbf{Q}_{\kappa_o}$ satisfies "∞ - AC" i.e. the set chain condition: that any class of mutually incompatible conditions is in fact a set. (Since any $p \in \mathbf{Q}_{\kappa_o}$ is in some $\mathbb{Q}_{\kappa_0}^{\alpha}$ which satisfies α^+-AC). Thus we may just prove:

<u>Claim 1.1</u> Let $\Delta \in M$ be predense in $\mathbf{Q}_{\kappa_0}$. Then

$\exists h \exists q \in G'_h$ s.t. q meets Δ.

To see that this is sufficient let Δ be any class as in the statement of Claim 1 and Δ' any predense set in M. Setting $\Gamma = \{p \mid p \text{ meets } \Delta'\}$; and if $\Gamma \not\supseteq \Delta$, then the members of $\Delta \backslash \Gamma$ are incompatible with those of Δ' and hence taking a maximal incompatible collection from $\Delta \backslash \Gamma$, which by the above remark is a set, we may add these to Δ' to obtain another predense set in M. Then showing Claim 1.1 will do. So suppose Claim 1.1 is false. Let Δ be a counterexample with $\Delta \subseteq \mathbb{Q}_{\kappa_0}^{\kappa_i}$ with i chosen minimally. W.l.o.g. we may suppose Δ is a maximal antichain in $\mathbf{Q}_{\kappa_0}$ which implies that it is one in $\mathbf{Q}_{\kappa_0}^{\kappa_i}$. Note that since U is an M-ultrafilter $M \models$ "κ_0 is weakly compact" and similarly for the other κ_i. It is easily seen that there exist a $\nu < \kappa_i$ such that Δ is a maximal antichain in $\mathbf{Q}_{\kappa_0}^{\nu}$ (see A.1) i.e. $\mathbf{Q}_{\kappa_0}^{\kappa_i}$ satisfies the κ_i-AC. Thus $\Delta \in H_{\kappa_i}$ and i must be $j+1$ for some j, for otherwise $\Delta \subseteq \mathbf{Q}_{\kappa_0}^{\kappa_n}$ for some $n < i$ contradicting the minimality of i.

Set:

$$\Delta^* = \{p \in \mathbf{Q}_{\kappa_j}^{\kappa_{j+1}} \mid p \text{ meets } \Delta \text{ or } \Delta^p = \{q \in \mathbf{Q}_{\kappa_0}^{\kappa_j} \mid q \cup p \text{ meets } \Delta\} \text{ is dense in } \mathbf{Q}_{\kappa_0}^{\kappa_j}\}.$$

Then Δ^* is dense in $\mathbb{Q}_{\kappa_j}^{\kappa_{j+1}}$. Hence there is $h < \omega$ s.t. $p_{h,j} \in \Delta^*$. But $p_{h,j}$ does not meet Δ, since $p_{h,j} \in G'_h$. Hence $\Delta^{p_{h,j}}$ is dense in $\mathbf{Q}_{\kappa_0}^{\kappa_j}$ and thus $\Delta^{p_{h,j}}$ is predense in $\mathbb{Q}_{\kappa_o}$. So, by the minimality of i again, there is an $r \in G'_k$ $(k \geq h)$ s.t. r meets $\Delta^{p_{h,j}}$. And therefore $r \upharpoonright \kappa_j$ meets $\Delta^{p_{h,j}}$, so we may assume $r \in \mathbb{Q}_{\kappa_o}^{\kappa_j}$. But then finally $r \cup p_{k,j} \in G'_k$ and $r \cup p_{k,j}$ meets Δ - a contradiction.

QED (Claim 1)

Let $A' = \{\xi \mid \exists h \exists j ,\ p_{h,j}(\xi) = 1\}$. By the result just proven A' is $\mathbb{Q}_{\kappa_o}$-generic over M . But κ_o^{+M} is countable and so there is a $G \subseteq \mathbb{Q}^{\kappa_o}$ which is $\mathbb{Q}^{\kappa_o}$-generic over M . Set:

$$A'' = \{\xi \mid \exists r \in G ,\ r(\xi) = 1\}$$

$$A = A'' \cup A' .$$

Then A is $\mathbb{Q}$-generic over M . Set:

$$N = \langle M[A],A\rangle .$$

<u>Claim 2</u> $N \models \phi(\vec{\xi}) \longleftrightarrow N \models \phi(\pi_{ij}(\vec{\xi}))$ for $\vec{\xi} \in [On]^{<\omega}$ and ϕ a Σ_1 formula.

Proof Let $N \models \phi(\vec{\xi})$. So $\exists r \in G \exists p \in G'_h$ s.t. $r \cup p \Vdash \phi(\vec{\xi})$. But $\pi_{ij} \upharpoonright \mathbb{Q}^{\kappa_o} = id \upharpoonright \mathbb{Q}^{\kappa_o}$ and $\pi_{ij}{}''G'_h \subset G'_h$. Hence $\pi_{ij}(p) \in G'_h$ and thus $r \cup \pi_{ij}(p) \Vdash \phi(\pi_{ij}(\vec{\xi}))$. Hence $N \models \phi(\pi_{ij}(\vec{\xi}))$.

QED (Claim 2)

But now we can extend π_{ij} to a $\tilde{\pi}_{ij}: N \xrightarrow[\Sigma_1]{} N$ by setting

$$\tilde{\pi}_{ij} f(\vec{\xi}) = f(\pi_{ij}(\vec{\xi}))$$

for $f: On^n \longrightarrow N$, $f \in \Sigma_1(N)$. This extension must be unique since any such must satisfy this equation.

QED (Lemma 9.3)

So let A be as in Lemma 9.3. Define a $V_i \subseteq P(\kappa_i) \cap M[A]$ by:

(1) $X \in V_i \longleftrightarrow \kappa_i \in \tilde{\pi}_{ij}(X) \quad (i < j < \infty)$.

(2) $M[A]$ is the $\Sigma_0(M[A])$-closure of $\mathrm{ran}(\tilde{\pi}_{ij}) \cup \{\kappa_\ell \mid i \leq \ell < j\}$ $(i \leq j < \infty)$.

But these two together yield:

(3) a) $\langle H^{M[A]}_{\kappa_i^+}, V_i \rangle$ is amenable.

b) V_i is normal on κ_i in $\langle H^{M[A]}_{\kappa_i^+}, V_i \rangle$.

c) $\langle M[A], A, V_0, \kappa_0 \rangle$ is iterable, yielding $\langle M[A], A, V_i, \kappa_i \rangle$ with iteration maps $\tilde{\pi}_{ij}: \langle M[A], A \rangle \xrightarrow[\Sigma_1]{} \langle M[A], A \rangle \quad (i \leq j < \infty)$.

So we may now assume:

(**) $M = L[A]$ where $A \subseteq On$ and

(I) (*) (b)-(d) of P64 hold.

(II) $\pi_{ij}: \langle M, A \rangle \xrightarrow[\Sigma_1]{} \langle M, A \rangle$.

Since if (**) does not already hold we may make it do so by invoking Lemma 9.3.

As before we shall initially assume that either $0^{\#} \notin M$ or M is closed under "#" choosing our conditions accordingly - later we indicate how to weld together the two processes as in the case of Theorem 0.1. It suffices to show:

Lemma 9.4 *There is* $D \subset [\omega,\infty)$ *s.t.*

a) $D \cap \alpha^+$ *is* $\mathbf{P}_\omega^{\alpha^+}$*-generic over* $\mathfrak{A}'_{\alpha^+}$ *for* $\alpha \in \mathrm{Card} \cap [\omega,\infty)$ [*cf. "pseudogeneric" of* §4.4].

Set: $L^D = \langle L[D], D\rangle$

b) $L^D \models \phi(\vec{\xi}) \longleftrightarrow L^D \models \phi(\pi_{ij}(\vec{\xi}))$ *for* $\xi \in [\mathrm{On}]^{<\omega}$ *and* Σ_1 *formulae* ϕ.

Before proving this we make some remarks and show how this yields the result. Firstly note that in a) above $\mathfrak{A}'_{\alpha^+}$ is to be read as $\mathfrak{A}^1_{\alpha^+}$ or $\mathfrak{A}'_{\alpha^+}$ depending on whether $0^\# \notin M$ or M is closed under "#" respectively. Similarly μ'_{α^+}.

Again cardinal properties are preserved in the extension to L^D since for all $\alpha \in \mathrm{Card} \cap [\omega,\infty)$ we have that the properties of τ are preserved in $\mathfrak{A}'_{\alpha^+}[D \cap \alpha^+]$ for $\tau \in \mathrm{Card} \cap \alpha^+$. Moreover for $\alpha \geq \omega$, $\mathfrak{A}'_{\alpha^+}[D \cap \alpha^+] = L_{\mu'_{\alpha^+}}[D_\omega]$, thus $L[D] = L[D_\omega]$. Also $D_\omega \in \mathbf{S}^+_\omega$. Since ω^+ is countable, there is an $a \subseteq \omega$ which is $\mathbb{P}^{D_\omega}$-generic over $L[D]$ (see 4.17). Then $L[a \cup D] = L[a]$. $\mathbb{P}^{D_\omega}$ has cardinality ω_1, and hence is a small set of conditions compared to the κ_i, satisfying the ω_1-AC in $L[D]$. Hence we continue to preserve the requisite properties and (b) yields:

(b') $L^a \models \phi(\vec{\xi}) \longleftrightarrow L^a \models \phi(\pi_{ij}(\vec{\xi}))$ for $\vec{\xi} \in [\mathrm{On}]^{<\omega}$ and Σ_1 formula ϕ.

The unique extendability of π_{ij} to a $\tilde{\pi}_{ij}: L^a \xrightarrow[\Sigma_1]{} L^a$ follows exactly as in Claim 2 of the last lemma.

Proof of Lemma 9.4 Let $g: \omega \longleftrightarrow \kappa^+$. Define g_j $(j < \infty)$ by: $g_j(u) = \pi_{oj}(g(u))$. So $g_j: \omega \longrightarrow \kappa_j^+$.

We shall define a sequence p^{hi}, s_{hi} for $h < \omega$, $i < \infty$ such that:

(a) $s_{hi} \in \mathbf{S}_{\kappa_i^+}$.

(b) $p^{hi} \in \mathbf{P}_\omega^{s_{hi}}$ is smooth.

(c) $p^{hi}_{\kappa_i^+} = s_{hi}$ for $i < j < \infty$.

(d) $\pi_{ij}(p^{hi}) = p^{hj}$ for $i \le j < \infty$.

(d) yields:

(e) $p^{hi} \restriction \kappa_i = p^{hj} \restriction \kappa_i$ for $i \le j < \infty$.

$\pi_{ij}(p^{hi}_{\kappa_i}) = p^{hj}_{\kappa_j}$ so clearly $\pi_{ij}(\mathfrak{A}_{p^{hi}_{\kappa_i}}) = \mathfrak{A}_{p^{hj}_{\kappa_j}}$.

Set $\pi' = \pi_{ij} \restriction \mathfrak{A}_{p^{hi}_{\kappa_i}}$, then

$$\pi' : \mathfrak{A}_{p^{hi}_{\kappa_i}} \xrightarrow[\Sigma_\omega]{} \mathfrak{A}_{p^{hj}_{\kappa_j}}$$

$\pi' \restriction \kappa_i = \mathrm{id} \restriction \kappa_i$; $\pi'(\kappa_i) = \kappa_j$.

Hence $\pi' = \pi_{\kappa_i}^{\mathfrak{A}_{p^{hj}_{\kappa_j}}}$ and an easy application of the collapsing lemma gives:

(f) $p^{hi}_{\kappa_i} = p^{hj}_{\kappa_i}$; $\dot{p}^{hj}_{\kappa_i} = \emptyset$ for $i < j < \infty$.

$\langle p^{hi} \mid i < \infty \rangle$, $\langle s_{hi} \mid i < \infty \rangle$ are defined by induction on h , in a similar fashion to §4.4. We define simultaneously

$\langle X^\gamma_{hi} \mid i < \infty , \gamma \in \mathrm{Card} \cap [\omega, \kappa_i^+) \rangle$ and

$\langle f_{hi} \mid i < \infty \rangle$ s.t. $\overline{\overline{X^\gamma_{hi}}} = \gamma$, $X^\gamma_{hi} \prec \mathfrak{A}'_{s_{hi}}$,

and $f_{hi} \in \mathbf{F}(p^{hi})$, as follows:

(1) $s_{hi} = p^{hj}_{\kappa_i^+}$ $(i < j < \infty)$.

(2) X^γ_{hi} = the smallest X such that $X \prec \mathfrak{A}'_{s_{hi}}$ with $\gamma \cup g_i{}''h \subseteq X$; $X^\nu_{ki} \in X$ for $\nu \leq \kappa_i$, $k < h$; $p_{ki} \in X$ for $k \leq h$.

(3) $f_{hi}(\gamma) = X^\gamma_{hi} \cap \mathfrak{A}'^{\,hi}_{p_\gamma}$ if $\gamma < \kappa_i$ and $\gamma^+ \in X^\gamma_{hi}$;

$= X^\gamma_{hi}$ if $\gamma = \kappa_i$; otherwise undefined.

(4) p^{oi} = the constant function $\langle \emptyset, \emptyset \rangle$ on $\mathrm{Card} \cap [\omega, \kappa_i^+)$.

(5) $p^{h+1,i}$ = the $\mathfrak{A}'_{s_{hi}}$ -least smooth $p \in \mathbb{P}^{s_{hi}}_\omega$ such that

$p \leq p^{hi}$ and $p \in \sum^{s_{hi}, p^{hi}}_{f_{hi}}$.

Claim 1 Given $\gamma \in \mathrm{Card}\setminus\omega$, let i be the least i s.t. $\gamma \leq \kappa_i$, then

$$H_{\kappa_i^+} \subseteq \bigcup_{h<\omega} X^\gamma_{hi}$$

Proof Set $X = H_{\kappa_i^+} \cap \bigcup_{h<\omega} X^\gamma_{hi}$.

Then $\pi_{oi}{}''\kappa_o^+ = \mathrm{ran}(g_i) \subseteq X$ and $\{\kappa_j \mid j < i\} \subset \gamma \subset X$. But $X \prec \langle H_{\kappa_i^+}, A \cap \kappa_i^+ \rangle$ and every $y \in H_{\kappa_i^+}$ is $\langle H_{\kappa_i^+}, A \cap \kappa_i^+ \rangle$ - definable from parameters in $\pi_{oi}{}''\kappa_o^+ \cup \{\kappa_j \mid j < i\}$.

QED (Claim 1)

Corollary $\mathfrak{A}'_{p_{\kappa_i^+}} \subseteq \bigcup_{h<\omega} X^\gamma_{hi}$.

Proof Every $y \in \mathfrak{A}'_{p_{\kappa_i^+}}$ is $\sum_1(\mathfrak{A}'_{p_{\kappa_i^+}})$ with parameters from $H_{\kappa_i^+}$.

QED (Corollary)

Definition $p^h = \bigcup_{i<\infty} p^{hi} \upharpoonright \kappa_i$.

Then $p^h \upharpoonright \kappa_i = p^{hi} \upharpoonright \kappa_i$ and $p^h_{\kappa_i} = p^{hi}_{\kappa_i}$, but $\dot{p}^h_{\kappa_i} = \emptyset$ and $\dot{p}^{hi}_{\kappa_i} \neq \emptyset$ for $h > 0$, by (3). Clearly $s_{hi} = p^h_{\kappa_i^+}$. Still following §4.4 we define:

Definition Let $\alpha \in \mathrm{Card}\setminus\omega$, $h < \omega$.

G^h_α = the set of $q \in \mathbb{P}_\omega^{p^h_{\alpha+}}$ s.t. for some finite $u \subseteq \mathrm{On}$

(i) $i \in u \longrightarrow \kappa_i \leq \alpha$.

(ii) $q(\kappa_i) = p^{hi}(\kappa_i)$ for $i \in u$.

(iii) $q(\gamma) = p^h(\gamma)$ in all other cases.

Then G^h_α is a set of mutually compatible conditions in $\mathbb{P}_\omega^{p^h_{\alpha+}}$. Using Claim 1, we imitate the proof of 4.14 to prove:

<u>Claim 2</u> Let $\Delta \in \mathfrak{A}'_{p^h_{\alpha+}}$ be predense in $\mathbb{P}_\omega^{p^h_{\alpha+}}$. Then $\exists k \geq h$ s.t. $\exists p \in G^k_\alpha$ with p meeting Δ.

As a straightforward corollary we get:

<u>Claim 3</u> Set $D_\gamma = \bigcup_{h<\omega} \tilde{p}^h_\gamma$, $D = \bigcup_{\gamma \in \mathrm{Card}\setminus\omega} D_\gamma$. Then $D \cap \gamma^+$ is $\mathbb{P}_\omega^{\gamma^+}$-generic over $\mathfrak{A}'_{\gamma+}$ for $\gamma \in \mathrm{Card}\setminus\omega$.

It remains only to show:

<u>Claim 4</u> $L^D \models \phi(\vec{\xi}) \longleftrightarrow L^D \models \phi(\pi_{ij}(\vec{\xi}))$ for $\vec{\xi} \in [\mathrm{On}]^{<\omega}$

and Σ_1 formulae ϕ, where

$L^D = \langle L[D], D\rangle$.

Proof It suffices to show

(1) $L^D_\eta \models \phi(\vec{\xi}) \longrightarrow L^D_\eta \models \phi(\pi_{ij}(\vec{\xi}))$ for arbitrarily large η.

Let $\eta = \kappa_\eta$ be such that all of $\vec{\xi}, \kappa_i, \kappa_j$ are less than η. Then $\pi_{ij}(\eta) = \eta$ and $\pi_{ij}(\mathfrak{A}'_{\eta^+}) = \mathfrak{A}'_{\eta^+}$.

Assume $L^D_\eta \models \phi(\vec{\xi})$. Then:

(2) $\mathfrak{A}'_{\eta^+}[D \cap \eta^+] \models$ "$L^D_\eta \models \phi(\vec{\xi})$".

Hence $\exists h < \omega$, $\exists q \in G^h_\eta$ s.t.

(3) $q \Vdash L^{\mathring{D}}_{\check{\eta}} \models \phi(\check{\vec{\xi}})$

forcing over $\mathfrak{A}'_{\eta^+}$ with $\mathbf{P}^{\eta^+}_\omega$. But π_{ij}"$G^h_\eta \subseteq G^h_\eta$,

hence $\pi_{ij}(q) \in G^h_\eta$ and thus

(4) $\pi_{ij}(q) \Vdash L^{\mathring{D}}_{\check{\eta}} \models \phi(\widecheck{\pi_{ij}(\vec{\xi})})$.

Hence

(5) $\mathfrak{A}'_{\eta^+}[D \cap \eta^+] \models$ "$L^D_\eta \models \phi(\pi_{ij}(\vec{\xi}))$"

Hence $L^D_\eta \models \phi(\pi_{ij}(\vec{\xi}))$.

QED (Lemma 9.4)

We now turn to the intermediate cases where M is not closed under $\#$. So then, $\exists \beta \in$ Card such that $\exists b \subseteq \beta$ with $\neg b^\#$ in M. This will be true of a $\beta < \kappa_0$ and we may take it to be the least such. By modifying the construction so far we can get:

Lemma 9.5 *Let* $\alpha \in \text{Card} \cap \kappa_0$ *s.t.* $\exists b \subseteq \alpha^+$ $b\neg^\#$ *in* M.

Then $\exists a \subseteq \alpha^+$ *such that*

(i) $M \subseteq L[a]$.

(ii) $H^{L[a]}_{\alpha^+} = H^M_{\alpha^+}$.

(iii) The cardinals and cofinalities of M *are preserved in* $L[a]$ *, as are the aforementioned large cardinal properties.*

(iv) $L[a] \models \phi(\vec{\xi}) \longleftrightarrow L[a] \models \phi(\pi_{ij}(\vec{\xi}))$ *for* $\vec{\xi} \in [On]^{<\omega}$ *and* Σ_1 *formulae* ϕ .

Consider the same three cases as the final proof of Theorem 8.12.

Case 1 $\beta = \omega = 0^+$. Then apply Lemma 9.5.

Case 2 $\beta = \gamma^+ > \omega$. Let $b \subset \beta$ such that $L[b]$ satisfies (i)-(iv) of Lemma 9.5. Note that β^+ is countable since $\beta^+ < \kappa_0$. We may simply carry out the further generic extensions of the previous "Case 2" getting an $L[a]$ with $a \subseteq \omega$, with $L[a]$ satisfying (i) and (iii) above. But $L[a]$ will also satisfy (iv), since the further extension was by a small set of conditions relative to κ_0 .

Case 3 β is a limit cardinal.

Again let $b \subseteq \beta^+$ s.t. $L[b]$ satisfies (i)-(iv) above. Since β^{++} continues to be countable we can still make the generic extensions constructed in Case 3 before, again getting an $L[a]$, $a \subseteq \omega$ satisfying (i),(iii). (iv) continues to hold for the same reason.

QED

A modification of the above proof yields:

Corollary 9.6 *Let* M *satisfy (a)-(c) of Theorem 9.1. Suppose moreover, that:*

(d)* $\kappa^+ = \bigcup_{i<\omega} Y_i$ *, where* $Y_i \in M$ *and* $\overline{\overline{Y_i}} \leq \tau^+$ *in* M

where $\tau < \kappa$ *. Then* $\exists\, a \subseteq \tau^+$ *s.t. (i)-(iii) of Theorem 9.1 hold, as does* $H^M_{\tau^+} = H^{L[a]}_{\tau^+}$.

Looking now at the proof of Theorem 9.2, we see that Theorem 9.1 can be used to establish this lemma:

Lemma 9.7 *Let* M *be as in Theorem 9.2. Let* $M \models V = L[A,U]$ *with* $A \subseteq \kappa$ *. Then the conclusion of Theorem 9.2 holds.*

Proof $M_i = L[A_i, U_i]$, where $A_i = \pi_{oi}(A)$. Hence $H = L[A^*]$, where $A^* = \bigcup_{i<\infty_M} A_i$. Clearly $A_i = A^* \cap \kappa_i$.

Now generically add (by a set of conditions of course) a function f such that $f\colon \omega \longleftrightarrow \kappa^+$, and argue in $M[f]$: by Theorem 9.1 there exists $a \subseteq \omega$ such that $L[a] = H[a]$, and $L[a]$ has the preservation properties with respect to H ; further $\pi_{ij}\restriction H$ extends (uniquely) to $\tilde{\pi}_{ij}\colon L[a] \longrightarrow L[a]$. Note that as f was adjoined by a set of conditions so can a be, thus establishing the following claim shows that we can reduce to $N = M[a]$.

<u>Claim</u> $M_i[a] = L[a^\#]$.

Proof ($\supseteq$) The iteration of the U_i yields the indiscernibles $\{\kappa_j \mid j \geq i\}$ for $L[a]$. Hence $a^\# \in M_i[a]$.

($\subseteq$) $A_i, U_i \in L[a^\#]$ since $U_i = V_i \cap H$ and $A_i \in H$, where $X \in V_i \longleftrightarrow \kappa_i \in \tilde{\pi}_{ij}(X)$ $(i < j < \infty_M)$. Hence $M_i = L[A_i, U_i] \subseteq L[a^\#]$.

QED

Thus if we bring an arbitrary M into the form $L[A,U]$ with $A \subseteq \kappa$ we shall be done. As a first step we show:

Lemma 9.8 *There is an M-definable, in U, class of conditions $\mathbb{P}$ such that if M^* is a $\mathbb{P}$-generic extension of M, then $M^* = M[D]$ with*

(i) $D \subseteq \kappa^{+M}$

(ii) $M^* \models ZF + V = L[D]$

(iii) $(On^\kappa)_M = (On^\kappa)_{M^*}$

The proof is implicitly that of Theorem 8.12. If, e.g., M is closed under $\#$, we perform the construction of Chapter 8, if not we use the two stage process also described there.

So let $M^* = M[D]$ be as in Lemma 9.8. (iii) implies that $H^{M^*}_{\kappa^+} = H^M_{\kappa^+}$. So U will still be normal on κ in M^*. It also implies that the iterates of $\langle M^*,U,\kappa\rangle$ are well-founded, yielding iteration maps π^*_{ij} between the $\langle M^*_i,U^*_i,\kappa^*_i\rangle$ where $U^*_i = U_i$, $\kappa^*_i = \kappa_i$, $M_i \subseteq M^*_i$, $(On^{\kappa_i})_{M_i} = (On^{\kappa_i})_{M^*_i}$, and the π^*_{ij} are the necessarily unique extensions of the π_{ij}. Set $H^* = \bigcap_{i<\infty_M} M^*_i$. Then $H^* = \bigcup_i H^{M^*}_{\kappa_i} = \bigcup_i H^M_{\kappa_i} = H$. Thus it suffices to prove Theorem 9.2 with M^* in place of M. So, we see that we may assume from now on that $M \models V = L[D]$ for a $D \subseteq \kappa^+$. Assuming this we have:

Lemma 9.9 *There is a definable set of conditions $\mathbb{P}$ such that if M^* is a $\mathbb{P}$-generic extension, then $M^* = M[A]$, where $A \subseteq \kappa$ and:*

(i) $M^* \models V = L[A,U]$.

(ii) *The cardinals and cofinalities of* H_κ^M *are preserved in* $H_\kappa^{M^*}$*, as are the aforementioned large cardinal properties.*

(iii) *If* $f: \kappa \longrightarrow On$ *in* M^**, then there is* $X \in U$ *such that* $f \restriction X \in M$.

Before giving the proof we shall show how this yields Theorem 9.2. (iii) easily gives that

(1) In M^* if we define U^* by

$$U^* = \{Y \subseteq \kappa \mid \exists Z \in U, Y \supseteq Z\},$$

then U^* is normal on κ in M^*.

Similarly

(2) $\forall f \in M^\kappa \, \exists g \in M \;\; f = g \bmod(U^*)$ in M^*.

(1) and (2) together imply that if we let $\langle M_i^*, U_i^*, \kappa_i^* \rangle$ be the iterates of $\langle M^*, U^*, \kappa\rangle$, with iteration maps π_{ij}^* then $\kappa_i^* = \kappa_i$, $M_i \subseteq M_i^*$ and π_{ij}^* extends π_{ij}. Let $A_i = \pi_{oi}^*(A)$, $D_i = \pi_{oi}(D)$. Then $H_{\kappa_i}^{M_i^*} = L_{\kappa_i}[A_i]$ will have the preservation properties with respect to $H_{\kappa_i}^{M_i} = L_{\kappa_i}[D_i]$.

Now set $H^* = \bigcap_i M_i^* = \bigcup_i H_{\kappa_i}^{M_i^*}$. The last remark implies that H^* has the preservation properties with respect to $H = \bigcap_i M_i = \bigcup_i H_{\kappa_i}^{M_i}$.

Then apply Lemma 9.7 to M^*. So $H^*[a] = L[a]$, and thus $H[a] = L[a]$, since $H \subseteq H^*$. $L[a]$ has the preservation properties with respect to H^* and hence, in turn, to H. We know (by Lemma 9.7) that $M_i^*[a] = L[a^\#]$.

<u>Claim</u> $M_i[a] = L[a^\#]$.

($\subseteq$) is trivial since $M_i \subseteq M_i^*$.

(⊇) $a^{\#} \in M_i[a]$, since the iteration of M_i yields the indiscernibles for $L[a]$, as in v and vi of Theorem 9.1. $\tilde{\pi}_{ij}$ is an extension of $\pi^*_{ij} \upharpoonright H^*$ and hence a (unique) extension of $\pi_{ij} \upharpoonright H$.

This completes the proof of Theorem 9.2 given Lemma 9.9.

Proof of Lemma 9.9 This will be a modification of an argument of Chapter 1, namely the proof of Theorem 1.5. We are assuming w.l.o.g. that $H_\tau = L_\tau[D]$ for $\tau \le \kappa^+$, $\tau \in \text{Card}$. We may also assume:

$$\xi \in [\kappa,\kappa^+) \longrightarrow \bar{\bar{\xi}}^{L[D \cap \xi]} = \kappa .$$

For if not, we may achieve this with a preliminary forcing as in Theorem 1.4.

Defining $L^U_\mu[D \cap \xi]$ as $\langle L_\mu[D \cap \xi, U], U \cap L_\mu[D \cap \xi, U]\rangle$ we define μ_ξ, $\mathfrak{A}_\xi$ for $\xi \in [\kappa,\kappa^+)$ as before and after Lemma 1.2, but replacing $L_{\mu_\xi}[A,B \cap \xi]$ by $L^U_{\mu_\xi}[D \cap \xi]$ and using in the definition for $\mathfrak{A}_\xi$ $L^U_{\mu_\xi}[D \cap \xi]$. We then proceed to define (for $\tau \le \kappa$) $X_{\xi\tau}$, $\rho_{\xi\tau}$, $\mathbb{P}$, $|p|$ (for $p \in \mathbb{P}$) , $\mathbb{P}_\tau$, $\mathbb{P}^\tau$ ($\tau \in \text{Card} \cap \kappa$) just as before. Similarly if $A \subseteq \kappa$ is $\mathbb{P}$-generic over M , then $A \cap [\tau,\kappa)$ is $\mathbb{P}_\tau$-generic, and $A \cap \tau$ is $\mathbb{P}^\tau$-generic over $M[A \cap [\tau,\kappa)]$. We then prove 1.6 on extendability of conditions and 1.7 on the (τ,∞)-distributivity of $\mathbb{P}_\tau$ (replacing b in the latter proof by $b = \langle L_{\kappa^{++}}[D],D,U\rangle)$. If A is $\mathbb{P}$-generic over M and A' codes A , $D \cap \kappa$, then consider $N = M[A']$. 1.9 is true for N , and $N \models V = L[A',U]$. The preservation of the large cardinal properties follows from the fact that $A' \cap \tau$ is $\mathbb{P}^\tau$-generic over $M[A' \setminus \tau]$ and $H_\tau^{M[A' \setminus \tau]} = H_\tau^M$ for $\tau \in \text{Card} \cap \kappa$.

We are left only with (iii) to prove, and this is done by imitating the construction of 1.7. So let $f: \kappa \longrightarrow \text{On}$, $f \in M^* = M[A']$. Let $\mathring{f}$ be a term in the forcing language for f. We may assume that $\mathring{f} \in H^M_{\kappa^{++}} = L_{\kappa^{++}}[D]$. Let $p \in G_A$, such that $p \Vdash \mathring{f}: \check{\kappa} \longrightarrow On$. Set:

$$\Delta = \{q \mid q \mid p \vee \exists x \in U\ q \Vdash \mathring{f} \restriction \check{X} \in \check{M}\}$$

We need to show that Δ is dense in $\mathbb{P}$. Let p_o be an arbitrary extension of p. We construct a $q \leq p_o$ with $q \in \Delta$. Set:

$$b = \langle L_{\kappa^{++}}[D], D, U, p_o, \mathring{f} \rangle$$

Define $X_i \prec b$ $(i \leq \kappa)$ as follows:

$$X_i = \text{the smallest } X \prec b \text{ s.t. } X_h \in X \text{ for } h < i .$$

(And so $X_\lambda = \bigcup_{i<\lambda} X_i$ for $Lim(\lambda)$.)

Let $\pi_i: b_i \overset{\sim}{\longleftrightarrow} X_i$ with b_i transitive. The

$$b_i = \langle L_{\delta_i}[D \cap \alpha_i], D \cap \alpha_i, U_i, p_o, \mathring{f}_i \rangle$$

where $\kappa^+ \cap X_i = \alpha_i < \delta_i < \kappa^+$. $U_i = U \cap L_{\alpha_i}[D \cap \alpha_i]$.

It is clear that $\pi_i \restriction \alpha_i = id \restriction \alpha_i$ and that $\pi_i(\alpha_i) = \kappa^+$. Thus α_i is a cardinal in $L_{\delta_i}[D \wedge \alpha_i]$ and hence, by virtue of the definition of μ_{α_i}, $\delta_i < \mu_{\alpha_i}$. Also $U_i \in \mathfrak{A}_{\alpha_i}$. Thus $b_i \in \mathfrak{A}_{\alpha_i}$. Now define $p_i, p'_i \in \mathbb{P}$ $(i \leq \kappa)$ as follows:

p_o is given

p'_i = the $L[D]$-least $p' \leq p_i$ such that

$\quad |p'| \geq \alpha_i$ and $\exists \xi\ p' \Vdash \mathring{f}(\check{\omega}_i) = \check{\xi}$

$$p_{i+1} = p_i \cup p'_i \restriction [\omega_i, \kappa)$$

$$p_\lambda = \bigcup_{i<\lambda} p_i \quad \text{if} \quad \bigcup_{i<\lambda} p_i \in \mathbb{P} .$$

As in 1.7 it follows that $p_i \in \mathbb{P} \cap \mathfrak{A}_{\alpha_i}$ and $|p_i| \geq \alpha_h$

for $h < i$. Hence $|p_\lambda| = \alpha_\lambda$ for $\mathrm{Lim}(\lambda)$ and also $\langle p'_h \mid h < i\rangle \in \mathfrak{A}_{\alpha_i}$.

Claim $\exists X \in U \cap \mathfrak{A}_{\alpha_\kappa}$ such that $\forall \alpha,\beta \in X$ if $\alpha \le \beta$. Then $\alpha = \omega_\alpha$, $\beta = \omega_\beta$ and $p'_\alpha \restriction \alpha = p'_\beta \restriction \alpha$.

Proof For $\nu < \kappa$, $m < 2$ set:

$$X^m_\nu = \{\alpha < \kappa \mid \nu < \alpha = \omega_\alpha \wedge p'_\alpha(\nu) = m\}$$

$$X^2_\nu = \{\alpha < \kappa \mid \nu < \alpha = \omega_\alpha \wedge \nu \notin \mathrm{dom}(p'_\alpha)\}$$

Then $X^0_\nu \cup X^1_\nu \cup X^2_\nu = \{\alpha < \kappa \mid \alpha = \omega_\alpha\} \in U$. Hence for each $\nu < \kappa$ $\exists m_\nu$ such that $X^{m_\nu}_\nu \in U$. Let X_ν equal that $X^{m_\nu}_\nu$ and $X = \{\alpha \mid \alpha \in \bigcap_{\nu<\alpha} X_\nu\}$. Then $X \in U$ and it is easily checked that it has the requisite properties, and that from the construction, it is in $\mathfrak{A}_{\alpha_\kappa}$.

QED (Claim)

Let $q = \bigcup_{\alpha\in X} p'_\alpha \restriction \alpha$. Then $q\colon |q| \longrightarrow 2$, $\overline{\overline{\mathrm{dom}(q) \cap \tau^+}} < \tau^+$ and $p_\kappa \subseteq q$, since $p_\kappa \restriction \alpha \subseteq p'_\alpha \restriction \alpha$ for $\alpha \in X$. Hence $p'_\alpha \subseteq (p'_\alpha \restriction \alpha) \cup p_\kappa \subseteq q$ for $\alpha \in X$.

Since $p_\kappa \subseteq q \in \mathfrak{A}_{\alpha_\kappa}$ and $|p_\kappa| = \alpha_\kappa$ we have that $q \in \mathbb{P}$ and $|q| = \alpha_\kappa$.

Since $p'_\alpha \subseteq q$ for $\alpha \in X$, there is a unique ξ_α such that $q \Vdash \mathring{f}(\check{\alpha}) = \check{\xi}_\alpha$. Set $g = \langle \xi_\alpha \mid \alpha \in X\rangle$. Then $g \in M$ and

$$q \Vdash \mathring{f} \restriction \check{X} = \check{g} .$$

Hence $q \in \Delta$.

QED

This completes the proof of Theorem 9.2.

Appendix

This appendix is independent of the rest of the book and contains properties of a model obtained by adding a class generic G which adds a Cohen generic to every successor cardinal. The results of the appendix are used in §4.2 which contains the definitions and elementary properties of the large cardinal properties used here. We shall prove Facts 4.2.13-15. Though much of the material presented here (if not all) is known it has never appeared in print.

We start with a model $M \models ZF + GCH$. Let $\mathbf{Q}$ be the set of conditions s.t.

$p \in \mathbf{Q}$ iff $p: \text{dom}(p) \longrightarrow 2$ where $\text{dom}(p) \cap [\omega,\infty)$ and

$\text{card}(\text{dom}(p \upharpoonright \kappa)) < \kappa$ for all regular cardinals κ.

We shall use the following notation:

For $p \in \mathbf{Q}$, $(p)^{\alpha} = p\upharpoonright\alpha$ and $(p)_{\alpha} = p\upharpoonright[\alpha,\infty)$.

$\mathbf{Q}^{\alpha} = \{p \in \mathbf{Q} \mid p = (p)^{\alpha}\}$, $\mathbf{Q}_{\alpha} = \{p \in \mathbf{Q} \mid (p)_{\alpha} = p\}$ and

$\mathbf{Q}^{\alpha}_{\tau} = \mathbf{Q}^{\alpha} \cap \mathbf{Q}_{\tau}$.

Hence $\mathbf{Q}^{\alpha^+}_{\alpha}$ are the standard Cohen conditions for adding a subset to $[\alpha,\alpha^+)$.

$\text{card} = \{\alpha \mid \alpha$ is an infinite cardinal$\}$.

For $\alpha \in \text{card}$, $\mathbf{Q}_{\alpha}$ is α-closed. $\text{card}(\mathbf{Q}^{\alpha}) = \alpha$ so in general $\mathbf{Q}^{\alpha}$ is α^+-AC. But for Mahlo cardinals, κ, $\mathbf{Q}^{\kappa}$ is κ-AC.

Lemma A.1 *Let* κ *be Mahlo, then* $\mathbf{Q}^{\kappa}$ *is* κ-AC.

Proof Let $\Delta \subseteq \mathbf{Q}^{\kappa}$ be a set of mutually incompatible conditions. By Mahloness there is a regular $\bar{\kappa} < \kappa$ s.t. $\Delta_{\bar{\kappa}} = \Delta \cap \mathbf{Q}^{\bar{\kappa}}$ is a set of mutually incompatible conditions. We claim $\Delta_{\bar{\kappa}} = \Delta$. If $q \in \Delta \setminus \Delta_{\bar{\kappa}}$ then $p = q/\bar{\kappa}$ must be mutually incompatible with all of $\Delta_{\bar{\kappa}}$ contradicting the maximality of $\Delta_{\bar{\kappa}}$.

Q.E.D.

Set $N = M[A]$, where A is $\mathbf{Q}$-generic over M. Set $A^{\alpha} = A \cap \alpha$ and $A_{\alpha} = A \cap [\alpha, \infty)$. Clearly $N \models ZF + GCH$ in which cardinals and cofinalities are preserved and universal choice holds in N (Easton [E]). By the comments in §2.1 we can assume w.l.o.g. that $H_{\alpha} = L_{\alpha}[A]$ for all infinite cardinals α (Levy [L]).

In general we will have to show that some property holds in N and many times it will be sufficient to show this for $N' = M[A^{\kappa}]$ since A_{κ} is κ-closed and any $f \in N'[A_{\kappa}] = N$ s.t. $f: \kappa \longrightarrow M$ is already in N' $(f \in N')$.

Lemma A.2 Let κ be Mahlo. Let $C \in N$ be cub in κ. Then $\exists D \in M$ $(D \subseteq C \wedge D$ is cub in $\kappa)$.

Proof $C \in M[A^{\kappa}]$. Let f be the monotone enumeration of C, $f \in M[A^{\kappa}]$. Let $p_0 \in \mathbf{Q}$ s.t.

$$p_0 \Vdash \text{" } \mathring{C} \text{ is cub in } \check{\kappa} \text{ and } \mathring{f} \text{ enumerates } \mathring{C} \text{ ".}$$

For $\tau < \kappa$ set $\Delta_{\tau} = \{q \in \mathbf{Q}^{\kappa} \mid \exists \eta (q \Vdash \mathring{f}(\check{\tau}) = \check{\eta}\}$.

Δ_{τ} is dense in $\mathbf{Q}^{\kappa}$ for $\tau < \kappa$. Let $\Delta^{*}_{\tau} \subseteq \Delta_{\tau}$ be a set of mutually incompatible conditions. Then $\forall \tau (\text{card}(\Delta^{*}_{\tau}) < \kappa)$. For $r \in \Delta_{\tau}$ set:

$$\eta_r = \text{that unique } \eta \text{ s.t. } r \Vdash \mathring{f}(\check{\tau}) = \check{\eta} .$$

Clearly $\{\eta_r \mid r \in \Delta_{\tau}\} = \{\eta_r \mid r \in \Delta^{*}_{\tau}\}$. For $\tau < \kappa$ set:

$$\eta_\tau = \sup_{r\in\Delta^*_\tau} (\eta_r) < \kappa .$$

Let $D = \{\alpha < \kappa \mid \forall \tau < \alpha(\eta_\tau < \alpha)\}$. D is cub in κ and $D \in M$.

$\forall \tau < \kappa \; (p_o \Vdash \mathring{f}(\check{\tau}) \leq \check{\eta}_\tau)$ so,

$\forall \alpha \in D \; (p_o \Vdash \mathring{C} \cap \check{\alpha}$ is unbounded in $\check{\alpha})$.

Hence $p_o \Vdash \check{D} \subseteq \mathring{C}$.

Q.E.D.

Corollary A.2.1 *Let* κ *be Mahlo. Let* $X \in M$ *be s.t.* $M \models$ "X is stationary". *Then* $N \models$ "X is stationary".

Proof Let $C \in N$ be cub in κ. Take $D \subseteq C$, $D \in M$ and D cub in κ. Then $D \cap X \neq 0 \longrightarrow C \cap X \neq 0$.

Corollary A.2.2 $M \models$ "κ is Mahlo" $\longrightarrow N \models$ "κ is Mahlo".

Lemma A.3 *Let* $X \subseteq \kappa$. $M \models$ "X is subtle" $\longrightarrow N \models$ "X is subtle".

Proof $\{\tau \in X \mid \tau$ is Mahlo$\}$ is subtle so we may assume w.l.o.g. that $\forall \alpha \in X\,(\alpha$ is Mahlo$)$. We shall use the following equivalent form of subtleness (Fact 4.2.6):

$X \subseteq \kappa$ is subtle iff for every sequence $\langle B_\alpha \mid \alpha \in X\rangle$ s.t. $B_\alpha \subseteq \alpha$ there are arbitrarily large α s.t. $B_\alpha = \alpha \cap B_\beta$ for arbitrarily large β.

It is sufficient to show that X is subtle in $N' = M[A^\kappa]$. Let $\langle B_\alpha \mid \alpha \in X\rangle \in N'$ s.t. $B_\alpha \subseteq \alpha$. Let $p_o \in \mathbb{Q}^\kappa$ s.t. $p_o \Vdash \forall \alpha \in \check{X}(\mathring{B}_\alpha \subseteq \check{\alpha})$ and $p_o \in G$, where G is the $\mathbb{Q}^\kappa$ generic that adds A^κ.

Let $\Delta^\alpha_\nu = \{r \in \mathbb{Q}^\kappa \mid r \,||\, \check{\nu} \in \mathring{B}_\alpha\}$ where,

$r \parallel \check{\nu} \in \mathring{B}_\alpha$ iff $r \Vdash \check{\nu} \in \mathring{B}_\alpha$ or $r \Vdash \check{\nu} \notin \mathring{B}_\alpha$.

$\forall\alpha\forall\nu < \alpha$ Δ_ν^α is dense in $\mathbf{Q}^\kappa$. By standard methods it can be shown (since $\mathbf{Q}_\alpha$ is α-closed) that

$H_\alpha = \{q \in \mathbf{Q}_\alpha^\kappa \mid \forall\nu < \alpha\ (\Delta_\nu^\alpha)^q$ is dense in $\mathbf{Q}^\alpha\}$ is dense in $\mathbf{Q}_\alpha^\kappa$, where $(\Delta_\nu^\alpha)^q = \{r \in \mathbf{Q}^\alpha \mid r \cup q \in \Delta_\nu^\alpha\}$.

For every $\alpha \in \kappa$ s.t. $\alpha > \sup(\mathrm{dom}(p_o))$ choose a $q_\alpha \in \mathbf{Q}_\alpha^\kappa$ s.t. $q_\alpha \in H_\alpha$. For every $\sup(\mathrm{dom}(p_o)) < \alpha < \kappa$ let $\Delta_\nu^{q_\alpha} \subseteq (\Delta_\nu^\alpha)^{q_\alpha}$ be a set of mutually incompatible conditions. Then by Lemma A.1 $\mathrm{card}(\Delta_\nu^{q_\alpha}) < \alpha$ since α is Mahlo. Partition $\Delta_\nu^{q_\alpha} = \Delta_{\nu_o}^{q_\alpha} \cup \Delta_{\nu_1}^{q_\alpha}$ where

$r \in \Delta_{\nu_o}^{q_\alpha} \longrightarrow r \Vdash \check{\nu} \notin \mathring{B}_\alpha$ and

$r \in \Delta_{\nu_1}^{q_\alpha} \longrightarrow r \Vdash \check{\nu} \in \mathring{B}_\alpha$.

We can code $\langle\langle \Delta_{\nu_o}^{q_\alpha}, \Delta_{\nu_1}^{q_\alpha} \rangle \mid \nu < \alpha\rangle$ by a $D_\alpha \subseteq \alpha, D_\alpha \in M$ for $\alpha \in X \setminus \sup(\mathrm{dom}(p_o))$. By the subtleness of X in M there are arbitrarily large α s.t. $D_\alpha = \alpha \cap D_\beta$ for arbitrarily large β . Set $E = \{q_\alpha \mid D_\alpha = \alpha \cap D_\beta$ for arbitrarily large $\beta\}$. E is predense in $\mathbf{Q}^\kappa$ since for $p \in \mathbf{Q}^\kappa$, $p \cup q_\alpha \leq q_\alpha \in E$ for $\alpha > \sup(\mathrm{dom}(p))$. So for arbitrarily large α, $q_\alpha \in G$ s.t. $D_\alpha = \alpha \cap D_\beta$ for arbitrarily large β .

<u>Claim</u> Let α be s.t. $D_\alpha = \alpha \cap D_\beta$ for arbitrarily large β and $q_\alpha \in G$. Then $B_\alpha = \alpha \cap B_\beta$ for arbitrarily large β .

Proof Let $\xi < \kappa$. Set:

$F = \{q_\beta \mid \beta > \xi$ and $D_\alpha = \alpha \cap D_\beta\}$. F is predense in $\mathbf{Q}^\kappa$.

Hence $\exists\beta > \xi\ (q_\beta \in F \cap G)$. But then $\Delta_{\nu_o}^{q_\alpha} = \mathbb{Q}^\alpha \cap \Delta_{\nu_o}^{q_\beta}$ and $\Delta_{\nu_1}^{q_\alpha} = \mathbb{Q}^\alpha \cap \Delta_{\nu_1}^{q_\beta}$ for $\nu < \alpha$.

$$\nu \in B_\alpha \longleftrightarrow r \in \Delta_{\nu_1}^{q_\alpha} \cap G(r \cup q_\alpha \Vdash \check{\nu} \in \mathring{B}_\alpha) \longrightarrow$$

$$\longrightarrow \exists r \in \Delta_{\nu_1}^{q_\alpha} \cap G(r \cup q_\beta \Vdash \check{\nu} \in \mathring{B}_\beta) \longrightarrow \nu \in B_\beta .$$

In a similar manner $\nu \notin B_\alpha \longrightarrow \nu \notin B_\beta$.

Q.E.D.

Corollary A.3.1 *Subtle cardinals are preserved in* N .

Fact A.4 Let κ be regular. Let ϕ be a Π_n formula. There is a Π_n^1 formula $\tilde{\phi}$ s.t. for $B \subseteq H_\kappa$: $H_{\kappa^+} \models \phi(B)$ iff $H_\kappa \models \tilde{\phi}(B)$ and the map $\phi \longrightarrow \tilde{\phi}$ is uniform for all regular κ .

Lemma A.5 *Let* κ *be (subtly)* Π_n^1 *indescribable* $(1 \leq n < \omega)$. *Let* $X \in H_{\kappa^+}$ *and let* $H_{\kappa^+} \models \phi(X)$ *, where* ϕ *is* Π_n . *Let* $X \in U$ *, where* U *is transitive,* $\mathrm{card}(U) = \kappa$. *Let* $f: \kappa \longrightarrow U$ *be surjective. Set*

$$D = \{\alpha \in \mathrm{card} \cap \kappa \mid f''\alpha \prec U \wedge \kappa \cap f''\alpha = \alpha \wedge X \in f''\alpha\} .$$

For $\alpha \in D$ *set* $U_\alpha = f''\alpha$, $\pi_\alpha: \langle \bar{U}_\alpha, \in \rangle \cong \langle U_\alpha, \in \rangle$ *, where* $\bar{U}_\alpha$ *is transitive and* $X_\alpha = \pi_\alpha^{-1}(X)$.
Then $\{\alpha \in D \mid H_{\alpha^+} \models \phi(X_\alpha)\}$ *is stationary (subtle) in* κ .

Proof Let $f^{-1}(X) = \nu \in \kappa$ set:

$$E = \{\langle f^{-1}(Y), f^{-1}(Z)\rangle \mid Y, Z \in U, Y \in Z\}$$

and let

$$\psi(E,\nu) \equiv \forall U \forall \pi (\pi: \langle \kappa, E\rangle \cong \langle U, \in\rangle \wedge U \text{ is transitive} \longrightarrow \phi(\pi(\nu))).$$

Then $\psi(E,\nu)$ is Π_n , $E \subseteq \kappa \times \kappa$ (E can be coded $\subseteq \kappa$) and by Fact A.4 there is a Π_n^1, $\tilde{\psi}$ s.t.

$$H_{\kappa^+} \models \psi(E,\nu) \text{ iff } H_\kappa \models \tilde{\psi}(E,\nu) .$$

By (subtly) Π_n^1 indescribability $\{\alpha < \kappa \mid H_\alpha \models \tilde{\psi}(E \cap \alpha^2, \nu)\}$

is stationary (subtle) in κ . But D is cub in κ and hence $B = \{\alpha \in D \mid H_\alpha \models \tilde{\psi}(E \cap \alpha^2, \nu)$ and α is regular$\}$ is stationary (subtle) in κ .

By Fact A.4 the map $\phi \longmapsto \tilde{\phi}$ is uniform for regular cardinals so $B = \{\alpha \in D \mid H_{\alpha^+} \models \psi(E \cap \alpha^2, \nu)\}$. By the definition of $\psi(E,\nu)$, $B = \{\alpha \in D \mid H_{\alpha^+} \models \phi(X_\alpha)\}$.

Q.E.D.

Fact A.6 Σ_o forcing with $\mathbf{Q}^\kappa$ is uniformly Δ_1 over any transitive U , $U \models ZF^-$ and $\mathbf{Q}^\kappa \in U$.

Lemma A.7 *(Subtly)* Π^1_n *indescribable cardinals are preserved for* $1 \leq n < \omega$.

Proof Let κ be (subtly) Π^1_n indescribable in M . It is sufficient to show that κ is preserved in $N' = M[A^\kappa]$. Let $B \subseteq \kappa$, $B \in N'$ s.t. $N' \models$ "$H_\kappa = \phi(B)$" , where ϕ is Π^1_n . We may assume w.l.o.g. that $\mathring{B} \in H_{\kappa^+}$ in M . Let $p \in G(p \Vdash \mathring{B} \subseteq \check{\kappa} \wedge \phi(\mathring{B}))$.

<u>Claim 1</u> There is a Π_n formula ψ s.t.

$$M \models \text{"}H_{\kappa^+} = \psi(\mathbf{Q}^\kappa, H_\kappa, \mathring{B}, p) \text{ iff } p \Vdash_{\mathbf{Q}^\kappa} (\mathring{B} \subseteq \check{\kappa} \wedge \phi(\mathring{B}))\text{"} .$$

Proof We will show ψ for the case where ϕ is Π^1_2 and the general case is analogous. Let $\phi(B) \equiv \forall X \exists Y\ \chi(X,Y,B)$ where χ is first order.

$$\psi(\mathbf{Q}^\kappa, H_\kappa, \mathring{B}, p) \equiv ((p \Vdash_{\mathbf{Q}^\kappa} \mathring{B} \subseteq \check{\kappa}) \wedge \forall \mathring{X} \exists q \leq p\ (q \Vdash \mathring{X} \subseteq \check{\kappa} \longrightarrow$$
$$\longrightarrow \forall q' \leq q\ \exists q'' \leq q'\ \exists \mathring{Y}\ (q'' \Vdash (\mathring{Y} \subseteq \check{\kappa} \wedge \check{H}_\kappa \models \chi(\mathring{X}, \mathring{Y}, \mathring{B})))))) .$$

By Fact A.6 it follows that ψ is Π_n .

Q.E.D. (Claim 1)

Let U be a transitive ZF^- model s.t. $\mathring{B}, \mathbf{Q}^\kappa, H_\kappa \in U$ ($H_\kappa \in U \longrightarrow p \in U$). Let $f: \kappa \longrightarrow U$ be surjective and set:

$$D = \{\alpha \in \mathrm{card} \cap \kappa \mid f''\alpha \prec U,\ \alpha = \kappa \cap f''\alpha \wedge p, \mathring{B}, \mathbf{Q}^\kappa, H_\kappa \in f''\alpha\}.$$

For $\alpha \in D$ let $\pi_\alpha: \langle U_\alpha, \in\rangle \cong \langle f''\alpha, \in\rangle$, where U_α is transitive. Then $\pi_\alpha: U_\alpha \prec U$, $\pi_\alpha(\mathbf{Q}^\alpha) = \mathbf{Q}^\kappa$, $\pi_\alpha(H_\alpha) = H_\kappa$, $\pi_\alpha(p) = p$ and $\pi_\alpha(\alpha) = \kappa$. Set $\mathring{B}_\alpha = \pi_\alpha^{-1}(\mathring{B})$ and $D^* = \{\alpha \in D \mid H_{\kappa^+} \models \psi(\mathbf{Q}^\alpha, H_\alpha, \mathring{B}_\alpha, p)\}$. By Lemma A.5, D^* is stationary (subtle) in κ.

For $\alpha \in D^*$, we have $p \Vdash_{\mathbf{Q}^\alpha} \mathring{B}_\alpha \subseteq \check{\alpha} \wedge \phi(\mathring{B}_\alpha)$, but A^α is $\mathbf{Q}^\alpha$-generic over M. Hence letting B_α be the $M[A^\alpha]$ interpretation of $\mathring{B}_\alpha$, we have:

$$M[A^\alpha] \models B_\alpha \subseteq \alpha \wedge H_\alpha \models \phi(B_\alpha).$$

$P(\alpha) \cap M[A^\alpha] = P(\alpha) \cap N'$ (where P is the power set operation), so $N' \models B_\alpha \subseteq \alpha \wedge H_\alpha \models \phi(B_\alpha)$. Thus it remains to show that $\forall \alpha \in D^* (B_\alpha = \alpha \cap B)$, since then $\forall \alpha \in D^*$ ($N' \models$ "$H_\alpha \models \phi(B \cap \alpha)$") and D^* is stationary (subtle).

<u>Claim 2</u> $B_\alpha = \alpha \cap B$ for $\alpha \in D^*$.

Proof Recalling the uniform definability of forcing (Fact A.4):

$$q \Vdash_{\mathbf{Q}^\alpha} \check{\nu} \in \mathring{B}_\alpha \text{ iff } q \Vdash_{\mathbf{Q}^\kappa} \check{\nu} \in \mathring{B}, \text{ for } q \in \mathbf{Q}^\alpha \text{ and } \nu < \alpha,$$

since $\pi_\alpha: U_\alpha \prec U$. Hence for $\nu < \alpha$,

$$\nu \in B_\alpha \longrightarrow \exists q \in G \cap \mathbf{Q}^\alpha (q \Vdash \check{\nu} \in \mathring{B}_\alpha) \longrightarrow$$
$$\longrightarrow \exists q \in G \cap \mathbf{Q}^\kappa (q \Vdash \check{\nu} \in \mathring{B}) \longrightarrow \nu \in B.$$

Similarly $\nu \notin B_\alpha \longrightarrow \nu \notin B$.

Q.E.D.

Lemma A.8 *Let* $X \subseteq \kappa$ *be stationary (subtle) in* κ *in* M. *Let* $\langle\langle a_\alpha, \delta_\alpha\rangle | \alpha \in X\rangle$ *s.t.* $\delta_\alpha \in (\alpha, \alpha^+)$ *and* $a_\alpha \subseteq \delta_\alpha$ *in* M. *Then* $X^* = \{\alpha \in X \,|\, a_\alpha = A \cap [\alpha, \delta_\alpha)\}$ *is stationary (subtle) in* κ *in* N.

Proof

Case 1 X is stationary.

We assume w.l.o.g. that $X \subseteq \text{card}$. Let $C \in N$ be cub in κ. By Lemma A.1 take $C' \in M$, $C' \subseteq C$ and C' cub in κ. For $\alpha \in X$ set $r_\alpha = \chi_{a_\alpha} \upharpoonright [\alpha, \delta_\alpha)$. $r_\alpha \in \mathbf{Q}_\alpha$ and $\{r_\alpha \,|\, \alpha \in X \cap C'\}$ is predense in $\mathbf{Q}^\kappa$, since $X \cap C'$ is unbounded in κ. Hence $\exists \alpha \in X \cap C'$ s.t. $r_\alpha \in G$ and $A \cap [\alpha, \delta_\alpha) = a_\alpha$.

Q.E.D. (Case 1)

Case 2 X is subtle.

We assume w.l.o.g. that $\alpha \in X \longrightarrow \alpha$ is Mahlo. Define r exactly as in Case 1. We repeat the proof of Lemma A.3 with a slight change. Define H_α as in Lemma A.3. Now define $q_\alpha \in H_\alpha$ and $q_\alpha \leq r_\alpha$. The proof is completed in the same way.

Q.E.D.

Lemma A.9 *Let* κ *be Mahlo in* M. *Let* $X \in N$, $X \subseteq \kappa$ *s.t.* $\forall \alpha < \kappa\ (X \cap \alpha \in M)$. *Then* $X \in M$.

Proof It is sufficient to prove the lemma for $X \in N' = M[A^\kappa]$. Let $p \in G \cap \mathbf{Q}^\kappa$ s.t. $p \Vdash \mathring{X} \subseteq \check{\kappa} \wedge \forall \alpha < \check{\kappa}\ (\mathring{X} \cap \check{\alpha} \in \check{M})$. Set $\Delta = \{q \leq p \,|\, \forall \nu < \kappa (q \,||\, \check{\nu} \in \mathring{X})\}$. Clearly it suffices to show:

Claim $\Delta^* = \{q \in \mathbf{Q}^\kappa \,|\, q/p \vee q \in \Delta\}$ is dense in $\mathbf{Q}^\kappa$.

Proof Assume by negation that,

$$\exists p' \le p\, \forall q \le p'\ \exists \nu < \kappa \sim (q \not\Vdash \check{\nu} \in \mathring{X}) .$$

Let $U \in H_{\kappa^+}$ (in M) s.t. $U \models ZF^-$, U is transitive and H_κ , $\mathbf{Q}^\kappa$, $\mathring{X} \in U$. As in Lemma A.7, Claim 1, there is a first order ϕ s.t.

$$M \models \text{"}U \models \phi(H_\kappa, \mathbf{Q}^\kappa, \mathring{X}, p') \text{ iff } \forall q \le p'\, \exists \nu < \kappa (q \not\Vdash \check{\nu} \in \mathring{X})\text{"}.$$

Let $f: \kappa \longleftrightarrow U$ be a bijection and set:

$$D = \{\alpha \in \text{card} \cap \kappa \mid f''\alpha \prec U \wedge \alpha = \kappa \cap f''\alpha \wedge H_\kappa, \mathbf{Q}^\kappa, \mathring{X}, p' \in f''\alpha\} .$$

D is cub in κ , hence there is a regular $\bar{\kappa} \in D$. Set $\sigma: \langle \bar{U}, \in\rangle \cong \langle f''\bar{\kappa}, \in\rangle$, where $\bar{U}$ is transitive. Then

$\sigma(H_{\bar{\kappa}}) = H_\kappa$, $\sigma(\bar{\kappa}) = \kappa$, $\sigma(\mathbf{Q}^{\bar{\kappa}}) = \mathbf{Q}^\kappa$ and $\sigma(p') = p'$. Set $\mathring{\bar{X}} = \sigma^{-1}(\mathring{X})$, then $\bar{U} \models \phi(H_{\bar{\kappa}}, \mathbf{Q}^{\bar{\kappa}}, \mathring{\bar{X}}, p')$. Hence $\forall q \in \mathbf{Q}^{\bar{\kappa}} (q \le p' \longrightarrow \exists \nu < \kappa (q \not\Vdash \check{\nu} \in \mathring{\bar{X}})$. Let $\bar{X}$ interpret $\mathring{\bar{X}}$ in $M[A^{\bar{\kappa}}]$. It follows that $\bar{X} \notin M$. Exactly as in Lemma A.7, Claim 2, we can show $\bar{X} = \bar{\kappa} \cap X$, hence $\bar{X} \in M$ and we have reached a contradiction.

Q.E.D.

Lemma A.10 *Let* κ *be (subtly)* Π^1_n *indescribable in* N . *Then* κ *is (subtly)* Π^1_n *indescribable in* M .

Proof Set $E_\kappa = \{A \subseteq \kappa \mid \forall \alpha < \kappa (A \cap \alpha \in H^M_\kappa)\}$. $N \models \text{"}E_\kappa = P(\kappa) \cap M\text{"}$ by Lemma A.9. Let $B \in M$, $B \subseteq \kappa$ and $H^M \models \phi(B)$, where ϕ is Π^1_n . Let $\phi_{E_\kappa}(B)$ be the relativisation of ϕ to E_κ . Set,

$\psi(B, H^M_\kappa) \equiv \text{"}\kappa \text{ is Mahlo} \wedge \phi_{E_\kappa}(B)\text{"}$. Then $H^N_\kappa \models \psi(B, H^M_\kappa)$ and ψ is Π^1_n . Set:

$$D = \{\alpha < \kappa \mid H^N_\alpha \models \psi(B \cap \alpha, H^M_\alpha \cap V_\alpha)\} =$$
$$= \{\alpha < \kappa \mid \alpha \text{ is Mahlo in } N \wedge H^N_\alpha \models \phi_{E_\alpha}(B \cap \alpha)\} =$$
$$= \{\alpha < \kappa \mid \alpha \text{ is Mahlo in } M \wedge H^M_\alpha \models \phi(B \cap \alpha)\} .$$

D is stationary in M since it is stationary in N .

Q.E.D.

Lemma A.11 *Let* κ *be regular. Let,*

$\mathscr{L}_\kappa = \{\phi(\mathring{X}_1, \ldots, \mathring{X}_n, \mathring{A}) \mid \phi$ *is* $\Sigma_0, \mathring{X}_1, \ldots, \mathring{X}_n \in H_{\kappa^+}$ *are terms of the forcing language for* $\mathbb{Q}^{\kappa^+}_\kappa$ *and* $\mathring{A}$ *is the canonical predicate for* $A \cap [\kappa, \kappa^+)\}$.

Then there is a relation $\rhd \subseteq \mathbb{Q}^{\kappa^+}_\kappa \times \mathscr{L}_\kappa$ *s.t.*

(a) $p \rhd \phi \longrightarrow p \Vdash_{\mathbb{Q}^{\kappa^+}_\kappa} \phi$.

(b) $\{p \mid p \rhd \phi \vee p \rhd \sim\phi\}$ *is dense in* $\mathbb{Q}^{\kappa^+}_\kappa$.

(c) $\rhd$ *is* $\Delta_1(H_{\kappa^+})$ *uniformly in* κ .

Proof For $p \in \mathbb{Q}^{\kappa^+}_\kappa$ set $a_p = \{\nu \mid p(\nu) = 1\}$. Set:

$C = \{p \in \mathbb{Q}^{\kappa^+}_\kappa \mid \exists \delta \in (\kappa, \kappa^+) \ (\text{dom}(p) = \delta \wedge \langle L_\delta[a_p], a_p \rangle \models ZF^-)\}$.

<u>Claim 1</u> C is dense in $\mathbb{Q}^{\kappa^+}_\kappa$.

Proof Let $p \in \mathbb{Q}^{\kappa^+}_\kappa$ and choose a $\mathbb{Q}^{\kappa^+}_\kappa$ generic $A \subseteq [\kappa, \kappa^+)$ s.t. $a_p \subseteq A$.

Set $D = \{\delta < \kappa^+ \mid \langle L_\delta[A \cap \delta], A \cap \delta \rangle \prec \langle L_{\kappa^+}[A], A \rangle\}$.

D is cub in κ^+, $\delta \in D \longrightarrow \langle L_\delta[A \cap \delta], A \cap \delta \rangle \models ZF^-$ and $\chi_{A \cap \delta} \leq p$ in $\mathbb{Q}^{\kappa^+}_\kappa$.

Q.E.D. (Claim 1)

For $p \in C$ set $U_p = \langle L_\delta[a_p], a_p \rangle$. For $\phi(\vec{\mathring{X}}) \in U_p$ set $p(\phi(\vec{\mathring{X}})) = \phi(p(\vec{\mathring{X}}))$, where $p(\vec{\mathring{X}})$ is the U_p-interpretation of $\vec{\mathring{X}}$. We set:

$$p \rhd \phi \text{ iff } p \in C \wedge \phi \in U_p \wedge U_p \models p(\phi) .$$

Then,

(i) If $p \in C$ and $\phi \in U_p$ then $p \rhd \phi \longleftrightarrow p \Vdash \phi$.

(ii) $\{p \mid p \in C \wedge \phi \in U_p\}$ is dense in $\mathbb{Q}_\kappa^{\kappa^+}$.

The conclusion is immediate.

Q.E.D.

Corollary A.11.1 *Let* ϕ *be* Π_n $(1 \le n < \omega)$. *There is a* Π_n *formula* ϕ^* *s.t. for all* $\mathbb{Q}_\kappa^{\kappa^+}$ *terms* $\mathring{X} \in H_{\kappa^+}$ *we have:*

$$[p \Vdash_{\mathbb{Q}_\kappa^{\kappa^+}} (\langle L_{\kappa^+}[\mathring{A}], \mathring{A} \rangle \models \phi(\mathring{X}))] \longleftrightarrow [H_{\kappa^+} \models \phi^*(p, X)] .$$

Proof As an example take ϕ to be Π_2^1.

$\phi \equiv \forall A \exists B\ \chi(A,B,X)$, where χ is first order. Then set ψ by:

$$\psi \equiv \forall p' \le p \forall \mathring{A} \forall q \le p' \exists q' \le q \exists \mathring{B}(q' \rhd \chi(\mathring{A}, \mathring{B}, \mathring{X})) .$$

Q.E.D.

The following is Fact 4.2.14.

Lemma A.12 *Let* κ *be (subtly)* Π_m^1 *indescribable in* M $(1 \le m < \omega)$. *Let* $\xi < \kappa^+$ *and let* ϕ *be a* Π_m *formula s.t.* $\langle L_{\kappa^+}[A], A \cap \kappa^+ \rangle \models \phi(\xi)$ *in* N. *For* $\alpha \le \kappa$ *set:*

$X_\alpha =$ *the least* $X \prec \langle L_{\kappa^+}[A], A \cap \kappa^+ \rangle$ *s.t.* $\alpha \cup \{\xi\} \subseteq X$,

$\sigma_\alpha : \langle L_{\delta_\alpha}[A'_\alpha], A'_\alpha \rangle \cong X_\alpha$, $\xi_\alpha = \sigma_\alpha^{-1}(\xi)$ *and*

$D = \{\alpha \in \text{card} \cap \kappa \mid \alpha = \kappa \cap X_\alpha \wedge A'_\alpha = A \cap \delta_\alpha \wedge \langle L_{\alpha^+}[A], A \cap \alpha^+ \rangle \models \phi(\xi_\alpha)\}$.

Then D is stationary (subtle) in κ.

Proof Set $N'' = N[A \cap [\kappa,\kappa^+)]$ and $N' = N''[A^\kappa]$. Then N'' is a $\mathbf{Q}_\kappa^{\kappa^+}$-generic extension of M and N' is a $\mathbf{Q}^\kappa$-generic extension of N''. Let $p \in G^\kappa \cap \mathbf{Q}^\kappa$ be s.t. (G^κ is the $\mathbf{Q}^\kappa$-generic that adds A^κ)

(*) $N'' \models [p \Vdash_{\mathbf{Q}^\kappa} (\langle L_{\kappa^+}[\mathring{A}],\mathring{A}\rangle \models \phi(\check{\xi}))]$ (forcing over N'').

Then as in Lemma A.7, Claim 1, there is a Π_m, ψ s.t.

(**) $N'' \models$ "$[\langle H_{\kappa^+}, A \cap [\kappa,\kappa^+)\rangle \models \psi(\mathbf{P},\mathbf{Q}^\kappa,\xi)]$ iff

$[p \Vdash_{\mathbf{Q}^\kappa} \langle L_{\kappa^+}[\mathring{A}],\mathring{A}\rangle \models \phi(\check{\xi})]$".

But then $\exists\, r \in \mathbf{Q}_\kappa^{\kappa^+} \cap G_\kappa^{\kappa^+}$ (where $G_\kappa^{\kappa^+}$ is the $\mathbf{Q}_\kappa^{\kappa^+}$-generic that adds $A \cap [\kappa,\kappa^+)$) s.t.

(***) $M \models [r \Vdash_{\mathbf{Q}_\kappa^{\kappa^+}} (\langle H_{\kappa^+}[\mathring{A} \cap [\check{\kappa},\check{\kappa}^+)], \mathring{A} \cap [\check{\kappa},\check{\kappa}^+)\rangle \models \psi(\check{\mathbf{P}},\check{\mathbf{Q}}^\kappa,\check{\xi}))]$.

This can be expressed in M by:

(****) $M \models$ "$[H_{\kappa^+} \models \chi(r,p,\mathbf{Q}^\kappa,\xi)]$ iff

$[r \Vdash_{\mathbf{Q}_\kappa^{\kappa^+}} (\langle H_{\kappa^+}[\mathring{A} \cap [\check{\kappa},\check{\kappa}^+)], \mathring{A} \cap [\check{\kappa},\check{\kappa}^+)\rangle \models \psi(\check{\mathbf{P}},\check{\mathbf{Q}}^\kappa,\check{\xi}))]$,

where χ is the Π_m formula given by Corollary A.11.1.

Let δ = the least δ, $\delta \in \Upsilon(\xi,\kappa^+)$ s.t.

$U = \langle L_\delta[A], A \cap \delta\rangle \prec \langle L_{\kappa^+}[A], A \cap \delta\rangle$.

Note that $U = X_\kappa = \bigcup_{\alpha<\kappa} X_\alpha$ since $X_\kappa \models \forall\alpha(\mathrm{card}(\alpha) \leq \kappa)$ and hence $X_\kappa \cap \kappa^+$ is transitive. $U \in M$ is a transitive ZF^- model.

For $\gamma \in [\kappa,\kappa^+)$ set $r_\gamma = \chi_A \wedge [\kappa,\gamma) \in \mathbf{Q}_\kappa^{\kappa^+}$.

Since $U \prec \langle L_{\kappa^+}[A], A \cap [\kappa,\kappa^+)\rangle$, there is a $\gamma < \delta$ s.t. $M \models (H_{\kappa^+} \models \chi(r_\gamma,p,\mathbf{Q}^\kappa,\xi)$. Let $f\colon \kappa \longrightarrow u$ be surjective.

Set:

$E = \{\alpha<\kappa \mid \alpha \text{ is regular} \wedge f''\alpha \prec U \wedge \alpha = \kappa\cap U \wedge \mathbf{Q}^{\kappa}, p, \xi, \gamma \in f''\alpha\}$.

For $\alpha \in E$ set $\pi_\alpha : \langle L_{\delta'_\alpha}[a_\alpha], a_\alpha\rangle \cong \langle f''\alpha, A\cap[\kappa,\kappa^+)\cap f''\alpha\rangle$.

Then $\pi_\alpha(\mathbf{Q}^\alpha) = \mathbf{Q}^\kappa$ and $\pi_\alpha(p) = P$. Set $\xi_\alpha = \pi_\alpha^{-1}(\xi)$, $\gamma_\alpha = \pi^{-1}(\gamma)$ and

$$E' = \{\alpha \in E \mid H_{\alpha^+} \models \chi(r'_{\gamma_\alpha}, p, \mathbf{Q}^\alpha, \xi_\alpha)\} \text{ in } M .$$

By Lemma A.5 E' is stationary (subtle) in M . Note that $r'_{\gamma_\alpha} = \chi_{a_\alpha} \restriction [\alpha,\gamma_\alpha)$. By Lemma A.6 we get:

$E'' = \{\alpha \in E' \mid a_\alpha = A\cap[\alpha,\delta'_\alpha)\}$ is stationary (subtle) in N .

Hence for $\alpha \in E''$, $r'_{\gamma_\alpha} = r_{\gamma_\alpha} \in G_\alpha^{\alpha^+} \cap \mathbf{Q}_\alpha^{\alpha^+}$ and retracing the steps (****) to (*) we have:

For $\alpha \in E''$, $\langle L_{\alpha^+}[A], A\cap\alpha^+\rangle \models \phi(\xi_\alpha)$. $\bigcup_{\alpha<\kappa} X_\alpha = U = \bigcup_{\alpha<\kappa} f''\alpha$ hence, $C = \{\alpha \in \text{card} \mid \alpha = \kappa\cap X_\alpha \wedge X_\alpha \cap U = f''\alpha \wedge \mathbf{Q}^\kappa, p, \gamma, \xi \in X_\alpha\}$ is cub in κ . Thus $C\cap E''$ is stationary (subtle) in N . But for $\alpha \in C\cap E''$ it is true that $\delta_\alpha = \delta'_\alpha$ and $A'_\alpha = A\cap\alpha = a_\alpha = A\cap\delta_\alpha$.

Q.E.D.

Lemma A.13 *Let* κ, ξ, X_α, σ_α *and* δ_α *be as in Lemma A.12. Then,* $D^* = \{\alpha \in \kappa \mid \alpha$ *is regular* $\wedge\ \alpha = \kappa\cap X_\alpha \wedge A'_\alpha = A\cap\delta_\alpha \wedge \wedge \langle L_{\delta_\alpha}[A], A\cap\delta_\alpha\rangle \prec_{\Sigma_m} \langle L_{\alpha^+}[A], A\cap\alpha^+\rangle\}$ *is stationary (subtle) in* κ .

Proof Let δ = the least $\delta > \xi$ s.t.

$$\langle L_\delta[A], A\cap\delta\rangle \prec \langle L_{\kappa^+}[A], A\cap\kappa^+\rangle .$$

(Hence $L_\delta[A] = X_\kappa$.) The statement $\langle L_\delta[A], A\cap\delta\rangle \prec_{\Sigma_m} \langle L_{\kappa^+}[A], A\cap\kappa^+\rangle$ can be expressed by: $\langle L_{\kappa^+}[A], A\cap\kappa^+\rangle \models \phi(\delta)$, where ϕ is uniformly π_m for regular κ .

Define X^*_α from δ as X_α was defined from ξ for $\alpha \le \kappa$. Set $\sigma^*_\alpha : \langle L_{\gamma^*_\alpha}[A^*_\alpha], A^*_\alpha\rangle \cong X^*_\alpha$ and $\delta^*_\alpha = \sigma^{*-1}_\alpha(\delta)$. By Lemma A.12 we get:

$D' = \{\alpha<\kappa \mid \alpha$ is regular $\wedge\ \alpha = \kappa \cap X^*_\alpha \wedge A^*_\alpha = A\cap\delta^*_\alpha \wedge \xi \in X^*_\alpha \wedge$
$\wedge \langle L_{\alpha^+}[A], A\cap\alpha^+\rangle \models \phi(\delta^*_\alpha)\}$ is stationary (subtle) in κ.

For $\alpha \in D'$ we have,

$\langle L_{\delta^*_\alpha}[A], A\cap\delta^*_\alpha\rangle \prec_{\Sigma_m} \langle L_{\alpha^+}[A], A\cap\alpha^+\rangle$ since,

$\langle L_{\alpha^+}[A], A\cap\alpha^+\rangle \models \phi(\delta^*_\alpha)$.

<u>Claim</u> $X_\alpha = X^*_\alpha \cap L_\delta[A]$, $\delta_\alpha = \delta^*_\alpha$ and $A^*_\alpha \cap \delta = A'_\alpha$, for $\alpha\in D'$.

Proof $\alpha \in D' \longrightarrow \xi \in X^*_\alpha$. Let $\xi^*_\alpha = \sigma^{*-1}_\alpha(\xi)$. By the isomorphism $\delta^*_\alpha =$ the least $\delta' > \xi^*_\alpha$ s.t.

$\langle L_{\delta^*_\alpha}[A^*_\alpha], [A^*_\alpha \cap \delta^*_\alpha]\rangle \prec \langle L_{\gamma^*_\alpha}[A^*_\alpha], A^*_\alpha\rangle$.

$\langle L_{\gamma^*_\alpha}[A^*_\alpha], A^*_\alpha\rangle \models$ "$\langle L_{\delta^*_\alpha}[A^*_\alpha], A^*\cap\delta^*_\alpha\rangle$ = the set of y definable from $\alpha \cup \{\xi^*_\alpha\}$.

$X^*_\alpha \cap L_\delta[A] = (\sigma^*_\alpha)^{-1}$ "$L_{\delta^*_\alpha}[A^*_\alpha]$ = the set of y definable in $\langle L_{\kappa^+}[A], A\rangle$ from $\alpha \cup \{\xi\}$.

Hence $X^*_\alpha \cap L_\delta[A] = X_\alpha$, $\delta_\alpha = \delta^*_\alpha$ and $A^*_\alpha \cap \delta_\alpha = A'_\alpha = A \cap \delta_\alpha$.

Q.E.D.

Lemma A.14 *Let* $\alpha < \kappa$ *and let* κ *be* α*-Erdos in* M. *Then* κ *is* α*-Erdos in* N.

Proof We will work with $\mathbf{Q}$ and our generic $A \subseteq [\omega,\infty)$. $\mathbf{Q}$ is ω-closed. It suffices to show that κ is α-Erdos in $N' = M[A^\kappa]$. Let $C, f \in N'$ s.t. C is cub in κ and $F: [C]^{<\omega} \longrightarrow \kappa$ is regressive. By Lemma A.2 we may assume that $C \in M$, since there is a cub $C' \in M$, $C' \subseteq C$ and it is sufficient to show that $f \restriction [C']^{<\omega}$ has a homogeneous set of type α.

Let $\mathring{f}$ be a term for f and let $p \in \mathbf{Q}$ s.t.

(*) $\quad p \Vdash (\mathring{f}: [C]^{<\omega} \longrightarrow \kappa$ is regressive).

Set $E = \{\langle \xi,q,u\rangle \mid u \in [C]^{<\omega}, q \leq p, q \Vdash \mathring{f}(\check{u}) = \check{\xi}\}$ and $\mathfrak{A} = \langle L_\kappa[B],\alpha,\mathbf{Q}^\kappa,E,C,p,B\rangle$, where $H_\kappa = L_\kappa[B]$.

Using the Corollary to Fact 4.2.12, let I be a good set of indiscernibles for $\mathfrak{A}$ of order type α s.t. $\gamma \in I \longrightarrow \mathfrak{A}\restriction\gamma \prec \mathfrak{A}$.

Claim There is a $q \leq p$ s.t.

$q \Vdash \check{I}$ is a set of indiscernibles for $\check{\mathfrak{A}}$.

Note Since the claim only assumes (*), its proof will give that the set of such q is dense in $\{r \mid r \leq p\}$. Hence, taking $p \in G$, there is such a $q \in G$ which will complete the lemma.

We note the following facts:

(1) $I \subseteq C$ and for $\gamma \in I$, $\langle L_\gamma[B] \ldots C \cap \gamma \ldots\rangle \prec \mathfrak{A}$, hence $C \cap \gamma$ is bounded in γ.

(2) Each $\gamma \in I$ inaccessible.

(3) If $p_\gamma \in \mathbf{Q}^\kappa$ is $\mathfrak{A}$-definable from $\gamma \in I$ and $\nu < \gamma$, then $p_\gamma \in \mathbf{Q}^{\gamma'}$ for all $\gamma' \in I \setminus (\gamma+1)$.

Proof $\gamma' \in I \setminus (\gamma+1)$ implies $\langle L_{\gamma'}[B], \ldots \mathbf{Q}^\kappa \cap L_{\gamma'}[B] \ldots\rangle \prec \mathfrak{A}$. But $\gamma \in L_{\gamma'}[B]$ and p_γ is γ definable, hence $p_\gamma \in \mathbf{Q}^\kappa \cap L_{\gamma'}[B] = \mathbf{Q}^{\gamma'}$.

(4) Let $p_{\gamma,\vec{\gamma}}$ be uniformly $\mathfrak{A}$-definable from $\langle\gamma,\vec{\gamma}\rangle \in I^{n+1}$ and $\nu < \gamma$ and $p_{\gamma,\vec{\gamma}} \in \mathbf{Q}^{\min(\vec{\gamma})}$.

(a) Then $p_{\gamma,\vec{\gamma}} = p_{\gamma,\vec{\delta}}$ for all $\langle\gamma,\vec{\delta}\rangle \in I^{n+1}$.

Proof First it is clear that for $\langle\gamma,\vec{\delta}\rangle \in I^{n+1}$, $p_{\gamma,\vec{\delta}} \in \mathbf{Q}^{\min(\vec{\delta})}$. $\gamma' = \min(I \setminus \gamma + 1)$ then $p_{\gamma,\gamma',\eta} \in \mathbf{Q}^{\gamma'}$ for $\langle\gamma,\gamma',\vec{\eta}\rangle \in I^{n+1}$. Hence $\exists\, \xi < \gamma'$ s.t. $p_{\gamma,\gamma',\vec{\eta}} \in \mathbf{Q}^{\xi}$. Since $I \setminus \gamma'$ is a set of indiscernibles for $\langle \mathfrak{A}, \nu(\nu<\gamma')\rangle$, it follows by standard methods that $p_{\gamma,\vec{\gamma}} = p_{\gamma,\vec{\delta}}$ for all $\langle\gamma,\vec{\gamma}\rangle, \langle\gamma,\vec{\delta}\rangle \in I^{n+1}$.

(b) Since $p_{\gamma,\vec{\gamma}}$ is independent of $\vec{\gamma}$ set $p_{\gamma,\vec{\gamma}} = p_\gamma$. Now assume that we have such p_γ for $\gamma \in I$ (i.e. all defined by the same definition). Then $(p_\gamma)^\gamma = (p_{\gamma'})^{\gamma'}$ for $\gamma,\gamma' \in I$.

Proof Clear.

(c) Let $\langle p_\gamma \mid \gamma \in I\rangle$ be as in (b) and $X \subseteq I$, then $p_X = \bigcup_{\gamma\in X} p_\gamma \in \mathbf{Q}^\kappa$.

Proof All we have to show is that $\mathrm{card}(\mathrm{dom}(p_X) \cap \lambda) < \lambda$ for regular λ . If β is a limit point of I then β is singular since $\mathrm{ot}(I) = \alpha$ and $\alpha < \min(I)$. Let $\lambda \le \kappa$ be regular. For $\gamma \in X \setminus \lambda$ we have by (b) $(p_\gamma)^\gamma$ is constant. $p_X \upharpoonright \lambda = (p_\gamma)^\gamma \cup \bigcup_{\delta\in X\cap\lambda} p_\delta$ for some $\gamma \in X \setminus \lambda$. Since λ is regular, λ is not a limit point of X . If $X \cap \lambda$ has a maximal element δ^* then $p_\delta \in \mathbf{Q}^{\delta^*}$ for all $\delta < \delta^*$. If $X \cap \lambda$ has no maximal element then $\delta' = \sup(X \cap \lambda) < \lambda$ and for $\delta \in X \cap \lambda$, $p^\delta \in \mathbf{Q}^{\delta'}$. In all cases $\mathrm{card}(\mathrm{dom}(p_X)\cap\lambda) < \lambda$.

Q.E.D. (4)

We now construct $\langle p^n_\gamma \mid \gamma \in I\rangle$ and $\xi_n < \min(I)$ for $n < \omega$ s.t. $p^{n+1}_\gamma \le p^n_\gamma \le p$ and

(a) $p^n_\gamma \in P^{\gamma'}$ for $\gamma' \in I \setminus (\gamma+1)$ is uniformly $\mathfrak{A}$-definable in $\gamma, \vec{\gamma}$ for $\langle \gamma, \vec{\gamma} \rangle \in I^{n+1}$. (For $n = 0$ this means p^o_γ is uniformly $\mathfrak{A}$-definable in γ.) By 4(a) p^n_γ is independent of the choice of $\vec{\gamma}$.

(b) $p^n_{\gamma_o} \cup \ldots \cup p^n_{\gamma_n} \Vdash \mathring{f}(\{\check{\gamma}_o, \ldots, \check{\gamma}_n\}) = \check{\xi}_n$ for $n < \omega$, $\langle \vec{\gamma} \rangle \in I^{n+1}$.

After we succeed to define such p^n_γ, set $p^n = \bigcup_{\gamma \in I} p^n$. By 4(c) $p^n \in \mathbf{Q}^\kappa$. $\mathbf{Q}^\kappa$ is ω-closed, so $q = \bigcup_{n<\omega} p^n \in \mathbf{Q}^\kappa$. But then clearly $q \Vdash \check{I}$ is homogeneous for $\mathring{f}$. Thus all we have to do is define $\langle p^n_\gamma \mid \gamma \in I, n < \omega \rangle$ and $\langle \xi_n \mid n < \omega \rangle$.

We define $\langle p^n_\gamma \mid \gamma \in I \rangle$ and ξ_n by induction on n.

p^o_γ = the $\mathfrak{A}$-least $q \le p$ s.t. $\exists \xi (q \Vdash \mathring{f}(\check{\gamma}) = \check{\xi})$.

By (3) $p^o_\gamma \in \mathbf{Q}^{\gamma'}$ for $\gamma' \in I \setminus (\gamma+1)$.

ξ^o_γ = that unique ξ s.t. $p^o_\gamma \Vdash \mathring{f}(\check{\gamma}) = \check{\xi}$.

$p^o_\gamma \le p \Vdash$ "$\mathring{f}$ is regressive", so $\xi^o_\gamma < \omega$. Hence we may set $\xi^o = \xi^o_\gamma = \xi^o_{\gamma'} < \min(I)$ for $\gamma, \gamma' \in I$.

Assume we already have defined $\langle p^m_\gamma \mid \gamma \in I \rangle$, ξ_m for $m \le n$.

Notation For $\vec{\nu} \in (On)^n$, $\vec{\nu} < \gamma$ implies $\max(\nu) < \gamma$.

For $\vec{\nu} \in (On)^n$, $\vec{\nu} < \gamma$ set,

$\Delta_{\vec{\nu},\gamma} = \{q \le p \mid \exists \xi (q \Vdash \mathring{f}(\check{\vec{\nu}}, \check{\gamma}) = \check{\xi}\}$. $\Delta_{\vec{\nu},\gamma}$ is obviously dense in $\{r \in \mathbf{Q}^\kappa \mid r \le p\}$. As a preliminary step to defining p^{n+1}_γ we define p^{*i}_γ for $i \le n$ as follows:

p^{*o}_γ = the $\mathfrak{A}$-least $q \le (p^n_\gamma)_\gamma$, $q \in \mathbf{Q}^\kappa_\gamma$ s.t. $\Delta^q_{\vec{\nu},\gamma}$ is dense in $\{r \in \mathbf{Q}^\gamma \mid r \le (p)^\gamma\}$ for all $\vec{\nu} < \gamma$, $\vec{\nu} \in (On)^n$.

This is possible because $\mathbf{Q}_\gamma$ is γ-closed.

p^{*1}_γ = the $\mathfrak{A}$-least $q \le (p^n_\gamma)_\gamma$, $q \in \mathbf{Q}^\kappa_\gamma$ s.t.

$(\Delta^{p^{*o}_{\gamma'}}_{\vec{\nu},\gamma,\gamma'})^q$ is dense in $\{r \in \mathbb{Q}^\gamma \mid r \le (p)^\gamma\}$ for all $(\vec{\nu}) \in (On)^{n-1}$ and any $\gamma' \in I \setminus (\gamma+1)$.

$(\Delta^{p^{*o}_{\gamma'}}_{\vec{\nu},\gamma,\gamma'})^q \subseteq \mathbb{Q}^\gamma \in L_{\gamma'}[B]$ for $\gamma' \in I\setminus(\gamma+1)$, so by the same argument as 4(a) it is independent of the choice of γ'.

If we already have p^{*i}_γ and $i+1 < n$ then

p^{*i+1}_γ = the $\mathfrak{A}$-least $q_\gamma \le (p^n_\gamma)_\gamma$, $q \in \mathbb{Q}^\kappa_\gamma$ s.t.

$(\Delta^{p^{*o}_{\delta_n}}_{\vec{\nu},\gamma,\vec{\delta}})^{p^{*1}_{\delta_{n-1}}}) \ldots {}^{p^{*i}_{\delta_{n-i}}})^q$ is dense in $\{r \in \mathbf{Q}^\gamma \mid r \le (p)^\gamma\}$

where $\vec{\nu} \in (On)^{n-(i+2)}$, $\vec{\delta} = (\delta_{n-i},\ldots,\delta_n) \in I^{i+1}$ and $\vec{\nu} < \gamma < \vec{\delta}$.

As before the above set is independent of $\vec{\delta} \in I^{i+1}$, $\gamma < \vec{\delta}$.

We have defined p^{*i}_γ for $i < n$.

p^{*n}_γ = the least $q \le p^n_\gamma$ s.t.

$q \in (\Delta^{p^{*o}_{\delta_n}}_{\gamma,\vec{\delta}}) \ldots)^{p^{*n-1}_{\delta_1}}$ for $\gamma \in I$ and $\delta = (\delta_1,\ldots,\delta_n) \in I^n$, $\gamma < \vec{\delta}$. Clearly the choice is independent of $\vec{\delta} \in I^n$, $\gamma < \vec{\delta}$.

Set $p^{n+1}_\gamma = p^{*o}_\gamma \cup \ldots \cup p^{*n}_\gamma$.. It is clear from the definitions that $p^{n+1}_\gamma \in \mathbf{Q}^{\gamma'}$ for $\gamma' \in I\setminus(\gamma+1)$ and p^{n+1}_γ is definable from $\nu < \gamma$, $\langle\gamma,\vec{\delta}\rangle \in I^{n+1}$ and is independent of the choice of $\vec{\delta} \in I^{n+1}$, $\gamma < \vec{\delta}$. Set:

$\xi^{n+1}_{\vec{\gamma}}$ = that ξ s.t. $p^{n+1}_{\gamma_o} \cup \ldots \cup p^{n+1}_{\gamma_n} \Vdash \mathring{f}(\vec{\gamma}) = \check{\xi}$, where $\vec{\gamma} = \langle\gamma_o,\ldots,\gamma_n\rangle \in I^{n+1}$.

This is tedious but it can be done since:

(0) $p^{*\gamma}_{\gamma_o} \in (\Delta^{p^{*o}_{\gamma_n}}_{\gamma_o,\gamma_1,\ldots,\gamma_n}) \ldots)^{p^{*n-1}_{\gamma_1}}$, hence so does $p^{n+1}_{\gamma_o} \le p^{*n}_{\gamma_o}$. But then

$p^{n+1}_{\gamma_o} \cup p^{*n-1}_{\gamma_1} \in (\Delta^{p^{*o}_{\gamma_n}}_{\gamma_o,\gamma_1,\ldots,\gamma_n}) \ldots)^{p^{*n-2}_{\gamma_2}}$ etc. until,

$p^{n+1}_{\gamma_o} \cup \ldots \cup p^{n+1}_{\gamma_n} \in \Delta_{\gamma_o,\gamma_1,\ldots,\gamma_n}$ and hence such a ξ exists. It is also clear that $\xi^{n+1}_{\vec{\gamma}} = \xi^{n+1}_{\vec{\delta}} = \xi^{n+1}$ is independent of the choice of $\vec{\gamma},\vec{\delta} \in I^{n+1}$ since f is regressive and $\xi^{n+1} < \min(I)$.

Q.E.D.

Lemma A.15 *Suppose* $M \models$ *"Every set of ordinals has a sharp", then the same is true of* N .

Proof Since in N we have $H_\kappa = L_\kappa[A]$ for every regular κ , it suffices to prove the lemma for $N' = M[A^\kappa]$ and to show $(A^\kappa)^\#$ exists. Let $\langle \iota_\tau \mid \tau \in On \rangle$ enumerate I , the canonical indiscernibles for $L[\mathbb{Q}^\kappa]$ ($\mathbb{Q}^\kappa$ considered as coded by a set of ordinals) and let $\mathring{A}$ be the canonical name for A^κ . Let $p \in \mathbb{Q}^\kappa$, $\vec{c}$, $\vec{c}\,' \in [I]^n$, then

$$p \Vdash_{\mathbb{Q}^\kappa} \phi(\vec{c},\mathring{A}) \quad \text{iff} \quad p \Vdash_{\mathbb{Q}^\kappa} \phi(\vec{c}\,',\mathring{A})$$

and the result follows.

Q.E.D.

Bibliography

[BA] J. Baumgartner, Ineffability Properties of Cardinals I, Colloquium Mathematical Society Janos Bolyai 10, Infinite and Finite Sets, Keszthely, 1973, pp 105-130.

[BO] W. Boos, Lectures on Large Cardinal Axioms, Logic Conference, Kiel, 1974, Ed. G.H. Muller, A. Oberschlep and K. Potthoff, Springer Lecture Notes, Volume 499, pp 25-88.

[BU] P. Burgess, Forcing, Handbook of Mathematical Logic, Ed. J. Barwise, North Holland, 1977, pp 403-452.

[DEV] K. Devlin, Aspects of Constructability, Springer Lecture Notes, 1973, Volume 354.

[DJ] K. Devlin and R.B. Jensen, Marginalia to a Theorem of Silver, Logic Conference, Kiel, 1974, Ed. G.H. Muller, A. Oberschlep and K. Potthoff, Springer Lecture Notes, Volume 499.

[DoJ] A.J. Dodd and R.B. Jensen, The Core Model, Annals of Mathematical Logic, to appear.

[DR] F.R. Drake, Set Theory and an Introduction to Large Cardinals, North Holland, 1974.

[EA] Easton, Powers of Regular Cardinals, Annals of Mathematical Logic 1 (1970), 139-179.

[F] H. Friedman, One Hundred and Two Problems in Mathematical Logic, Journal of Symbolic Logic, 40 (1975), pp 113-129.

[JO] R.B. Jensen, Coding the Universe by a real, handwritten Manuscript, 1975.

[J1] R.B. Jensen, The Fine Structure of the Constructible Hierarchy, Annals of Mathematical Logic, 4 (1972), pp 229-308.

[J2] R.B. Jensen, Forcing which preserves Ramsey Cardinals, hand-written manuscript, 1966.

[J3] R.B. Jensen, More Coding by a real, hand-written manuscript, 1978.

[J4] R.B. Jensen, Preserving Measurable Cardinals, AMS Proceedings of Symposia in Pure Mathematics Vol.XIII, 2

[J5] R.B. Jensen, Some Applications of K, hand-written manuscript, 1978.

[JS] R.B. Jensen and R. Solovay, Some Applications of Almost Disjoint Sets, Y. Bar-Hillel Ed., Mathematical Logic and the Foundations of Set Theory, North Holland, Jerusalem, 1968, pp 84-104.

[K] K. Kunen, Some Applications of Iterated Ultrapowers in Set Theory, Annals of Mathematical Logic 1 (1970), pp 179-227.

[LE] A. Levy, A Generalization of Gödel's Notion of Constructability, Journal of Symbolic Logic, 25 (1960), pp 147-155.

[SA] G.E. Sacks, F-Recursiveness, Logic Colloquium 1969, Ed. R.O. Gandy and C.M.E. Yates, North Holland, 1971, pp 289-303.

[SH] J. Shoenfield, Unramified Forcing, AMS Proceedings of Symposia, Volume XIII, Part 1, Ed. D. Scott, pp 357-381.

[SI] J. Silver, Some Applications of Model Theory in Set Theory, Annals of Mathematical Logic, 3 (1971), pp 45-110.

[SO] R. Solovay, A non-constructible Δ^1_3 set of integers, Transactions of the AMS, 127 (1967), pp 58-75.

[Z] A. Zarach, Forcing with proper classes, Fundamentica Mathematica, 81 (1973), pp 1-27.

Notational Index

In the following index we have collected together the principal notational definitions. Where two page references coexist, this usually represents an item that has been redefined for the purposes of Chapter 8.

By an *inner model* we mean a transitive class M that contains all ordinals and is a model of ZF.

A structure $\langle M,A\rangle$ is *amenable* if $x \in M \rightarrow x \cap A \in M$. We use $\text{card}(x)$ or $\overline{\overline{x}}$ to denote the cardinality of a set x, and $\text{ot}(\langle A,<\rangle)$ [or $\text{otp}(\langle A,<\rangle)$] to represent the order type of A as ordered by $<$.

We define $0^{\#}$ as the complete theory of

$$\langle L,\in,\nu_0,\nu_1 \dots \nu_n, \dots\rangle_{n<\omega}$$

where the ν_i enumerate the first ω canonical indiscernibles for L.

We say a class of forcing conditions $\mathbf{Q}$ satisfies the κ-anti-chain condition (κ-AC) when every set of mutually incompatible elements of $\mathbf{Q}$ has cardinality less than κ. We use "$p \quad \psi$" as an abbreviation for "$p \Vdash \psi$ or $p \Vdash \neg\psi$". We say $\Delta \subseteq \mathbf{Q}$ is *predense* if the class of extensions of members of Δ is dense. $\mathcal{P}(x)$ denotes the power set of x.

Index